One Legacy of Paul F. Brandwein

CLASSICS IN SCIENCE EDUCATION

Volume 2

Series Editor:

Karen C. Cohen

For further volumes:
http://www.springer.com/series/7365

Deborah C. Fort

One Legacy of Paul F. Brandwein

Creating Scientists

Springer

Deborah C. Fort
3706 Appleton St. NW
Washington, DC 20016
USA
deborah.fort@starpower.net

ISBN 978-90-481-2527-2   e-ISBN 978-90-481-2528-9
DOI 10.1007/978-90-481-2528-9
Springer Dordrecht Heidelberg London New York

Library of Congress Control Number: 2009926887

© Springer Science+Business Media B.V. 2010
No part of this work may be reproduced, stored in a retrieval system, or transmitted in any form or by
any means, electronic, mechanical, photocopying, microfilming, recording or otherwise, without written
permission from the Publisher, with the exception of any material supplied specifically for the purpose
of being entered and executed on a computer system, for exclusive use by the purchaser of the work.

Printed on acid-free paper

Springer is part of Springer Science+Business Media (www.springer.com)

Foreword

Once again, our nation has a powerful need for a revolution devoted to creating scientists. As we face the challenges of climate change, global competitiveness, biodiversity loss, energy needs, and dwindling food supplies, we find ourselves in a period where both scientific literacy and the pool of next-generation scientists are dwindling. To solve these complex issues and maintain our own national security, we have to rebuild a national ethos based on sound science education for all, from which a new generation of scientists will emerge. The challenge is how to create this transformation. Those shaping national policy today, in 2009, need look no further than what worked a half-century ago.

In 1957, Sputnik circled and sent a clarion call for America to become the world's most technologically advanced nation. In 1958, Congress passed the National Defense Education Act, which focused the national will and called for scholars and teachers to successfully educate our youth in science, math, and engineering. It was during this time period that Paul F. Brandwein emerged as a national science education leader to lay the foundation for the changes needed in American education to create the future scientists essential to the nation's well-being. Paul's hands-on teaching experiences at George Washington and Forest Hills high schools (both in New York) and at universities and colleges, including—among others—Columbia Teachers College and other postsecondary institutions, particularly in Colorado; his seminal writings (especially *The Gifted Student as Future Scientist* [1955/1981]); and the strategies he employed as a top editor and administrator at Harcourt Brace Jovanovich uniquely prepared him to help create the paradigm shift in science education that allowed us, as a nation, to meet the challenge to become the leading society of the 20th century in the excellence of our science education. Today, a half-century later, we continue to reap the benefits of this national focus in all fields of science, in technology, and in medicine.

The cultural and scientific revolutions of the 1960s, coupled with Paul's strong understanding of the interconnectedness of science and society, brought a new focus and intensity to Paul's work and strengthened his and his wife Mary's commitment to environmental conservation. It was during these years, while Paul served as the codirector of the Pinchot Institute for Conservation Studies at Grey Towers, in Milford, Pennsylvania, that his work in the newly emerging field of ecology inspired a growing national environmental awareness. Paul also fostered greater understanding of the role that the natural world can play in cultivating the curiosity and culture of investigation necessary to create new scientists. Paul understood from his research that the best way to encourage the young in science was to help them early to do original work investigating scientific questions for which the answers were still unknown. What resulted was the introduction of inquiry-based, ecologically focused science education into classrooms across the nation.

Paul spent his professional life as a scientist, educator, author, and publisher focused on the deep question of how we as a nation can create the scientist within. Through varied contributions from his former students, his colleagues, and his friends, *One Legacy of Paul F. Brandwein: Creating Scientists* explores how one man's teachings and philosophies on science, education, and environmentalism both laid the groundwork for the first great science education revolution in our nation's history and prepared the way for the one so necessary today. Many of the essays in this book offer firsthand reflections by former students and colleagues of Paul's during the 1950s and beyond that record the impact of, and inspiration that resulted from, their encounters. Paul's insights highlighted in this book[1] illuminate a path forward for us today, as we work to create the second American science education revolution.

As president of the Paul F-Brandwein Institute, I wish to extend on behalf of its board of directors heartfelt congratulations and thanks both to Deborah C. Fort, the contributing editor of this book, and to all the other individuals who provided their own generous and invaluable offerings. Deborah's commitment to and perseverance during this project are overshadowed only by the timeliness of this publication in insuring the vital and continuing impact of Paul F. Brandwein's legacy in creating the next generation of scientists.

Keith A. Wheeler
Unionville, New York

---

[1] Some of them in his own words (see Part II).

Preface

My Path to This Book

I first learned of the existence of Paul Franz Brandwein in 1986, when, working as a freelance editor at the National Science Teachers Association (NSTA), I was asked to edit a manuscript on teaching science to talented students. The head of the publications department and I agreed that the decidedly ungifted manuscript had to be rejected, and she suggested that I pull together materials for another book on the subject. I started by compiling a bibliography and discovered, to my astonishment—these were the days when a future devastating shortfall of research scientists was widely (but it turned out falsely) predicted—that only one significant book existed on the subject. The sole volume was Paul's *The Gifted Student as Future Scientist* (first published in 1955 by Harcourt Brace and republished in 1981 by the National/State Leadership Training Institute on the Gifted and the Talented). In the intervening years, Paul had become copublisher of Harcourt Brace Jovanovich and, coincidentally, he had been that year's recipient of NSTA's prestigious Robert H. Carleton Award, which "recognizes one individual who has made outstanding contributions to, and provided leadership in, science education at the national level. . . ." (2007). He seemed to me to be the perfect editor of the projected NSTA volume on encouraging "gifted and talented"[1] students.

My bibliographic research also turned up the name of a man active in fostering education for the gifted, one A. Harry Passow, then president of the World Council for Gifted and Talented Children. I promptly wrote to both men in hopes that they might be persuaded to coedit the book. Harry, the Jacob H. Schiff Professor at Teachers College of Columbia University, was easy to find. But my letter to Paul at Harcourt went unanswered. More research revealed that he had recently retired to a farm in Unionville, New York. There I wrote to him again, and he replied by telephone days later that only if Harry, who described himself as a "defrocked science teacher," would serve as coeditor, would he agree. Since Harry had made Paul's coeditorship his provision for joining the project, I had my editors.

Although I had been making my living as a freelance editor for more than 10 years at that point, Paul had much to teach me, not only about editing but more importantly about the business of living responsibly, kindly, and significantly.

---

[1] On the definition of these tricky terms, more later.

ix

I became one of a long list of people over the course of nearly half a century lucky enough to have Paul as a mentor. This was a task he took seriously, and he had a bigger job in teaching me than he could possibly have anticipated. For the next 8 years, we had several lengthy phone conversations each week, touching not only on the progress of the book on which we were working, now titled *Gifted Young in Science: Potential Through Performance*,[2] but also on many, many other subjects.

In an essay I cowrote with a colleague in 2005, the two of us not only pondered the question of why terrifically busy people choose to make available time to be mentors[3] but also defined two classes of mentors—those who confine their advice to professional matters and those who also guide their protégé(e)s personally (Haseltine and Fort). Both my coauthor, a reproductive endocrinologist who founded the Society for Women's Health Research and who directs the Center for Population Research at the National Institute of Child Health and Human Development at the National Institutes of Health, and Paul saw their roles as mentors as comprising aid to both aspects of their protégé(e)s' natures. Much as he taught me professionally, Paul was much more important to me as a personal adviser in the business of living than as a role model in writing, editing, and publishing.

Once, he explained, as if it were the simplest concept on earth, "When I see help is needed, I give it," adding,

Never take advantage of anyone.
Never humiliate anyone.
Never harm anyone.
And, if you can, lend a hand.
Hardest of all, forget yourself.

And he continued, "I look at people and ask myself, 'How on earth did they get themselves into that mess?' and then, 'What can I do to help?'" He had his work cut out for him with me—stubborn, proud, recalcitrant, sure of myself (in spite of contrary evidence), impolitic, always in trouble. "You must learn patience ...," he cautioned, suggesting some avenues down which he had successfully walked:

I am able to lose graciously.
I don't want to win in all things.
I do want to learn to write better.
I want certain enemies.
I would like to have the friendship of all human beings who can stand fast.
Defeat is as important as victory, and neither is worthy of being noticed.

---

[2]Then-NSTA President Gerald Skoog would later join Paul, Harry, and me as a contributing editor.
[3]Perhaps Stephen Jay Gould has an answer:

What do mentors get in return?... The answer, strange as this may sound, is fealty in the genealogical sense. The work of graduate students is part of a mentor's reputation forever, because we trace intellectual lineages in this manner. I was Norman Newell's student, and everything that I ever do, as long as I live, will be read as his legacy (and if I screw up, will redound to his detriment—though not so seriously, for we recognize a necessary asymmetry: errors are personal, successes part of the lineage). (1990, p. 140)

Preface

And, he added, on numerous occasions,

"Grace
Silence
Unfailing courtesy
have worked for me."

Between that first talk in 1986 and his death in 1994, between our frequent phone discussions, face-to-face meetings were rare, because of the serious illnesses that plagued but did not stop his last years' contributions. Conversations with Paul sparkled with quotations from literature from many nations, in many languages, over many centuries, as well as with laughter and jokes, often the worse the better. "You," he said (correctly, alas) on numerous occasions, "are the east end of a westbound horse." Paul in conversation was stunning in a different way than he was in writing or even in formal speech (though those lucky enough to have heard him have said he was wonderful in that medium as well). After the publication of *Gifted Young in Science: Potential Through Performance* in 1989, I occasionally did the research for him that his condition made impossible. The Internet was in its infancy then, and Paul originally did not use a computer at all, though he was pleased to present me with a disk version of his last, posthumously published, book for the National Research Center on the Gifted and Talented.

Without question, except for the members of my immediate family, Paul Brandwein was the most important teacher, friend, and mentor of my life. I am delighted to be able to gather here materials from others—former students, colleagues, friends—who felt similar impacts.

Acknowledgments Without the editorial aid of my longtime friend and colleague Suzanne Lieblich, this book would not be as finished as it is. Without the cooperation of my friends, Brandwein alumni all, James P. Friend, Richard Lewontin, and the late Walter G. Rosen, Part III would not exist. Thank you all.

References

Gould, Stephen J. (1990). *Wonderful life: The Burgess Shale and the nature of history*. New York: Norton.

Haseltine, Florence P., with Fort, Deborah C. (2005). Why be a mentor? In Deborah C. Fort (Ed.), *A hand up: Women mentoring women in science* (2nd ed., pp. 353–366). Washington, DC: Association for Women in Science.

National Science Teachers Association (2007). Retrieved June 7, 2007, from http://www3.nsta.org/awardees

Deborah C. Fort
Washington, District of Columbia

Contents

Part I Remembering Paul F. Brandwein: Essays

Brandwein Alumni

Turning a Dream (Deftly, Subtly, and Effectively) into Reality   5
Andrew M. Sessler

How Dr. Paul Brandwein's Mentorship and Guidance Affected My Scientific Interests and Career   9
Josephine Baron Raskind von Hippel

How to Win Converts and Influence Students   15
Richard Lewontin

Paul F. Brandwein's Influence on My Life: The Essential Spark   21
James P. Friend

Paul Brandwein's Influence on My Life   29
Barbara Wolff Searle

Brief Encounters, Lasting Effects

One Year   37
Richard Goodman

Research at a Tender Age   39
Tom Schatzki

Intellectually Exciting Years at Forest Hills High School   41
Lisa A. Steiner

Encouraging the Uncertain   43
Herbert L. Strauss

Saturdays at the Brooklyn Botanic Gardens   45
Walter G. Rosen

Question Everything   47
Naomi Goldberg Rothstein

xiv

One Class Was Enough 51
Joanne Gallagher Corey

Science Education and Beyond: Colleagues

Paul F. Brandwein—A Personal Reflection 55
Sigmund Abeles

Paul Brandwein and the National Science Teachers Association 61
Marily DeWall

**Dr. Paul F. Brandwein: Messages on Teaching and Learning for
All Educators** 69
E. Jean Gubbins

Some Reflections on Paul F. Brandwein's Impact on Science Education 81
Robert W. Howe

Paul F. Brandwein and Conservation Education 91
Charles E. Roth

Environmental Education and Paul F. Brandwein's *Ekistics* 101
Rudolph J. H. "Rudy" Schafer

Watershed Education for Sustainable Development 105
William B. Stapp

**The Paul F-Brandwein Institute: Continuing a Legacy
in Conservation Education** 119
Keith A. Wheeler, John "Jack" Padalino, and Marily DeWall

**Part II   Paul F. Brandwein in His Own Words—Reprints
1955–1995**

The Gifted Student as Future Scientist 135
Paul F. Brandwein

Science Talent: In an Ecology of Achievement 215
Paul F. Brandwein

**Science Talent in the Young Expressed Within Ecologies of
Achievement: Executive Summary** 243
Paul F. Brandwein

Part III   The Surveys

Remembrances from More than a Half-Century Back: The Surveys 263
Deborah C. Fort

Part IV   Appendixes

Appendix A: The Survey   289

Appendix B: Bibliographies of the Works of Paul F. Brandwein   293
Deborah C. Fort and Suzanne Lieblich

Index   305

Introduction

Deborah C. Fort

Born an Austrian with Spanish and Basque roots,[1] Paul Franz Brandwein (1912–1994) emigrated to America before World War II. Though his family had suffered during World War I, he once laughingly reported one action taken to avoid unnecessary carnage in that pointless exercise: "The Italians," said Paul, "like to stay in the kitchen, away from the fighting. . . . My father," Paul continued, "fought at Bolzano against the Italians. One day the Austrian artillery advanced, and the Italians retreated; then the Italian artillery advanced and fired, and the Austrians retreated. Then they met to make sure that no one got hurt. Then a German battalion arrived to help the Austrians. My father warned the Italians."

The horrors of World War II (which Paul called "my war") followed the mindless slaughter of the first great war of the last century, which Paul said was "my father's." After emigrating in the 1930s, during the summer, when he was not teaching, Paul traveled to Europe to work with British and American intelligence against the Nazis. "In spite of self-esteem, integrity, kindness, and love," he mourned, years later, "there are wars. Wars are not within the human sphere but the nonhuman," observing further, "Men wear uniforms to hide weakness and gather strength from others in like costumes." Paul once told me that he taught so that he would not have to look at young men dead on battlefields anymore.

"Hitler couldn't be said to be mad," he told me. "He was inhuman, and madness is human. He treated people like cattle. Still, it took 18,000,000[2] people to defeat Hitler in World War II."

In 1991, during the first Gulf War, Paul said, sadly, "Bombs are falling on Iraq. Young men [and women] are dying again. If Bush had only known about the Iraqis'

---

[1] Neither his birthplace nor his Basque and Spanish roots appear in any of his official biographies, which are bare-boned summaries of his American career that conclude by referring the reader to his biographies in *American Men and Women in Science, Who's Who in the Humanities,* and *National Leaders of American Conservation.* My sources: Besides discussions between Paul and me (confidentiality respected, of course), his former student, the late Eleanore Berman, provided some confirming information.

[2] I'm sure Paul had a reason for this figure. Most sources include fighting Japan, making the number much higher.

terrible pride, he would not have gone in."[3] (Paul would have told George Herbert Walker Bush: "Let's give them life. You and I are old men. Let's let the young men have their lives.") With deep sadness and gentleness in his voice, he added, "The simple art of communication has been lost, but the brutality remains."

A profoundly modest and private man in his personal life,[4] Paul came nonetheless to be professionally at home at podiums and in laboratories, in classrooms, in the boardrooms of the publishing industry, in scientific societies, and in education associations. During the course of his long, distinguished, varied career, he worked productively as a scientist, an author, an educator (grade school to graduate school), an editor and publisher, and an environmentalist (a noun he preferred to the less inclusive term "conservationist"). In the service of science and education, he delivered hundreds of speeches to audiences worldwide.

Author

---

[3] Paul's alarm at Bush 41's invasion would have been exponentially greater had he lived to see the carnage of Bush 43's.

[4] Speaking at the first meeting of the Brandwein Institute, on the subject of "Remembering Paul Brandwein," his acquaintance of 30 years Calvin W. Stillman (professor emeritus, Rutgers University) noted that "Paul was a folk hero to science teachers. As with most folk heroes, his origins are obscure." While regretting that his contact with Paul had been limited, Stillman summarized, "I have the utmost admiration for Paul. He was one of the greatest people in my life and a very great man of our time" (quoted in Fort, 1998, p. 7).

Introduction xix

Paul's wide-ranging publications concern the humanities, science, and education. His first published essay of hundreds, as well as more than 50 books and textbooks in several languages, appeared in 1937. His last book, *Science Talent in the Young Expressed Within Ecologies of Achievement,* was published posthumously in 1995 by the National Research Center on the Gifted and Talented. Over the course of his 82 years, he was author and/or coauthor of many papers and books (and textbooks) in science and science education, particularly in relation to the science shy, the science prone, the science talented, the gifted, and the disadvantaged. He also published widely in the humanities and the social sciences. (See Part IV, Appendix B, for bibliographies.) According to a tribute from the National Science Teachers Association on the occasion of Paul's receipt of its highest award,

His eloquence as a platform speaker, delivering powerful ideas with grace, to teachers, supervisors, administrators, as well as superintendents of schools, has conveyed an urgency for competence and compassion in providing opportunities for all youngsters, however diverse their talents and special needs. His qualities of selfless quiet service and commitment are well-known in the community of science educators .... (Calvin W. Stillman, quoting a National Science Teachers Association document, in Fort, 1998, p. 8)

Scientist

Although childhood arthritis cut short Paul's formal piano studies, as many of his listening audiences would later testify, he never gave up the piano as avocation. He often broke up his talks by sitting down to the keyboard to emphasize analogies between music, science, art, and teaching and learning.

As a young man, while being treated in hospitals, Paul became interested in science, and his professional focus changed from music to biology. Shortly thereafter, a hospital chemist sponsored him as an assistant in the Littauer Pneumonia Research Laboratory (New York). While working at Littauer during the 1930s, Paul completed his BS, Phi Beta Kappa, from New York University, in night, afternoon, and summer classes. During those years, he was cited as author or coauthor on several research papers.

Thus, before beginning his doctoral studies, Paul spent 4 years observing and assisting in research on the biochemistry of pneumococcus. His practical experience at Littauer in what he would come to call "the well-ordered empiricism of research" (including the processes and protocols of problem finding and solving) focused on the microecology of protists and the ecology of host plant–fungus relationships. By the time he earned his master's degree (1937) and doctorate (1940), both also from New York University, he had come to a belief based in his own experience that would be lifelong: The best way to encourage the young in science was to help them early to do original work. And the best way to help that happen was through mentoring.

Philosopher

Paul took his obligation to serve as a mentor to those who chose him and whom he chose as seriously as his many other duties. Those he advised ranged from any vulnerable child whose path crossed his, to the high school students he taught at mid-century (some of whom went on to become scientists and some of whom are contributors to this volume), to his colleagues in publishing, to *anyone*, at any age, he found worthy and in need of help. When I was working with him, he was not only my mentor but also mentor, among others, to a man of over 60 and to a struggling, illiterate youth who worked on a nearby farm.

I will be focusing here on the principles to which he introduced me during our conversations. For example, here is his version of "act locally; think globally": "Try to change the world; maybe the universe will follow. If you would be on the cutting edge, you will screw up, but each time you will screw up less." He also advised me,

> Don't expose your back.
> We cannot undo the past, even if we didn't want to.
> Accuracy is the prime reason for embarrassment.
> Adulthood is by definition the ability to identify the consequences, find them, and bear them in silence.

"Write this down," he once commanded. *"Conflict resolution requires firmness and gentleness in alternating doses—firmness permeated by gentleness and gentleness permeated by firmness."* He also suggested that I follow "the sage advice" of John Dewey, "Between impulse *or desire* and action, we are to interpose first evidence; failing this, reason and judgment; failing this, compassion; *failing this, fair play"* (quoted in McKenzie, 1984, p. 98 [Brandwein additions in italics]). Essentially an optimist, he frequently quoted a favorite statement from the late UN Secretary-General Dag Hammerskjöld (1964):

> The night is nigh.
> For what has been, *thanks.*
> For what will be, *yes.*

After the war, in the late 1940s, Paul and his wife, Mary, moved into the historic Orange County farmhouse where they lived for most of the rest of their lives. Although professional demands would in the future frequently take him away from Sun Hill Farm, it was to this haven of pastoral and forested lands to which he would always return. There, on the surrounding acres, Paul established extensive gardens and an arboretum and continued his botanical research on rusts and smuts.

Introduction   xxi

Teacher and Mentor

> None of us got where we are solely by pulling ourselves up by our bootstraps. We got here because somebody—a parent, a teacher, an Ivy League crony, or a few nuns—bent down and helped us pick up our boots.
>
> —Thurgood Marshall, 1991

Paul taught first at George Washington High School (New York) and then, between the early 1940s and the mid-1950s, first as a member and later as chair of the science department, at Forest Hills High School (also New York). There, Paul instituted a program where students could select themselves to do original work in science. According to John Curtis Gowan and George D. Demos (1964), more of Paul's early students—who studied in a heterogeneous American public school,[5] not a specialized one training mathematicians and scientists—won the Westinghouse Science Talent Search than those of any other teacher. Gowan and Demos wrote that

> His record of National Science Talent Search winners is unchallenged in the nation, but so modest is he that this fact could never be deduced from his writings. While his own character and personality have much to do with the successes of his students, and his method cannot be communicated completely to less able teachers, it will still pay us handsomely to examine in detail his procedures as seen in his excellent book *The Gifted Student as Future Scientist* .... (1964, p. 118)

The research Paul completed on students' progress toward science came out in 1955 as *The Gifted Student as Future Scientist: The High School Student and His Commitment to Science* and was republished in 1981, without the subtitle. Until the National Science Teachers Association in 1989 published *Gifted Young in Science: Potential Through Performance,* Paul's 1955/1981 volume was the only one devoted to suggesting means to encourage students to develop gifts in science.

A stunningly gifted teacher and mentor himself, Paul was convinced by William Jovanovich that he could make a bigger difference in American education by writing, editing, and publishing both textbooks and other volumes than by working with individual students in individual classrooms, and he therefore turned to publishing. Before and during his development of curricula and instructional materials, Paul's publishing career was flourishing. He became president of Harcourt Brace Jovanovich's Center for the Study of Instruction (San Francisco) and its director of Research in Curriculum and Instruction; later, he was director and editor-in-chief of the School Division; finally, he was copublisher of Research-Based Publications.

Among his most widely distributed books were those making up the series *Concepts in Science,* best-selling, grade-specific texts published by Harcourt, Brace,

---

[5] His own precollege education was based on the elimination of almost all other children from access to his opportunity, he noted with both regret and anger: "In Austria, all children take, at a very young age, a brutal test. Then, all but 15 percent are eliminated from high school. Those 15 percent go to *Gymnasium.* I went to something even more selective, called *Realgymnasium.* ... The others, the vast majority, don't have a chance. It's better here in America, but inequalities still abound."

and World (and later by Harcourt Brace Jovanovich) that transformed the teaching of science in US schools. (See Part IV, Appendix B, "Textbooks and Series," and the essay by E. Jean Gubbins in Part I.)

Paul was well aware of the limitations of the lecture-textbook, laid-out laboratories process, which his many classroom visits taught him were the norm in American science education. Partially because of this concern, in the late 1950s and early 1960s, he joined other scientists and educators nationwide on the Sputnik science project, which worked to change science education in response to that "educational crisis"—one, Paul dryly noted, of a continuous and recurrent series. He served on the Steering Committee of the Biological Sciences Curriculum Study (BSCS), as chair of its Gifted Student Committee, and as consultant to the Physical Science Study Committee (PSSC). These committees developed programs of what Paul called "originative" inquiry, designed to interest high school students in science.

Profoundly committed to the American vision of education for all and considering the equal treatment of unequals both unfair and absurd, Paul worked especially to improve circumstances for the two groups of children whose needs he felt were most neglected—the disadvantaged and the gifted. "We do pretty well for the 80

Introduction xxiii

percent of the students in the middle," he once said. "But the 10 percent at the top and the bottom: We grind them under our feet!" The cornerstone of his philosophy was deeply democratic: He believed all young should be given equal access to opportunity, so that those *who freely chose, not those who tested high,* could select themselves for the original work through which—in science as in any field—they could discover their talent. "One is not gifted," Paul believed, "until one has given a gift. A creative person does a work; a gifted person may never do one," continuing, "A gifted person is a promissory note*; he*[6] *must give a gift, do a work, to become creative.* He is a cashier's check when he becomes creative."

Believing, as he put it, that "the value of a person's advice about teaching is inversely proportional to the square of the distance he or she is away from the classroom," Paul visited, in the course of 30 years, classes in about 600 schools and interviewed some 120 administrators and some 2,000 teachers in America and on four other continents to observe teaching and learning firsthand. Paul's sense of humor and humility informed his refusal to lecture rather than to teach, but did not undercut his gentle but unwavering commitment to his fundamental mission: The best education takes place when an "ecology of achievement" results as "the school-community ecosystem acts in mutualism with cultural and university ecosystems" (1995, p. 115).

On the possibility of the federal government unifying the U.S. educational "system," first of all, Paul was clear that there *was* no American educational system.[7] Paul said that when President Kennedy asked if Paul could help create one, "I told him of what happened when the Nazis set up an 'educational system' that controlled texts and, therefore, gained the ability to perpetuate—for a time—lies," and added, "American schooling is the best. (The 'disorder' works.)"

Educational Researcher and Publisher

Paul's research in this area culminated in *Memorandum: On Renewing Schooling and Education* (1981). There, and throughout his career, Paul emphasized that "education" is made up of much more than "schooling," a distinction that goes back as far as Plato, whose

> ... view of education gave only passing attention to schools; he did not equate schooling with education. Indeed, for him, and for scholars in the centuries following, schooling was a useful part, but certainly not all, of education. It was, Plato insisted, the community that educated, that shaped mind and character, vocation and avocation. (1981, p. 10)

---

[6] A man of the last century, Paul used the masculine pronoun, often over my objections, to mean human, men *and women.* Indeed, as several of his former students commented, his egalitarian approach to women's capabilities was far ahead of his time. So were his passionate attacks on racism.

[7] Any more than there is a "scientific method."

xxiv D.C. Fort

Paul noted that he rarely found an effective school in a poor neighborhood or a severely flawed one in an affluent area. Failed educational ecologies, he noted, should not be blamed on teachers, noting,

"We ask of them. . .everything.
"We pay them. . .nothing.
"And we give them. . .dark and dreary scoldings."

Paul's visits to schools, where he observed classes and met with science teachers and administrators again and again, saddened him in that the teachers had so much to do—run lunchrooms, monitor halls, oversee recess, fill out forms—that they had almost no time to live, much less to do their own work. In the welter of demands, they rarely had the time to give especially interested students (or especially neglected ones) the care they needed. In spite of the odds stacked against teachers, however, many succeeded splendidly: Teaching remains, he said, "a personal invention. It is a performing art. It is a mercy. Teachers help students cross their Rubicons. We are the last line of defense against meaninglessness, against chaos."

Though not formally professionally affiliated with a single college or university, Paul was no stranger to higher education, speaking, consulting, and teaching at many graduate and undergraduate institutions nationwide and internationally. His awards include several honorary degrees and many citations and honors from organizations devoted to science, the humanities, and teaching.

Environmentalist

Out of his lifelong interest in conservation, Paul became education director and, later, codirector of the Pinchot Institute for Conservation Studies at Grey Towers, in Milford, Pennsylvania (1954–1966). Its proximity to the Brandwein home and property in nearby Greenville, New York, probably influenced Paul only minimally to take these posts. He was one of the world's first long-distance commuters, traveling nation and worldwide as a matter of course, generations before such mobility became the norm.

He and his wife long planned to bequeath their property (as the Rutger's Creek Wildlife Conservancy) to an organization committed to students, teachers, scientists, historians, staff, and volunteers interested in the environment and natural systems. The Brandwein-[Evelyn] Morholt[8] Trust, a 501(c)(3) nonprofit organization, was

---

[8]Evelyn Morholt (1914–1995), a former science teacher with Paul at Forest Hills High School and an old friend of both Mary and Paul, bequeathed her residence to the Brandweins in 1994. After her death, the Institute turned her house, which is close to the Brandwein residence, into a laboratory.

Over the course of her long career, Morholt served as editor of *The Teaching Scientist* (Federation of Science Teachers, New York City), chair of a New York City high school science department, and acting examiner for the New York Board of Education. She wrote nine books used in the schools; the most recent (in 1986, in collaboration with Paul), *A Sourcebook for the Biological Sci-*

Introduction

formed shortly before Paul's death. Paul's widow and Morholt decided to ask the Pocono Environmental Education Center (PEEC) in Dingmans Ferry, Pennsylvania, to administer the conservancy. This collaboration led to the establishment in 1996 of the Paul F-Brandwein Institute, of which Mary Brandwein was the first chairwoman and John "Jack" Padalino (of PEEC), the first president. The Institute continues its work to this day. (See the chapter by Wheeler, Padalino, and DeWall in Part I.)

One Legacy of Paul F. Brandwein: Creating Scientists

Besides this introductory material, this book comprises four parts:

- "Remembering Paul F. Brandwein" includes essays by 12 former students (mostly now scientists), essays by 9 educators (mostly but not exclusively science educators), and one essay from an alumna who did not choose science or teaching as a career.
- "Paul F. Brandwein in His Own Words": reprints of three seminal publications, spanning 40 years.
- The 26 surveys filled out by people Paul influenced and gathered by James P. Friend, Richard Lewontin, the late Walter G. Rosen (three Brandwein alumni), and me are analyzed here.
- Two appendixes, which contain a copy of the survey and several bibliographies of Paul's publications in English—first, essays and books (except for textbooks); second, textbooks and series; and third, audiovisual contributions.

Remembering Paul F. Brandwein

Five of the Brandwein student alumni—former director of the Lawrence Berkeley National Laboratory Andrew M. Sessler; California psychiatrist Josephine Baron Raskind von Hippel; Alexander Agassiz Research Professor (Harvard University) Richard Lewontin, a population geneticist; Professor Emeritus of Atmospheric Chemistry (Drexel University) James P. Friend; and former ombudsman at the World Bank Barbara Wolff Searle—contributed their memories of Paul and his influence upon them at some length. Seven other alumni were more concise. I have gathered their short contributions into a subsection—"Brief Encounters, Lasting Effects." These alumni comprise Professor Emeritus of Geological Engineering Richard Goodman (University of California, Berkeley); U.S. Department of Agriculture Researcher Tom Schatzki; Professor of Immunology Lisa A. Steiner (Department of Biology, the Massachusetts Institute of Technology); Herbert L. Strauss, professor

---

ences (3rd ed.), which is highly regarded by science teachers in spite of its age (now over 20 years old). Out of print, it still fetches between $40 and $100 on the Internet, and it may be digitized for the use of the many high school teachers who find it valuable (some call it "the Bible").

in the graduate division, University of California, Berkeley; Walter G. Rosen, among whose many hats was one dedicated to science education; and precollege teachers Naomi Goldberg Rothstein, who taught all subjects at all levels; and award-winning middle school science teacher Joanne Gallagher Corey.

Several of Paul's colleagues contributing essays to Part I focused on his interest in conservation of the environment, as well as attending to other concepts. These writers include Audubon Scientist and Environmentalist Charles E. Roth, the late Professor of Resource Planning and Conservation in the School of Natural Resources and Environment Emeritus William B. Stapp (University of Michigan), and the late Environmental Education Director for the California Department of Education Rudolph J. H. "Rudy" Schafer. Writers looking at Paul's contributions to science education in a broader sense include Robert W. Howe, former chair, Department of Biology, University of Wisconsin—Green Bay, and Sigmund Abeles, former science consultant to the Connecticut State Department of Education and the New York State Education Department. Two essays examine Paul's cooperation with science education institutions: Marily DeWall, science education consultant, secretary to the Paul F-Brandwein Institute, former director of professional development at the JASON Foundation, and former NSTA associate executive director, writes about Paul's role in NSTA and other science education organizations. DeWall also joins John "Jack" Padalino, president emeritus of both the Pocono Environmental Education Center and the Brandwein Institute, and current Institute President Keith A. Wheeler, who also serves as president of the board of directors of the Foundation for Our Future, to discuss the work of the Paul F-Brandwein Institute. Finally, E. Jean Gubbins, associate director of the National Research Center on the Gifted and Talented and associate professor of educational psychology at the University of Connecticut, looks at Paul's philosophy of teaching and learning in general.

Paul F. Brandwein in His Own Words

The next section of this book reprints three of Paul's important works, spanning four decades. The first, *The Gifted Student as Future Scientist* (1955/1981), is the seed from which this book originally sprang as James P. Friend, Richard Lewontin, the late Walter G. Rosen, and I tried to check Paul's theories about what can be done to encourage the young to turn their talents to science. Paul's data were gathered during his years teaching science at Forest Hills High School between 1944 and 1955. Ours were based on surveys filled out by his surviving students (see Part III for analysis of these data).

The next reprint, "Science Talent: In an Ecology of Achievement," from *Gifted Young in Science: Potential Through Performance* (1989), jumps ahead about a third of a century to trace the evolution of his theories from a vantage point with the benefit of hindsight. Those young who freely choose to do science, no matter what the commitment of time and, on occasion, discomfort, are the ones who will likely make a similar decision in adulthood. But it is *through* work, be it experimental or theoretical, that their commitment is discovered.

Introduction   xxvii

The third reprint, the executive summary from *Science Talent in the Young Expressed Within Ecologies of Achievement* (1995), offers Paul's culminating conception of both what can hinder and what can facilitate the development of science talent in the young. He defines, among other phenomena, three essential ecologies necessary to this process: "Perhaps the family-school-community, college-university, and cultural ecosystems would contribute to the brilliance of the world if, in their interconnectedness, they would lend their collaborative resources to *all* young who aspire and are capable of achieving" (p. xii).

The Surveys: Protégé(e)s and Colleagues Reply

In Part III of this book, I analyze the data gathered on the 26 surveys that originally inspired this book. They include 18 from Brandwein alumni who became scientists, 6 from colleagues in education, and 2 from alumnae who went into nonscientific fields.

The Bibliographies and the Survey

Appendix A, in Part IV, reprints the original survey.

The bibliographies in Appendix B, the last section of *One Legacy of Paul F. Brandwein,* fall into three categories: The first lists almost a hundred of Paul's essays, articles, and books, excluding textbooks; the second summarizes—insofar as it is possible since the dissolution of Harcourt Brace Jovanovich in the late 1980s—his textbook contributions, including *Concepts in Science, Science and Technology,* and others; the last categorizes audiovisual contributions. Because my collaborator, Suzanne Lieblich, and I could not physically examine all the books and other materials listed, we could not always determine whether Paul was author, editor, or both of the textbooks and similar materials. We have made no effort to find Paul's publications in languages other than English.

References

Brandwein, Paul F. (1955/1981). *The gifted student as future scientist: The high school student and his commitment to science.* New York: Harcourt, Brace. (Reprinted in 1981, retitled *The gifted student as future scientist* and with a new preface, as Vol. 3 of *A perspective through a retrospective*, by the National/State Leadership Training Institute on the Gifted and the Talented, Los Angeles, CA)

Brandwein, Paul F. (1981). *Memorandum: On renewing schooling and education.* New York: Harcourt Brace Jovanovich.

Brandwein, Paul F. (1989). Science talent: In an ecology of achievement. In P. F. Brandwein and A. Harry Passow (Eds.), Gerald Skoog (Contributing Ed.), and Deborah C. Fort (Association Ed.), *Gifted young in science: Potential through performance* (pp. 73–103). Washington, DC: National Science Teachers Association.

Brandwein, Paul F. (1995). *Science talent in the young expressed within ecologies of achievement* (Report No. EC 305208). Storrs, CT: National Research Center on the Gifted and Talented. (ERIC Document Reproduction Service No. ED402700)

Fort, Deborah C. (1998). *Science in an ecology of achievement, November 13–16, 1997.* Dingmans Ferry, PA: Paul F-Brandwein Institute.

Gowan, John Curtis, and Demos, George D. (1964). *The education and guidance of the ablest.* Springfield, MA: Thomas.

Hammerskjöld, Dag (1964/1973). *Markings.* New York: Alfred A. Knopf.

McKenzie, Floretta Dukes (1984, Spring). Education, not excuses. *The Journal of Negro Education, 53*(2), 97–105.

Part I
Remembering Paul F. Brandwein: Essays

Brandwein Alumni

Turning a Dream (Deftly, Subtly, and Effectively) into Reality

Creating Scientists

Andrew M. Sessler

From the earliest times that I can remember, I had a dream to become a scientist. My parents were schoolteachers, so they instilled in me at an early age a respect for scholarship. My father taught biology and, consequently, often called my attention to natural phenomena.

In the fifth grade, I had a wonderful teacher who allowed me to instruct the class, for perhaps a half hour each week, on some scientific project. Often my "lectures" were tied to a demonstration using a set of apparatuses that my father had brought home that week. In the later years of elementary school (grades seven and eight), I studied high school texts in science.

I had a childhood friend who dreamed of becoming a medical physician as I dreamed of becoming a scientist. We thought that scientists did not do well financially (which actually was true during those Depression years), but physicians did, and he agreed to help support me in future years. As it developed, he did not get into medical school, and I have enjoyed a financially rewarding career in science.

Completing elementary school, my dream still strong, I wanted to go to one of the science magnet schools in New York City (Stuyvesant or Bronx High School of Science). My arithmetic teacher refused to write me a letter of recommendation because, she said, I was too poor a student ever to become a scientist. Stimulated by this put-down, I learned algebra on my own, passed the entrance examination, and was admitted to one of the magnet schools. Having "shown the teacher," I elected to go to the local high school, for it called for a shorter commute and was coed. Forest Hills High School was rather new in those days. (I attended from the fall of 1942 until June of 1945.) In those war years, it was possible to study on one's own, take the final exam, and get credit if one passed. Thus, I was able to finish in 3 years. There were many excellent teachers in the school, probably better ones than in subsequent years, for, during the Depression, teaching offered a secure job with a reasonable salary. In those special years, many who in later years went on to graduate school and far better paying positions became teachers. Those of us who were students just after the Depression experienced a quality of teaching that has never since been matched.

Dr. Brandwein took me under his wing, perhaps because he knew my father. For some reason, however—perhaps jealousy—they were professional colleagues, but never friends. I do not actually remember how it happened, but soon I was working—during my free period and before and after school—preparing the demonstrations used in various science classes as well as preparing material (such as 30 microscopes) for individual students to use. This work—wonderful training for a beginning student interested in science—went on throughout the time I was in high school, always with new kids involved in the work, but kids under the supervision of "old hands," that is, students in their second year.

At the end of my freshman year, Dr. Brandwein took me aside and gently, subtly, deftly, told me that my grades (in the 70s and 80s) would not allow me to become a scientist. It was just one brief conversation, but it was enough. The following semester I got three 98s, a 99, and a 95. I had learned that getting good grades was necessary for success (even if you thought the subject was uninteresting, beneath one, or just stupid). I never forgot the lesson and from then on always tried to do well in my courses.

In my second year, Dr. Brandwein suggested a victory garden soil-testing project. Soon a number of us were out collecting samples of soil (typically arriving by bicycle), analyzing them, and sending a written report to the owner as to what he/she should do so as to improve production of vegetables. At first we used a commercial kit, but we needed to test many samples, and so it was economically prudent to buy the reagents in bulk. We had to learn how to find out what the various chemicals were, how to order them, and how to decant from large bottles to small ones. The whole project was excellent training in sample gathering, laboratory work, and writing a report. Once again, wonderful training for budding scientists.

Those of us under Dr. Brandwein's tutelage, perhaps 20 kids, became a social group. We had parties, spent time together, and enjoyed being with each other. Most of us were too young to go on dates, so the relaxed atmosphere, associated with just hanging out, was attractive to us. Besides learning science, we were acquiring social skills.

At the end of my sophomore year, Dr. Brandwein suggested to me a science project that would soon become my Westinghouse[1] project. This was the first time someone from Forest Hills even entered that competition. Certainly, I did not even know of its existence. The project involved pill bugs, and my job was to learn how to raise them in the laboratory and then (but I never got this far) to see if I could use them for some biological study. So, now, training in actual research.

In the competition, which consisted of a written examination and evaluation of one's research project, I did well and became a finalist. Subsequently, many Forest Hills kids—I suspect all under Dr. Brandwein's influence—won Westinghouse Awards, but mine—being the first one—showed that it was within our grasp and, I believe, made it easier for those who followed.

Because of the Westinghouse, good grades, and probably good recommendations, I gained admission to Harvard. So, at age 16, with an advanced understanding of what was needed for scientific research, I entered the school immediately (i.e., in June 1945). I was going to major in biology, but a first course in botany turned me

---

[1] Now the Intel Science Search.

off (certainly not the proper way to choose—or not to choose—a lifetime career). I turned to mathematics. My honors teacher (a very fine mathematician) who worked individually with me for a great many sessions (comparable to what we often do with graduate students) gave me the impression that I was not very good in mathematics.

Consequently, when I graduated from Harvard in 1949 I went into physics (the nearest thing to mathematics). I did my graduate studies at Columbia University, where I had many fine mentors. For example, I had courses from nine professors who had, or would receive, Nobel Prizes. What an education!

Many years later, I met the Harvard honors professor of mathematics and told him how I was doing "okay" in physics, and it was very good that I had left mathematics, for he had shown me that I would not have done well in that field. He was surprised by my comment and responded that I was one of the best students he ever had. The contrast with Dr. Brandwein's advice need not be emphasized.

Following my student days, I have taught at the university level, been active in science all these many years (having written more than 400 research papers), been director of the Department of Energy's Lawrence Berkeley National Laboratory (1973–1980), been president of the American Physical Society (1998), and been elected to membership in the National Academy of Sciences (1990). I have also served on many committees, been active in human rights matters, worked with various physics and society matters, and, along the way, won a number of prizes. It has been—and is being—a full and rewarding scientific life.

My debt to Dr. Brandwein is too great for me to ever have paid back, but then he never asked for, or desired, "pay back." His reward was helping students succeed, and that he did with a masterful, gentle, subtle touch—a gesture that needed no reward. But his manner of teaching should be studied and transmitted to others so that they may learn what true mentoring and teaching is all about. I believe that is what Dr. Brandwein really wanted.

A youthful dream was turned into reality. And it probably would not have happened without Dr. Brandwein's influence. Actually, I am sure it would not have happened without Dr. Brandwein's influence.

Andrew M. Sessler (AB, Harvard University, MA and PhD, Columbia University) spent 7 years on the faculty of Ohio State University, during which time he interacted with the Midwestern Universities Research Association, and, in 1961, joined the Lawrence Berkeley National Laboratory, which he later directed and where he still works. He has published over 400 scientific papers, for which he has received a number of awards, including the Lawrence Award and the Wilson Prize. He is a member of the National Academy of Sciences and a former president of the American Physical Society. He has served on many national committees and has been active in arms control and human rights. For the latter, he was the first winner of the Nicholson Award. He is active in the Union of Concerned Scientists and Amnesty International.

How Dr. Paul Brandwein's Mentorship and Guidance Affected My Scientific Interests and Career

Josephine Baron Raskind von Hippel

Dr. Paul Brandwein was my science teacher and mentor when I was a student at Forest Hills High School, in Queens, New York, from 1942 to 1946. I learned long after graduating from Forest Hills that our school had apparently been chosen by the New York City School System to receive some especially excellent teachers, perhaps as a demonstration project to determine whether being taught by unusually qualified teachers would lead to increased student achievement. One of these teachers was Dr. Paul Brandwein, an unusual high school science teacher in many ways, in part because he had earned a PhD. There were then few people with doctorates in science working as teachers in New York City's public high schools.

Dr. Brandwein had the ability to recognize students of above-average intelligence who were hardworking and independent, several of whom, including myself, were also friends with one another. Most of these students, like the majority of those at Forest Hills High, were from lower-middle-class families, and their parents often did not have college educations. My background was unusual in that both of my parents were physicians. My father was an orthopedist; my mother, a psychiatrist and psychoanalyst. They were both intelligent and well educated. My father graduated from Yale University, and my mother from Bryn Mawr College.

I skipped the second grade and was in Opportunity Terman (a class for gifted students) when I was in the sixth grade at P.S. 150 in Queens. I think that my verbal skills were superior from a very young age, and my sensory and neuromuscular controls were more than adequate. (I am a good athlete.) I didn't labor too much. Luckily, my achievements came easily. I was, however, taught that work is a necessary part of life, and good grades were expected. I am more an idea person than a persistent working-through one. I was lucky that my home environment did offer financial comforts with no worries about money and sustenance. My parents were not only supportive; they expected academic success. They were not unrealistic in their expectations in my case. I don't recall any specific unfulfilled emotional needs. My warm, kind, loving parents worked hard at professions that they enjoyed.

I developed an early interest in nature because I went to a summer camp, where we were taught the names of trees, flowers, ferns, and so forth. I loved it! I have always loved nature and biology. When I took biology and chemistry at Forest Hills, Dr. Brandwein turned me on to research and research methodology. He encouraged me to work as a teacher's aide and lab assistant. It was the job of the student aides to

D.C. Fort, *One Legacy of Paul F. Brandwein*, Classics in Science Education 2, DOI 10.1007/978-90-481-2528-9_2, © Springer Science+Business Media B.V. 2010

bring science equipment for lectures and experiments to the teacher's classroom and to return the equipment to the lab after the lecture. Dr. Brandwein worked with us after school in the lab and in the school museum. I recall one especially memorable occasion when I went to pick up a demonstration from a classroom to return it to the lab and the teacher turned and placed a 3- to 4-foot snake in my arms. It was expected that I take it back to the lab—so I did! It was, however, somewhat of a surprise.

I suppose Dr. Brandwein must have recognized something about me that indicated to him that I was interested and could succeed in a scientific career, because he certainly encouraged me. There were a few of us whom he encouraged to do original experiments and to apply for the Westinghouse Science Talent Search competition. He suggested that I research the embryology of the *Physa* (a pond snail), because that study had apparently not previously been done. I raised the snails and observed their eggs and developmental stages under the microscope. My observations were accompanied by drawings of what I saw. Luckily (and amazingly!), along with Richard Lewontin,[1] also from our class, I was one of 40 national finalists of the Science Talent Search. It was a fantastic experience. We went to Washington, D.C., and met some prominent scientists. We had tea at the White House and met some very famous people, including Lisa Meitner and Edward Teller. Mrs. Harry Truman was also in the receiving line; I politely shook her hand but had no idea who she was until someone later identified that nice little old lady as Bess Truman. We also visited the Smithsonian Institute and other interesting places in the city.

Being a finalist in the Westinghouse (now Intel) Science Talent Search was probably the most exciting and moving experience of my high school years, and it would not have happened to me were it not for Dr. Brandwein. I still remember the whole experience in great detail. I made friends with some of the finalists from other high schools in New York City.

Dr. Brandwein's PhD research was on oat smut, a disease of oat grass, and this fact may have influenced my desire to go to Cornell University to study agricultural science, even though I eventually chose to go to Bryn Mawr College in Pennsylvania (my mother's alma mater) instead. One of my good friends in high school, who had interests in writing and poetry rather than science, wrote a short poem about Dr. Brandwein that included these lines:

I do not love thee Sir Oat Smut.
Your dimples fascinate me, but
I do not love thee Sir Oat Smut.

Becoming a Westinghouse finalist was obviously a boost to my self-esteem, and I knew then that I really did want to do research. This interest continued in college, where I majored in biology and did a research project in my biology class. At Bryn Mawr, my science teachers were all excellent and inspiring and tough as nails in their expectations. They included biologists Jane Oppenheimer (whose

---

[1] See his essay in this volume.

major interest was embryology) and Mary Gardiner, chemist Ernst Berliner, and physicist Walter Michels, all PhDs.

Another teacher at Forest Hills who was inspiring and a mentor to me was George Schwartz, our physics teacher. A kind and warm person, he also influenced me to choose a scientific career.

After graduating from Bryn Mawr, I was torn between research and going to medical school. As luck would have it, I didn't get into any of the three medical schools to which I applied (Harvard, Yale, and Columbia), and so I ended up doing research in biochemistry as a special graduate student at the Massachusetts Institute of Technology. (In those days most medical schools admitted only one or two women. It is different today. Most medical school classes enroll 50 percent or more women.) The Massachusetts Institute of Technology required that each student do an original piece of research and write a thesis in order to get a master's degree in biochemistry. My research in enzymology, under the direction of Dr. Irwin Sizer, dealt with "The Effects of Peroxidase on Invertase." When my lab mates began writing "Visiting Hours: 2 to 4 PM" on my lab door, however (because so many other students liked to talk with me), I realized that I was probably headed in the wrong direction and decided medical school was right for me after all.

After my first 2 years of medical school at the Women's Medical College of Pennsylvania I transferred to Harvard Medical School (8 women in a class of 120), did my internship and residency training in medicine and psychiatry, and became a psychiatrist. I suppose Dr. Brandwein would have called that "self-identifying." In retrospect, if I had received better mentoring in graduate school at the Massachusetts Institute of Technology, I might well have opted for a career in research or at least in academic medicine. Frankly, however, I think that I made the right choice for my career. I do think that it is more difficult for women to combine research and raising a family than to combine a medical career with family duties. Obviously, there are a few women who can wear research and domestic hats simultaneously, but the stretch is not easy, and if your research career is postponed, you quickly fall far behind your peers in knowledge and technology. In my day, women followed their husbands because mostly their careers came first. Today priorities can be different, and often the woman's career decides where the couple will live.

I did my internship and residency at George Washington University Hospital, in Washington, D.C. At the time, my husband, Pete, was enrolled in Officers' Candidate School in Newport, Rhode Island. He returned as a naval officer to do research at the Naval Medical Research Institute in Bethesda, Maryland.

Circumstances determined many of my lifetime decisions. While I was standing in a lunch line at the George Washington Hospital one day, the assistant chair of medicine offered me the job of chief resident in medicine in the outpatient department of the hospital. Oddly, given medicine's bias against women doctors at mid-century, he probably picked me in part because I was a young married woman. In that job, he said, I could start having a family and wouldn't need to take emergency calls with 24 hours on and 24 hours off. I immediately said, "Yes!—OK—I'll do it." For the next year, while pregnant with my firstborn (now a freelance energy consultant), I taught fourth-year medical students at George Washington University

Medical School and ran the outpatient clinics. In the late 1950s, medicine was much more thorough than it is today, with all of our low-income patients receiving EKGs, sigmoidoscopies, lab work, and chest X-rays as regular parts of their yearly medical checkups.

In large city hospitals, low-income patients were treated and examined gratis by fourth-year medical students, interns, and residents, under the supervision of the attending doctors, who were largely unpaid for their work. This kind of service is rarely offered in private practice, and there are fewer hospital clinics today than in the past.

After my medical residency, our family moved to Hanover, New Hampshire, where Pete became an assistant and then an associate professor of biochemistry at Dartmouth College. I got a job at the Hitchcock Clinic doing part-time psychiatry with personal supervision from the head of the psychiatry department. I was able to raise our three children at the same time, since I did not have to take calls 24/7, and my job was part-time. Eight years later we moved across the country to Eugene, Oregon, where we both decided that we would stay forever, if possible. Pete had about 28 job offers, so his choice of the University of Oregon's Molecular Biology Institute is something of an honor both for that facility and for Eugene. After 6 months of part-time work in general medicine at the University of Oregon's Student Health Service, I opened my own office in the solo practice of psychiatry in 1968. After 30 years of private practice, outpatient *and* hospital work, *and* taking emergency calls, I retired in 1998. I would have continued to work in private practice were it not for having to take emergency calls night and day, which—unlike psychologists, counselors, and scientists—doctors *have* to do. In a small town like Eugene, we are responsible for our patients 24 hours a day, 365 days a year, unless we get someone to cover for us, and then we have to cover for them if they go on vacation or out of town.

I loved my work and was used to being super-scheduled all my life, so I immediately made a new schedule for myself when I retired. For the last 10 years, I have been working as a nature guide for schoolchildren at our local arboretum. We guide kids from kindergarten to fifth grade and take 1,500–3,000 school children from our county through the arboretum during the spring and fall of each year. We teach topics like life cycles, habitats, trees, water cycles, and ecology. We get kids thinking about biology, chemistry, and physics as they all apply to the above topics. It is lots of fun for us mentors, and the kids really enjoy it too. Maybe it helps start some of them on the road to science and reasoning as well, something that would have delighted Dr. Brandwein.

Also, every Monday morning, I work with a group at a local mountain park called Mt. Pisgah, doing habitat restoration and trail clearing. My original interests in biology and botany have blossomed again, and I am a member of the Native Plant Society, the National Butterfly Association, the Lane County Audubon Society, and the Eugene Tree Foundation (we plant trees). These groups all have meetings and field trips

I'm still learning. It's great! I hike almost daily and play tennis both socially and in competitions through tournaments organized by the United States Tennis Association.

My greatest professional and personal contribution was (and is) to help people to understand themselves, to increase their potential for work, and to help them enjoy life fully. These processes also strengthen their abilities to relate to and help their families and friends, because they themselves are stronger and more self-confident.

I have worked as a physician for 42 years, 37 of them in the practice of psychiatry and psychosomatic medicine. I think that I have helped a great many people during those years, and former patients still call and write to me. At present I do some psychiatric consultation as a volunteer at our local Volunteers in Medicine Clinic. This work, too, is particularly satisfying, because the patients at this clinic are people who are not financially able to access *any* health care under our present system. They are the working poor.

I thank Dr. Brandwein for his scientific influence on my life and hope that his example will help other science teachers to mentor their students equally well.

How to Win Converts and Influence Students

Richard Lewontin

In the creation of a work of pottery, not only the intent of the artist and the actions of the potter's hands but also the properties of the material from which the object is being made must be taken into account. Those properties extend beyond the immediate state of the clay as it is being molded as well as the way in which that material changes as it is fired and as it ages. There is a reciprocal interaction between the materials and the artist. The potter chooses different clays for different purposes and alters his/her pressure on the material and the time and temperature of its firing to suit the properties of the particular clay. The wrong method for the wrong clay does not succeed. But beyond the success or failure of the production process, the shape that is made constrains only weakly what the object's contents will be. From the look of a bowl one cannot predict with any accuracy the history of what it will contain during its lifetime.

If we are to understand both the immediate and long-term effects that Paul Brandwein had on students, we need to consider not only his technique of teaching, the content of what he taught, his behavior toward students, and the image he projected but also the nature of those who were acted upon. Brandwein knew and understood with whom he was dealing. So, to begin, we must look at the properties of the materials.

The most important—but easily overlooked—characteristic of most of the students with whom Brandwein dealt was that they were high school students, that is, *adolescents*. It would be a mistake to suppose that the methods he employed and the persona that he projected could be transferred to either elementary school or university contexts.[1] As adolescents, students are in the process of conscious self-formation in a social context. In the environment of the high school, groups form that constitute the main social milieu in which students spend their daily lives and that will have a major impact on their future attitudes and ambitions. Adolescents are in search of role models and are particularly susceptible to the influence of charismatic personalities, susceptibilities reinforced in their social interactions within the group in which they find themselves.

---

[1] Brandwein also worked with elementary school young and with postsecondary students, but it is teenagers about whom I am writing. Of his interactions with younger and older groups I have no knowledge.

D.C. Fort, *One Legacy of Paul F. Brandwein*, Classics in Science Education 2, DOI 10.1007/978-90-481-2528-9_3, © Springer Science+Business Media B.V. 2010

An extremely important influence on the attitudes and ambitions developed by individuals in these adolescent groups is the social class from which the majority of their members come. Paul Brandwein spent the major and most successful part of his teaching career at Forest Hills High School. The school, newly founded just before the outbreak of World War II in Europe, was originally characterized by having a large number of middle- and upper-middle-class families. It drew its students partly from similar neighborhoods, such as Kew Gardens, but it also served a lower-middle-class population from other adjacent neighborhoods and included black students from working-class families within its district boundaries.

This demographic distribution changed around 1943,[2] when the neighborhoods from which most of the black and working-class students came were excluded from the school district, and Forest Hills High School became more homogeneous both on the basis of race and class. This segregation by social class was mirrored within Forest Hills High by a further subdivision into three internal schools distinguished by whether their members were to be, for the most part, recipients of "academic," "general," or "commercial" diplomas. Within the (academic) largely middle-class student body were a number of families of European refugé(e)s from Nazism. For the most part, these families were originally well-to-do German and Austrian professionals or upper-middle-class entrepreneurs, whose cultural and educational levels were quite advanced. It was a standard joke that many of them had escaped from Germany with nothing but their Rembrandts. Although these families provided only a minority of the students influenced by Paul Brandwein, their impact on the social groups to which they belonged was considerable. It was typical, for example, that when the group to which my future wife and I belonged went to the local soda fountain to pass the time, we talked about Freudian psychology, and we often spent our weekends at the Museum of Modern Art or in the medieval collection at the Cloisters.

Although it would be impossible at this time to demonstrate objectively, the teaching staff at Forest Hills High School appeared to be elite compared to that of other New York schools. Several teachers, like Brandwein, had PhDs in a variety of academic disciplines, including biology and mathematics, and had gone into high school teaching rather than into their scholarly professions as a result of the economic depression of the 1930s. The principal of the school, Dr. Michael Lucey, had previously led another elite New York high school, Julia Richmond.

This teaching situation—both the nature of the educational environment and the presence of a large group of students destined for higher education and professional life—provided Paul Brandwein with the opportunity to carry out his unusual educational program simultaneously in the classroom and in the social group of students that formed around him.

What made Brandwein's approach to education possible was a quality that is easy to name but difficult to describe or analyze. He had *charisma*, that quality

---

[2]Brandwein taught at Forest Hills from 1944 to 1954.

of being at the same time far above us and intimately close and approachable, of being at once a distant model, yet one of us. He had great charm, not the sycophantic charm of someone anxious to please, but the charm radiated by a superior being whose presence was inspiring. There were many excellent teachers at Forest Hills in English, foreign languages, history, mathematics, and science, teachers who approached their subject matter in an intellectually stimulating and revelatory manner. Chief among these was George Schwartz, an extremely intelligent, decent, and knowledgeable biology teacher, who took a fatherly interest in my own education. He was, however, a modest person who did not fill the space around him with his aura, and so, for an impressionable adolescent, not intellectually seductive. Modesty and charisma are not easily combined in the eyes of a 16-year-old.

Paul Brandwein was able to deliver a program of education that reflected an intellectual hauteur, even hubris, a program that he could carry off despite its unconventionality. He was able to use his position as chairman of the science department to appropriate resources for the coterie of students who surrounded him and to tailor his classroom practice to reflect his own view of what was important in biology. That program consisted of five elements.

First, in his formal classroom instruction he created his own curriculum and informed the class of what he was doing. He assured the students that at the end they would have all the knowledge required for their performance on the standard New York State Regents examination that all students would have to pass in order to graduate as well as the facts they would need to perform well on the College Board examinations. His announced intention to ignore the rules and teach us the important truths appealed directly to our adolescent urges to rebel against the conventional wisdom of our parents. Moreover, it created in us a feeling of original intellectual work and the sense that we happy few were privileged sharers in an important enterprise under the aegis of a great man. He then brushed aside the official requirements as a basis for teaching and reconstructed biology as a coherent, intellectually challenging, and, most remarkably, socially relevant discipline.

One of the major changes he made in the curriculum was the introduction of Darwinism as a central concept in biology. In the early 1940s and before, the subject of evolution was not part of the American biology curriculum. Presumably this was a reflection of the antievolutionary ideology of religious fundamentalism that was then widespread—and is having a renaissance now—in the United States, which the outcome of the Scopes trial had done nothing to change. It was not until the work of the Biological Sciences Curriculum Study in 1958, in which Brandwein was intimately involved, that Darwinism became part of the standard American high school curriculum. As an irony of personal history, after Brandwein left high school teaching and worked as a senior officer in the publishing house of Harcourt Brace, he was forced by the exigencies of the market to keep evolution out of textbooks published by his employer so that their books would be adopted in states, like Texas and Florida, where fundamentalism remained politically powerful. The power he had as a high school teacher in New York to ignore contradictions between the official curriculum and his intellectual convictions could not survive the power of capital.[3]

Brandwein's second major change in the older content of biology teaching, a change shared in by other biology teachers at Forest Hills, was to examine the claims of biological racism. One consequence of the revelation of the effect of Nazism's racial ideology was a general revulsion in the 1940s against the widely accepted belief in the biological superiority of one race over another and a reexamination of the generally accepted claims about the biology of race. This had particular local resonance because of the significant number of students from German refugee families and the high proportion of Jews among the faculty. The biology curriculum at Forest Hills included the famous pamphlet *The Races of Mankind*, by the cultural anthropologists Ruth Benedict and Gene Weltfish, which had been originally written to distribute to American troops as part of the ideological war against Nazi racism. The point of this pamphlet and of the biology teaching at Forest Hills was that the physical biological differences between races were purely superficial and that mental and cultural differences were the consequence of historical cultural events.

The third element in Brandwein's approach to education, reflecting his own understanding of the science of biology, was a general skepticism of received wisdom. Science, and especially biology, was filled with claims that had either no concrete evidence to back them up or evidence gathered from experiments and observations with weak methodological standards. More than any other feature of his educational method, this demand for methodological rigor was the one that had the most influence on my own intellectual formation and professional practice. Up to the present, biology, and especially evolutionary biology, is filled with claims for which there is no acceptable level of proof or demonstration. This is a particular weakness of theories of human biological and cultural evolution, as exemplified by sociobiology, evolutionary psychology, and biological theories of evolution of culture. Brandwein's caution to students was that just because something was in a book, even a textbook, it could not be assumed to be true, a priori. This questioning of assertions made without adequate proof finds echoes in the adolescent's resistance to parental claims that base their validity only on the famous retort, "Because I say so, that's why!"

The fourth element of Brandwein's program stemmed from his recognition of the social group formation that was integral to students' maturation and socialization. He facilitated the formation of a student group with a social and intellectual identity by providing the group with a place provided with equipment linked to an intellectual purpose. Part of the physical facilities for teaching at Forest Hills was an inventory of supplies and equipment and a large preparatory laboratory, managed by a full-time staff member, where a variety of demonstration experiments were assembled for use in the classroom. Despite the understandable reluctance of

---

[3] By the early1980s, Brandwein's middle school biology texts included information on Darwin and evolution. Whether, if, and when it made it into the elementary school *Concepts in Science* series, I do not know.

the very able preparator, Bessie Lumnitz, Brandwein used his power as chairman of science to sequester one of the rooms in the laboratory as a kind of clubhouse for the exclusive and unsupervised use of a student biology club, of which he was the adviser. The room was equipped with laboratory benches, storage cabinets, and utilities. In addition to full-time reservation of the space and physical access to it from the opening until the closing of each school day, the members of the group had free use of the inventory of material and apparatus for their experiments. Each member of the group had his/her own bench and cabinet space, so that the "laboratory" became, in fact, a home base for the students during the day. Brandwein would visit the laboratory from time to time and engage in informal exchange, not only about the experiments being done but also about any subject that seemed interesting at the time.

The experiments were never mere laboratory exercises. They arose from discussions between the students and Paul Brandwein on particular unanswered questions in biology to which, at our level of knowledge, physical capability, and available time, we could provide a nontrivial contribution. In this context, Brandwein's principle of skepticism about poorly demonstrated claims translated into a physical reality.

My own experiment, which brought me to Washington as a finalist in the Westinghouse Science Talent Search, was a clear example. Brandwein had done his graduate work in plant pathology, and he was familiar with the textbooks on fungi. A leading text of the time, in its discussion of the formation of diploid sexual spores (zygospores) from the fusion of plus and minus strains of haploid mycelia, illustrated the germination of such a spore in a species of the mold *Phycomyces*. The standard theory of the time, still accepted in 2008, was that sexual spores were an evolutionary adaptation to adverse environmental conditions. Sexual recombination between strains that could not themselves survive the bad conditions would allow new genetic combinations to arise, some of which would enable the organisms to cope. The zygospores could indeed be induced to form in the laboratory when environmental conditions, were unfavorable for growth, for example on a nutrient-deficient medium culture. But the theory also required that these sexual spores would then germinate if subjected to a harsh environment. The drawing in the textbook showed such a germinating zygospore. Brandwein, however, said that he had never actually seen a zygospore of *Phycomyces* germinate under any of the usual unfavorable conditions and he was not at all sure that they ever did. He set me the problem of producing such spores and attempting to germinate them by subjecting them to a variety of conditions usually said to induce germination, such as high temperatures, acid or alkaline medium, desiccation, etc. Despite all my efforts, no zygospore that I produced could be induced to germinate.

Where the author of the textbook got the figure remained a mystery.

As an extension of the laboratory, the students also were a social group, walking home from school together, having parties on weekends, and going on outings and excursions, fictionally "field trips." More than one eventual marriage began as a friendship in our laboratory. One cannot imagine a more complete integration of social and school life.

Finally, quite aside from interactions with students in the classroom and in the biology laboratory group, Brandwein acted consciously as a mentor, a life model, and a father figure. We would sometimes walk home from school clustered around him, listening to his views and asking him questions as he went to his nearby apartment. He would, on occasion, accompany us on one of our "field trips," ostensibly as a source of natural historical information but primarily as someone we wanted to be with. We had access to advice from him on any subject whenever he was free. He was a surrogate parent and a source of great wisdom for adolescents who were often in the throes of rebellion and disenchantment with their own families. At least one member of the group left home and arrived on Mary and Paul's doorstep, asking to be taken in. The Brandweins behaved sensibly.

Paul Brandwein was conscious of his role as a mentor, as the master of a band of disciples. No doubt he was seduced to some extent by his own persona. Our idolization of him surely must have been a source of ego gratification to him. Yet his influence was always constructive. However much his self-image was increased and reinforced by his success as a teacher and model, he always used his skills to advance the interests of his students and their development into maturity.

Richard Lewontin is Alexander Agassiz Research Professor in the Museum of Comparative Zoology at Harvard University. After graduating in 1946 from Forest Hills High School, where he studied with Paul Brandwein, he obtained an AB at Harvard (1951) and a PhD from Columbia University (1954), having done his research on population genetics in the laboratory of Theodosius Dobzhansky. His laboratory research has centered around the molecular description of genetic variation in natural populations of organisms. His theoretical work has consisted of a variety of mathematical and statistical studies of the genetic changes in population genetic structures as a result of various evolutionary and genetic forces.

Paul F. Brandwein's Influence on My Life: The Essential Spark

James P. Friend

From my first encounter with Paul F. Brandwein as a sophomore in high school, he deeply affected my life and career as a scientist. For the first 13 years of my life, I had *no* formal educational exposure to *any* scientific field. Then, I took a routine and not very memorable introduction to science class in eighth grade. Another 3 years passed before Dr. Brandwein astonished and encouraged me in his biology class.

My Early Interest in Science

My earliest recollection of encountering science was in a volume of a set of books of knowledge my parents had bought about 1935, when I was 5 years old. The subject of one yellow volume, eventually much read and much battered, was science, and I have a vivid memory of a drawing depicting the attack of leukocytes on bacteria. I was most intrigued by the story of this dramatic battle carried on regularly within a human body, even though I could not understand many details. About 2 years later, in response to my pleading, my parents provided me with a chemistry set. Again, I was fascinated by the reactions that produced gases, precipitates, and colored solutions. I also discovered the value of following directions to achieve reproducible results.

Because of my parents' divorce when I was 8 years old, I lived for 2 years on a farm in a small New Hampshire town where I could find no books on science—even in the library—and I had no available materials designed to engage a youngster in science. I observed much of the workings of nature, however, through roaming the woods and fields and learning about how a dairy farm works. I learned to identify many species of plants, birds, and local wild animals. Observing nature was one of my favorite pastimes there.

In my remaining years before my freshman year in high school, I lived in rural areas of Massachusetts, where I continued my nature explorations in the woods, lakes, and swampy riverine regions nearby. I took the usual eighth-grade science course, which helped to reinforce my interests, though the teaching was peremptory. Just for my own enjoyment, however, I remember building a DC electric motor using a 3-volt dry cell and pieces of hardware and wire I found around my house.

D.C. Fort, *One Legacy of Paul F. Brandwein*, Classics in Science Education 2, DOI 10.1007/978-90-481-2528-9_4, © Springer Science+Business Media B.V. 2010

In my freshman year of high school, I enjoyed all my courses—algebra, English, ancient history, and Latin—no science. Nonetheless, I continued to tinker around the house—making various electrical and mechanical gadgets—but without any particular direction. For most of my 14th summer, I worked as a stock clerk for Polaroid Corporation in Cambridge. There, I learned many aspects of handling and working with technological materials. I became fascinated with the making of plastic optics and three-dimensional photographs. The summer ended with my departure from suburban Boston, because of the death of my father.

Then, I moved to Rego Park, in Queens, New York, to live with my mother and stepfather. There, I went to Forest Hills High School. The urban environment, the local culture, and the large high school all made major changes in my life. And I was finally in the right place to satisfy my quest for scientific knowledge. Forest Hills had a science department comprising teachers of courses in physics, chemistry, and biology, all of which I took. I first met up with Paul's teaching in the second term of my sophomore year, when he taught the honors biology class.

Encounters with Paul

This one-term course was one of the most intense learning experiences of my life. The intensity resulted because I was so eager to gain knowledge of science and because Paul had such a strong personality, which, with impeccable teaching technique, he used to inspire me and other aspiring scientists. Behind his authoritative and commanding presence, I sensed his true love of his subject and of his teaching profession, as well as deep affection for his students. To me, every class in that course was an exciting learning experience. There was never a dull, routine, or boring moment in Paul's class. Here are some remembered incidents from those hours.

On the second day of class, Paul called on a girl to say what she had learned from the reading assignment. When she began talking in vague generalities and her discussion wandered from the topic, Paul stopped her, saying merely, "Please sit down." Audible gasps came from the students, and then dead silence. Paul went on to admonish the students about saying only what you know to be true about a topic, by being able to explain how you know it to be true (by citing references and/or data), and, finally, by being obliged to tell why you know what you are saying is true (the quality of the citations). Despite the awe that overcame the class on that day, Paul subsequently displayed great understanding and affection for his students, and we all flourished.

Toward the end of the term, after we had studied and discussed heredity, evolution, and the races of mankind, Paul spoke often and passionately about the terrible tragedy of racism. It was the first I had ever even considered the topic, not having encountered any examples of it in my life up to that time. His emphatic discourse impressed me so powerfully that, when I did encounter racism, I always remembered Paul's urgings for tolerance and understanding and tried to fight it.[1]

---

[1] Though, at 15, I was unaware of the problem of gender discrimination, Paul was ahead of his time on that issue, too. He was mentor to both girls and boys.

I think it was around the last day of class that he said to us, "When you have something to say to someone, ask yourself, 'Is it true? Is it kind? Is it necessary?'" I cannot imagine any better advice for a teacher to give; it illustrates just what a fine, moral, and decent person Paul was.

As a result of my early interest in chemistry, I remained focused on gaining as much knowledge in that area as possible. So, unlike those students of Paul's who continued to work in the biology laboratory and to receive close guidance and support from him, I decided to work in the chemistry laboratory. By the time I had finished my two-term biology requirement,[2] I had read the entire chemistry textbook, and I asked Paul to allow me to take the second term of chemistry and skip the first. He readily agreed.

Later, remembering this episode, I realized that Paul was interested in helping my science education progress with as few obstacles as possible. As it turned out, even the second term of chemistry was unchallenging, because I was so familiar with the material. And for two and a half years, my work in the chemistry laboratory required me to help set up all the demonstrations and the student experiments for the current chemistry class.[3] Paul's freeing me from taking the first term of chemistry enabled me to take the course in solid geometry and thereby complete every science and math course offered in the academic curriculum, something that most of my peers could not do.

My last interactions with Paul involved the Westinghouse Science Talent Search. He encouraged all of the students in his courses who he thought might become scientists to take the Search's examination and to put together a project with a report. I must admit that my project was not very good, in part because no faculty member, including Paul, could provide much mentoring in chemistry. In my senior year, about five of us took the exam, and another student and I received honorable mention. Paul told me that I had one of the highest scores on the exam in the country.[4] His encouragement helped me immensely to do so well and solidified my determination to make my career science.

Learning Science

My learning in science throughout high school took place not only in the classroom (experiences that, for the time, were in themselves rich) but also beyond it. That richness was strongly influenced by Paul as the head of the science department. Paul, of course, was the master teacher, but my physics teacher was also talented, and his classes were rich in content and afforded enjoyable learning experiences. Chemistry

---

[2]Before I enrolled in Paul's class, I had taken another, unremarkable biology class in the fall.

[3]Editor's note: In this task, he was performing much the same work Andrew M. Sessler did for Paul's biology classes. (See Sessler's essay in this volume.)

[4]Editor's note: Compare the contrasting results of Paul's encouragement of Friend in chemistry and a mathematics teacher's downplay of Sessler's mathematics efforts.

did not have a master teacher then, but I augmented my learning with trips to the public library, where I spent many hours reading and trying to understand advanced ideas in the field beyond what high school curriculums could offer. Also, my work in the chemistry laboratory gave me ample opportunity to discuss chemistry with the young teacher in charge of it.

When I left Forest Hills High School to go to the Massachusetts Institute of Technology (MIT), I was more than adequately prepared in math and chemistry to meet the challenges of freshman year. Physics I found to be more fascinating and challenging than the other science courses I took. In retrospect, I believe it was Paul's course in biology that opened my eyes to the sheer joy of learning science. That feeling is what drove me to succeed through years of good, bad, and indifferent teaching, first during my undergraduate years at MIT and then in graduate school at Columbia University. My own ability to learn was more important than my teachers' ability to teach. A really gifted, caring teacher, however, can be a mentor and inspiration to catalyze the learning process. Paul was that.

To summarize my higher education, I found MIT a challenging place filled with many brilliant and talented students and faculty, and I worked hard there for all 4 years. That I could take several graduate-level courses in my senior year was enjoyable and exciting. As a result, I was able to bypass some first-year graduate courses at Columbia and begin taking several advanced courses, as well as some courses in the physics department. Generally, I found that my hard work at MIT helped make graduate school easier for me than for my peers from other schools. I started doing research for my PhD at the beginning of my second year. My thesis was to be in the field of microwave spectroscopy and molecular structure of gases, burgeoning fields at the time. I spent that year learning the relevant science and making measurements on my compounds (cyclopropyl chloride and formic acid). At the beginning of the third year (1953), however, I was drafted into the US Army and spent the next 2 years at the Army Chemical Center in Edgewood, Maryland, working on infrared spectroscopic detection of nerve gases, which turned out to be another learning opportunity, though at the time I did not know how much this work was going to determine the direction of my career.

After 2 years in the army, I returned to Columbia University, where I found that my research adviser would be leaving in 9 months to go to England on a Fulbright Fellowship. By working virtually day and night, I was able in those 9 months to redo every measurement I had made earlier on cyclopropyl chloride, to make new measurements on cyclopropyl cyanide, to combine the data to derive a model structure for the two compounds, and to write and successfully defend a thesis. At the time, these were the most complex molecules whose structures were determined by microwave spectroscopy. A key factor here was that I had time while I was in the army to decide on the course of research necessary to complete the work for the thesis, but I was able to accomplish this because of the self-confidence, in some sense instilled by Paul's encouragement years back, that I had in my knowledge and ability.

My Scientific Career

Industry

After graduating in 1956 from Columbia University with a PhD, because of contacts I had made with their senior scientist while I was in the army, I worked for 1 year for Perkin Elmer Corporation. During that year, I worked with an endlessly inventive original thinker, a genius from Switzerland named Marcel Golay. I learned much from working closely with him on the invention and development of capillary gas chromatography. I built the first instrument and obtained the first capillary gas chromatographs. Golay was the guiding mind behind it, but I helped make it practical.

After leaving Perkin Elmer, I went to work for a small company called Isotopes, Inc., which later became part of Teledyne Corporation. I worked there for nearly 10 years as a senior scientist on radiochemical analyses, carbon-14 dating, radioisotope technology, and, most importantly, the measurement of stratospheric radioactivity from nuclear weapons testing. I worked collaboratively with meteorological consultants and other senior scientists on the interpretation of the data collected. Also, while at that company, I

- managed a radiocarbon dating laboratory
- developed a new radiochemical analytical method for analyses of plutonium in soil and air filter samples
- managed two programs for the sampling and analyses of hundreds of samples of plutonium in soil and air filters
- participated in the design of a very large government micrometeorological experiment involving the plutonium sampling and analyses mentioned above
- managed the Stardust High Altitude Sampling Program (under a US Department of Defense contract), involving thousands of radiochemical analyses for various fission products and other tracer radionuclides in filter samples of stratospheric air collected by U-2 and other high-altitude aircraft
- collaborated in the interpretation of the data amassed in the above program
- collaborated in the design of an impactor sampler for collection of stratospheric particles using U-2 aircraft
- designed and supervised a program for the collection and analyses of the composition and size distributions of stratospheric particles
- identified for the first time an ultrafine particle component in stratospheric aerosols
- collaborated in the development of one of the earliest computer models (two-dimensional in space) for calculating the dispersion and global distribution of stratospheric trace materials
- designed and supervised the performance of a micrometeorological experiment to delineate the dispersion of a plume of radiokrypton released at the surface over rough terrain
- designed the sampling equipment and methodology for radiochemical assay for the above experiment

When I attended the exhilarating first meeting of the International Commission on Atmospheric Chemistry and Radioactivity in 1963, in Utrecht, the Netherlands, I was delighted to find that almost all of the Europeans attending knew of my work in stratospheric radioactivity. My work with meteorological professionals active worldwide led me to identify myself professionally as an atmospheric chemist. Participants felt they were present at the establishment of a new interdisciplinary field where chemists and meteorologists could exchange ideas and stimulate new thinking. Two of the scientists who attended that meeting (Sherwood Rowland and Paul Crutzen) years later were the first atmospheric chemists, along with Mario Molina, to be awarded Nobel Prizes.

Academia

After about 10 years in industry, I applied for a position at New York University. Because of my contact with two faculty members with whom I had consulted at Teledyne about isotopes, and because of my international reputation, I was quickly appointed associate professor of atmospheric chemistry in the department of meteorology and oceanography, thereby becoming the first professor of atmospheric chemistry in the United States. Many others now enjoy similar titles.

I spent 7 delightful years at New York University, where I established an atmospheric chemistry laboratory devoted to the study of formation of sulfate aerosols in the atmosphere. Beginning to give back the gifts I received from Paul and Marcel Golay, I mentored three graduate students through their PhD degrees and six through their master's. I was also coadviser to two other PhD students. In addition, I directed the interdisciplinary (with mechanical engineering) air resources program, which gave fellowships to master's candidates in air pollution science, analyses, and engineering control. During my time at New York University, I developed an early comprehensive model of the global biogeochemical cycling of sulfur and became one of the first scientists to identify volcanoes as major sources of the global stratospheric sulfate aerosol.

When financial problems forced New York University to close my department, I ultimately accepted a position as professor of atmospheric chemistry in the Department of Chemistry at Drexel University, in Philadelphia. There, I built another laboratory to continue my research into aerosol formation. It was often difficult to find graduate students in chemistry who were willing to take a daring step into a strange, young, and interdisciplinary field. Over several years, however, I managed to find five graduate students with whom to collaborate in my experiments. Eventually, two of them completed theses and obtained their PhDs. Through these collaborations, we were able to elucidate the processes by which gaseous compounds react in the atmosphere to form liquid and solid aerosols.

After I had been at Drexel for about a year, I became the R. S. Hanson Professor of Atmospheric Chemistry. Much of my time in my early years at Drexel addressed problems of national and international importance in the area of stratospheric pollution and its effect on climate. Because of my knowledge of stratospheric chemical and meteorological properties, I served on several committees of the National Academy of Sciences, including ones devoted to the stratospheric flight problem (1975–1980), the chlorofluorocarbon problem (1981–1982), and the nuclear winter problem (1983–1985).

In the early 1980s, I also directed a multi-investigator program to use aircraft to obtain samples of volcanic plumes of emitted particles and gases for future analysis. Planes flew into the plumes from Mount St. Helens and five recently erupting Central American volcanoes, including El Chichon in Mexico. The projects proved successful but expensive, so further studies of volcanic plumes have collected this information through remote sensing techniques.

Throughout my 25 years at Drexel University, I taught undergraduate courses in general chemistry and physical chemistry and graduate courses in atmospheric chemistry, community air pollution, environmental chemistry, and aerosol science and technology. At the end of my active career, I coadvised a graduate student in atmospheric science who chose a difficult interdisciplinary project for her PhD thesis—the interaction between atmospheric convection and the chemical kinetics of ozone formation. For 3 years, straddling the date of my retirement, the candidate, her coadviser, Carl Kreitzberg, and I met weekly for discussions of findings and planning future actions. The results were a fine thesis and a PhD for the student. Beyond that, however, both Carl and I found the interdisciplinary interaction stimulating and wished that we had found such collaborative opportunities earlier. Sadly, Carl died of a brain tumor shortly after he retired.

Of course, an academic career in science is multifaceted, and I found much satisfaction in teaching, research, advising students, attending and giving seminars, addressing and attending meetings of scientific societies, serving on professional committees, serving on editorial boards of scientific journals, advising government agencies, consulting for government and private organizations, and traveling to meet scientific colleagues from around the world.

Interdisciplinary research and interactions with colleagues with different scientific backgrounds were among the hallmarks and joys of my career, one always informed by the scientific and personal lessons I learned from my work in early high school with the generous and flexible guidance of Paul F. Brandwein. Aside from his marvelous teaching, all he really did, concretely, was to encourage me, to praise my performance on a chemistry test, to allow me to skip Chemistry 1, and to suggest that I enter what turned out to be an unsatisfactory project in the Westinghouse Science Search.

That was enough.

Paul Brandwein's Influence on My Life

Barbara Wolff Searle

Two people strongly influenced my intellectual (and, ultimately, professional) development during my early years—my father, Walter Wolff, and my biology teacher, Paul Brandwein. These influences were not independent of each other. As I learned recently (although perhaps I knew it when I was in high school), Paul Brandwein's first job as a high school teacher of biology was in the department my father chaired. The two of them created such a united front that it never occurred to me at the time to ponder what *I* might like to do with my life.

Dr. Brandwein influenced my life powerfully in four areas:

- His appreciation of what he saw as my strengths contributed mightily to my self-confidence.
- His presentation of the field of biology was sophisticated and powerfully motivating.
- He selected a "doable" project for my entry into the Westinghouse Science Talent Search and coached me for the examination (I won the top prize for those students who in those days [1947] were called "girls").[1]
- He selected a college for me—Swarthmore.

I will discuss these influences below, but first I will address two general topics: my impressions about what Forest Hills High School was like during the late 1940s and Paul Brandwein's teaching style. Both these sections are enriched by the recollections of my brother, Robert Paul Wolff, who graduated in 1950, 2 years after me. Then I will close with a few words about my life since I ceased to be a bench scientist in 1960.

Forest Hills High School in the 1940s

Forest Hills was a new school, designed to serve several rapidly developing communities in and around Forest Hills, in Queens. (We lived in a completely new

---

[1] From 1942 to 1948, the Search divided participants by gender, choosing two top winners, one from the "girls" group and one from the "boys" (Phares, 1990).

D.C. Fort, *One Legacy of Paul F. Brandwein*, Classics in Science Education 2, DOI 10.1007/978-90-481-2528-9_5, © Springer Science+Business Media B.V. 2010

area, Kew Gardens Hills, built adjacent to the grounds that had been used for the 1939 World's Fair.) I think the school was built during the war; in any case, the class of 1948 was one of the early graduating classes.[2] The school was large (3,000 students), but was divided into three "schools"—academic, commercial, and vocational—which significantly reduced the number of students with whom I interacted.

The surrounding communities, and hence the student body, were almost completely white and predominately Catholic and Jewish; the teaching staff reflected the students. (My third-semester English teacher said she knew what it was like to be in the minority—she was one of only a handful of Protestant teachers in the school.) To the extent that I was aware of such things, my memory of the students in the academic track is that we were middle- and upper-middle-class, well-behaved, smart, ambitious, and virtually all college bound.

The board of education had carefully selected the faculty, and, for that reason and because of the lack of alternative opportunities for academics in those years, we had more than the usual number of PhDs, in all fields but especially in math and science. My memory of the faculty (although certainly not my impression at the time!) is that they were quite young. They were like the student body—energetic, intelligent, open, and, especially, funny. They loved practical jokes, playing to an appreciative audience that, of course, included all the students who were part of the "in" group. So, while lots of serious learning took place, going to school was *fun!*

Paul Brandwein's Teaching Style

I remember Dr. Brandwein entering the honors biology classroom, shutting the door, and saying, conspiratorially, "Today we're going to ignore what's in the textbook—it's hopelessly out of date. I'm going to give you the *real scoop*." And he would proceed to explain the topic of the day, having ensured an attentive audience. My brother, Bob, has a slightly different recollection: "Brandwein used to perform a pedagogical sleight of hand that utterly mystified all of us. He would walk into class and say, 'Well, what would you like to talk about today?' We would hit on a topic, and a free-form discussion would ensue. About two-thirds of the way through the class, we would somehow come to a point at which Brandwein would pause, reach under his lab desk, and produce a prepared demonstration that illustrated the point we had been discussing. None of us could ever figure out how he did it."

Bob notes that "Brandwein was one of those old-time high school teachers who seemed not to have a very clear notion of a teaching load. He would take us on weekend field trips. Once, he took a group of us to Manhattan and to an ethnic restaurant of a type none of us had ever visited. This was the late forties, when even

---

[2]The building and its amenities reflected the austerity of the war years. For example, we had no swimming pool—a circumstance that I remember vividly because we had to take "dry land swimming" to satisfy New York State regulations.

middle-class Jewish teenagers had not been to Europe and had probably never eaten with chopsticks."

The biology preparation laboratory was a hangout for all of Dr. Brandwein's "special" students, and we were free to go there whenever we had spare time. In my freshman year, I was given the task of maintaining the ameba and paramecium cultures that were used for classroom demonstrations. Bob recalls, "When I took intro bio with him, he immediately spotted me as the son of his old department chair and the younger brother of his star student, so he decided that simply sitting in the biology class was a waste of time for me. He gave me a copy of a college genetics text by Laurence Snyder that was then, I think, quite new and cutting-edge. (This is almost 60 years ago, of course.) I can still remember reading about alleles and meiosis and mitosis and such. The point was not what I was learning but the sense that I was being introduced to *real* science." He cultivated in all of us the sense that biology was a wonderfully exciting subject and that we were smart enough and sophisticated enough to be treated as much more advanced students than we really were.

The Westinghouse Science Talent Search

I am not sure how it happened, but we quickly understood that the chief goal of Brandwein's special students was to prepare for the Search. The contest had two parts—the written examination and the project. Both Bob and I (at the appropriate time—junior year, I think) were given projects that did not work, followed by a second (or third) that did. My first project concerned coleus leaves. I do not remember the details, but no doubt it concerned the origin of the variegated foliage. Then Dr. Brandwein proposed that I investigate the production of phenocopies in *Drosophila.* I studied the papers that had been published by the geneticist Richard Goldschmidt (given to me by Dr. B., of course) and set to work. Note that half a century ago, almost no one expected high school students to do original work. (These days, the winners and finalists of the [now Intel] Search often make original contributions![3]) It would be enough if I could repeat previous work and understand what I was doing well enough to make a coherent presentation. This seemed possible.

Brother Bob reports, "As Barbara's little brother, it went without saying that I would enter the Search. Brandwein first put me on to studying the fauna of a pond in the Botanical Garden. Then, because of my interest in physical anthropology, he had me make a pair of slide calipers in metalworking shop, with the intention of measuring the skulls of first- and second-generation Chinese-Americans to see what effect environment had on such things as one's ascending ramus and zygomatic width. Eventually, I did a rather trivial math project and managed to eke out an honorable mention."

---

[3] Many current Search participants not only make such contributions as students but also go on to make important scientific breakthroughs as adults.

I passed through the first hurdles and found myself one of the 40 finalists. The pressure of the next elimination round was greatly reduced for me, since the woman who had won top prize the previous year was Rada Demerec, who had also worked with fruit flies. It was inconceivable to me that the powers that be would choose another fruit fly project for top prize. (Rada was the daughter of an eminent geneticist—altogether out of my league, I was sure.) It therefore was a complete shock when I won the top prize, but I remember being utterly delighted at all the attention it brought me. I have kept a scrapbook, which includes such treasures as a comic-book presentation of my project and a letter from a man who informed me that he could see neutrons.

After High School

Dr. Brandwein made it clear to me that the best possible college for me to attend was Swarthmore. I have no recollection on what he based his decision, but it was a wonderful one. Because of the gender-blindness of my father (he seemed never to expect different things from me or my brother) and the special support of Dr. Brandwein, I knew nothing of discrimination against women, and Swarthmore continued that kind of atmosphere. The biology department was very traditional—the department chair was a naturalist—but the men and women students were equally talented and treated alike, and we all flourished. In fact, in 1952, many of us took the first examination for National Science Foundation predoctoral fellowships, and Swarthmore's five winners were all women. The situation changed completely when I went to Radcliffe as a graduate student, but that is another story (see below).

I participated in the honors program at Swarthmore, which meant that I had seminars instead of classes, no examinations during the junior and senior year, and then eight written and eight oral exams at the end of the 2 years, given by outside examiners. It was a wonderful, heady, challenging experience, and, despite what I considered failing performance on my organic chemistry exam, I was awarded highest honors.

Then I went to Harvard (well, to Radcliffe, officially). I suffered from several culture shocks. The first was the substitution of lecture classes of 50 and 100, with lab sections taught by graduate students, for the freewheeling, intellectually challenging seminars of 5 or 6 students with a professor that I had experienced at Swarthmore. Second was the severe gender imbalance. I took not only the standard biology classes but also subjects like physical chemistry and thermodynamics. All the men seemed to have access to old lab reports and exams and, eventually, to mentors. The very few women were too intimidated even to seek each other out. It was a very lonely experience.

Finding a thesis adviser was an almost insurmountable challenge. I finally found a young professor in the medical school who was doing work that interested me (in biochemical genetics, although the field did not have a name back then). The major drawback was that he had a medical degree, not a PhD, and I was his first graduate student. So, smart and as capable as he was, he knew little about research. After a year, he was invited to be department chair at another medical school. By that time, I had been at Radcliffe for 3 years and had passed my qualifying exams and fulfilled the residence requirement, so I went with him.

By the time I completed my thesis, I was having grave doubts about my suitability for and interest in being a bench scientist. With the help of a Swarthmore classmate, who was then (and subsequently) much more successful than I, I was awarded a postdoctoral fellowship, which I spent at a wonderful biochemistry laboratory. Through that experience, it became clear that my dissatisfaction was deep and abiding, and I resigned my fellowship and set off in new directions. In the years that followed, I taught mathematics at a state college, joined a research institute (doing work on how children learn mathematics), became the director of a successful project financed by the U.S. Agency for International Development, and then joined the staff of the World Bank, first as a specialist in education and, ultimately, as the ombudsman of the Bank.[4]

The irony of this tale is that the substantial pushes I received from my father and Dr. Brandwein (not to speak of my own successes) sent me off in the wrong direction. But I cannot imagine how an alternative trajectory could have led to such a fascinating and varied professional life.

I have no regrets.

References

Phares, Tom K. (1990). *Seeking—and finding—science talent: A 50-year history of the Westinghouse Science Talent Search.* Pittsburgh, PA: Westinghouse Corporation.

Teitelbaum, Michael S. (2003, Fall). Do we need more scientists? *The Public Interest, 153,* 40–53.

Tobias, Sheila, and Sims, Leslie B. (2006, August). Training science and mathematics professionals for an innovation economy: The emergence of the professional science master's in the USA. *Industry & Higher Education, 20*(4), 263–267.

---

[4]Editor's note from the 21st century: The Wolffs' rejection of traditional scientific careers is not unusual. Science PhDs (often with significant postdoctoral work as well) who were trained in the hope of obtaining faculty positions at PhD-granting universities are often not finding such positions. The widely accepted (and as it turns out inaccurate) prediction some years back of an upcoming *shortfall* of American scientists destined for such positions turned instead into a glut (see "Do We Need More Scientists?" [Teitelbaum, 2003]).

If they are willing to take jobs in non-research-oriented colleges, or in industrial or governmental laboratories—or in the Wolffs' cases and those of a few other of Brandwein's high school science students, different fields altogether—their rigorous scientific habits can become valuable assets indeed.

Because of their difficulty in finding jobs suitable to their scientific training (bachelor's level engineering and computer science students excepted), some science majors choose a professional master's program instead of an advanced research degree to combine "advanced science/mathematics high-level communication and technical skills and a working knowledge of business principles" (Tobias & Sims, 2006).

Barbara Wolff Searle's brother, Robert Paul Wolff, also a Brandwein student in high school, did not become a scientist either. First a professor of philosophy and then of Afro-American Studies (now emeritus) at the University of Massachusetts Amherst, he is the author of 21 books, including *In Defense of Anarchism, The Poverty of Liberalism*, and *A Critique of Pure Tolerance* (with Herbert Marcuse and Barrington Moore, Jr.). His survey is analyzed in this volume.

Brief Encounters, Lasting Effects

One Year

Richard Goodman

My colleague Herb Strauss, professor of chemistry at the University of California, Berkeley, and also a Brandwein alumnus, told me about the Springer project. I was a student in Dr. Brandwein's biology class at Forest Hills High School (with Herb Strauss) in 1950 and 1951.

After my sophomore year in high school I went directly to the University of Wisconsin as a Ford Foundation Scholar. Subsequently, I transferred to Cornell, where I earned a BA and, after a year in industry, went on for a master's in engineering science, finally getting a doctorate at Berkeley in 1964. I was on the faculty at Berkeley, as professor, until retirement in 1994 and now still participate as professor emeritus of geological engineering.

I do not know if I was ever termed "gifted" (other than in music). I started with the piano,[1] became a geologist, and moved from the piano to opera (a baritone, I founded the Berkeley Opera in 1979 and am now artistic director emeritus and also sing). Now, in "retirement," I have returned to the piano.

I have achieved some success in my academic and engineering careers. I have won a number of honors, including a Guggenheim Fellowship as an assistant professor, a number of medals from civil and mining engineering associations, various lectureships deemed prestigious, the University of California's Berkeley Citation, prizes for publications, and appointment to an endowed chair at Berkeley. I was elected to the National Academy of Engineering in 1993 and received an honorary doctorate from the Technical University of Graz, Austria, in 2004. I have written five books and the usual large number of articles and so forth.

I remember Dr. Brandwein (we always called him "Doctor") as an unusually devoted teacher with a mission to influence young minds in a strongly positive way. He also had a great sense of humor—for example, telling a boy who was acting up that he [Dr. Brandwein] would "stunt the kid's growth" if he did not behave.

He taught us about chromosomes and genes and encouraged me to read a college biology text, which I loved. Every student's public utterance in class became a lesson in public speaking. No "uhhs" allowed in his classes! And Dr. Brandwein

---

[1] Editor's note: Like Paul himself, who as a child prodigy pianist hoped to become a professional, many Brandwein alumni love and play music.

added this stipulation. Before speaking, everyone should ask himself or herself these three questions: "Is it kind?" "Is it necessary?" "Is it true?"[2] If the answer to two of those questions was "yes," the person should speak. He also ordered us to dress well—including ties for the lads—and encouraged us to walk proud and do our best in every way.

He remains memorable, a hero, although I knew him for only a year.

---

[2]Editor's note: A number of Paul's former students, including James P. Friend, remembered this advice, with variations (see, for example, the preceding essay). When Paul quoted it to me, he added these questions as introduction, "How do we know it? How well do we know it? ..."

Research at a Tender Age

Tom Schatzki

I was a junior and senior at Forest Hills from 1943 to 1945. Those were the school's heydays (as they still might be). It was the last New York City high school opened before the United States entered World War II, and, according to the tradition of those days, faculty could choose where they wanted to teach. And, indeed, a great faculty gathered at Forest Hills.

But their real prize was Dr. Brandwein, a PhD, head of the science department. The school was proud of him. Dr. Brandwein taught biology, taught it very well indeed. But he taught more than that. He taught research science.

The *real* science club, which he probably initiated, was something I had (and have) never seen anywhere at the high-school level. Its maybe 20 members did experiments, but not the usual ones, where the outcome was predicted and success meant getting the right answer. Dr. Brandwein suggested research experiments where the outcome, if repeated and verified, actually advanced science. The members were more like graduate students in a professor's lab than high school kids. That sort of approach is more common today, but Brandwein was ahead of current teaching, in both scope and timing. Dr. Brandwein showed that the method worked.

He catapulted some of the brightest students into internationally acknowledged positions. The best examples, at least in my day, were probably Dick Lewontin and Andy Sessler (whose essays appear elsewhere in this volume). But there were a number of others. Dr. Brandwein left Forest Hills well after I did, to teach his teaching method to science teachers in general. He is justly renowned for that. A teacher of teachers is held in high regard in many cultures.

While teaching methods were his main love, his interests were far broader than that. He had a solid foot in the science community as well. I experienced that personally. When I applied to enter the Massachusetts Institute of Technology (MIT) as an undergraduate, I was rejected. There really was nothing on my record to justify my rejection. MIT probably decided that it had too many Jewish applicants from New York. Dr. Brandwein was a personal friend of Dr. Karl Compton, then president of MIT. A phone call from Dr. Brandwein to Dr. Compton was enough to reverse the decision. I did not go to MIT as an undergraduate because I wanted no special favors, choosing the University of Michigan instead.

I did go to MIT as a graduate student, however, where I earned my degree in chemical physics. The climate had changed, and phone calls were no longer required.

But viewing the development in science, maybe I should have joined the science club rather than the physics club, run by another fine teacher, Mr. Lazarus. I love physics and math, but my choice was a great disappointment for Dr. Brandwein, and I lost the chance to study under his guidance, to receive direct help from one of the great teachers.

After a couple of Midwestern postdocs in a hybrid field best called chemical physics, I went to do hard scientific research at the Shell Oil Company, then located in Berkeley, California. When Shell moved to Texas and basically discontinued its basic research, I—like most of its scientific staff—quit and moved on, in my case to a research branch of the U.S. Department of Agriculture in Berkeley, where (though retired) I still spend most of my time.[1]

---

[1] This chapter gathers memories over 60 years old. The gist is solid, but certain details may be misremembered.

Intellectually Exciting Years at Forest Hills High School

Lisa A. Steiner

I was a student at Forest Hills High School in the late 1940s, graduating in January 1950. I did not take a class from Paul Brandwein; my biology teacher was George Schwartz, whose class I enjoyed greatly.

I do not recall exactly how I met Brandwein, but I knew about the Science Talent Search and was friendly with some previous finalists, Barbara Wolff Searle and Ursula Victor Santer. Richard Lewontin[1] (whose essay appears elsewhere in this volume) was a senior when I entered Forest Hills, and I remember looking up to him. I must have expressed interest in participating, and Brandwein suggested a project, which I set up on the kitchen table at home. I do not remember specific details, but he must have provided guidance for the project as well as suggestions for how to prepare for the written test. Becoming one of 40 finalists in this national competition and the trip to Washington, where we met the president (Truman) and listened to J. Robert Oppenheimer give the after-dinner speech at the banquet, was certainly a high point in my young life.[2]

Forest Hills was an exciting place at that time, with many excellent teachers as well as students. Some of our teachers attended and graduated from the city colleges, especially the City College of New York, during the Depression: Teaching was a stable profession in uncertain times. Many of the students, me included, were Jewish refuge(e)s from Germany and Austria, and the family environment was highly supportive of education. The interaction between dedicated, well-educated teachers and bright, highly motivated students was synergistic. I remember feeling like a sponge that could soak up any amount of learning. In addition to our classroom work, we continued our education outside of class.

---

[1] Editor's note: Barbara Wolff Searle and Richard Lewontin completed the surveys and contributed essays to this volume; Ursula Victor Santer, who became a research biologist, died in 2003 before she could write an essay or fill out the survey, but her widower spoke warmly of her relationship to Paul Brandwein.

[2] The poster I prepared, describing the project—*The Effects of Light on Ascorbic Acid*—was carefully preserved by my mother and evidently transferred when we moved from Kew Gardens to Boston; its remnants were recently discovered, mixed with other effects, in a spare room in our house.

The father of Barbara and Robert Wolff,[3] a high school principal, invited us to their home for discussions, initially of semantics, but soon broadened to other subjects, including listening to music. We went to occasional evening lectures at Cooper Union and to concerts and standing room at the opera (the "old Met"). It was not dangerous in those days to ride the subways and buses into the small hours.

My chief interest during those high school years, extending into college, was mathematics, and I received much encouragement from several of my math teachers, especially Wallace Manheimer and David Frank. They provided additional problems, more challenging than those in the regular curriculum (the "math problem of the week," for example). I was on the math team, competing with students from specialized schools, like Stuyvesant. I have always felt indebted to the school and to teachers like Brandwein for providing such a rich intellectual environment and opportunities for personal growth.

I recently, and unexpectedly, reconnected to the Brandwein era, through Paul's colleague and friend in the publishing world Deborah Fort, the contributing editor of this volume. With the help of high school teachers in the Boston area, we have been trying to organize an update of his and Evelyn Morholt's A *Sourcebook for the Biological Sciences* (3rd ed.; San Diego: Harcourt Brace Jovanovich, 1986). In spite of its age, a number of teachers have told us that the book is a treasure, carefully handed down from teacher to teacher. If the revision process fails, we hope to extend its useful life by making it available on the Internet.

---

[3] Editor's note: On the joint influence of her father and Paul Brandwein on her development, see Barbara Wolff Searle's essay in this volume.

Encouraging the Uncertain

Herbert L. Strauss

Although I was too young to remember particular events occurring during our escape from Europe and arrival in New York, I shared my family's feelings of dislocation and fear. We barely made it safely out of Germany—by the skin of our teeth, in 1939, to England, thence to America in 1940. It took me years to be comfortable, especially in school, but by the upper grades in P.S. 99, I found myself near the top of the class in some subjects, like mathematics, science, and grammar, though somewhere near the bottom in subjects like spelling.

My first year of Forest Hills High School remains a bit of a blank in my memory, but the second year stands out because of Dr. Brandwein's class in biology. The class was full of bright students, some of whom remain good friends to this day. Dr. Brandwein encouraged us to achieve, and he made it clear to everyone (including my mother) that he had high expectations for us. Many of us entered the Westinghouse Science Talent Search: I started a project to investigate the properties of red cabbage as a natural pH indicator.

I am afraid I made a total mess. I boiled up red cabbage at home, and my parents put up with a refrigerator full of slowly molding red cabbage and red cabbage juice.[1] The juice turned a satisfying array of colors as the acidity was changed, but I never did figure out the basic ideas of buffered solutions and so never understood the results. I spent many happy hours at the Jamaica and Manhattan public libraries, however, and came to understand a bit of the structure and properties of cabbage pigments.

Dr. Brandwein's interest in us continued through our school careers and beyond. For example, he suggested I apply to Harvard. I remember feeling that I had let him down badly when I was not accepted, but I did get into Columbia, which turned out to be a perfect fit for me. Dr. Brandwein's challenge and encouragement played a crucial role in enabling me to become the scientist that I am today. Perhaps Dr. Brandwein had envisioned that from the beginning. I will always be grateful.

---

[1] Editor's note: In this, they were more willing to tolerate chaos and decay than the parents of another of Dr. Brandwein's students, whose tidy mother disposed neatly of the petri dishes full of rotting material that were supposed to become part of *his* Westinghouse entry.

Herbert L. Strauss is currently a professor in the graduate division at the University of California, Berkeley. While in high school, he was fortunate enough to have a biology class with Paul Brandwein (1950–1951). He attended Columbia College and did research on electron spin resonance spectroscopy. He went on to do postdoctoral research at Oxford. Joining the faculty at Berkeley in 1961, he became a professor in 1973. His research program has emphasized the use of vibrational spectroscopy to elucidate the dynamics of a wide variety of flexible molecules. He has won awards for spectroscopy and for service to the university.

Saturdays at the Brooklyn Botanic Gardens

Walter G. Rosen

At 17, Dick Lewontin[1] and I worked on Saturdays in 1946 at the Brooklyn Botanic Gardens in the lab/greenhouse of a plant pathologist, who worked on oat smut (the connection with PFB, who did his PhD on this plant pathogen) and on hybridized iris. PFB arranged for us to get this job. I do not recall exactly what we did for the scientist: Maybe we washed glassware, cleaned flower pots, and the like. My clearest memories are of getting there and back on the subway, accompanied by Mary Jane Christianson, then Dick's girlfriend and soon to be his wife.

Some of the attempted hybrid embryos would not develop in situ and had to be cultured. PFB must have explained this to me. Was something coming from the endosperm that inhibited embryo development? If so, it would be necessary to culture the embryo away from the endosperm.

I took the problem back to the high school lab. I worked there with oat seeds, which I think we had around, to make infusions for the growth of protozoa for use in biology classes. It was easy to remove the embryos from oat seeds that had been moistened for a few days, but if one wished to grow them free of endosperm influences, they must be separated from the dry, hard seed before the transfer of stuff from the endosperm could occur. I found I could do this by embedding them in the surface of paraffin in petri dishes to hold them. I could apply pressure with a scalpel at the side of the embryo, and it would pop out.

How I then got them to grow in sterile culture on agar slants, to which I transferred them in a sterile chamber I built, is another story. For here, it is enough to say that the problem, and the challenge it presented, was enough to hook me on the notion that this was the sort of thing I wanted to do, and this meant studying botany.

As a *B* student, I was rejected by Cornell, Bowdoin, and Duke and accepted by New York University and Queens College. I wanted badly to go *away* to college and not have to live at home and commute. PFB suggested that the University of Iowa had a good botany department. He gave no particulars, and I did not ask for any. I applied, was accepted, and off I went.

---

[1] See his essay elsewhere in Part I.

Walter G. Rosen (March 3, 1930–April 19, 2006), PhD in botany from the University of Wisconsin, began his academic career as a professor of biology at the State University of New York at Buffalo. In addition to teaching and research in plant and cellular physiology, he was active in environmental education and in science curriculum development.

From Buffalo, he moved to the University of Massachusetts Boston, then to Washington, D.C., first to the Environmental Protection Agency, then to the National Academy of Sciences/National Research Center. In 1985, the National Forum on Biodiversity, which he organized, was credited with publicizing the need to fight species extinctions.

Before his death, he did some work on the section of this book devoted to surveys of Brandwein's former students. Although Rosen's encounter with Brandwein was not brief but lifelong, his essay is short, because its completion was interrupted by his death. Hence, I have placed it with the others briefly describing large impacts from quick meetings.

Question Everything

Naomi Goldberg Rothstein

I graduated from Forest Hills High School in 1948 and did not meet with Dr. Brandwein after that. He encouraged me to enter the Westinghouse Science Talent Search (which I definitely would not have known about or done without his interest). I won an honorable mention.

I recall his personal interest in me and every other student. He once praised me for questioning something he said, after I had looked it up after class. I was not assertive and wondered how he would react, but he beamed at me for questioning him.

I still remember the pride I felt when he told me I should continue to question everyone and everything; that this was a good thing. I wrote to him after graduation and he sent me back two encouraging letters.[1]

I graduated from Radcliffe in 1952 with a major in social relations (a combined field of psychology, anthropology, and sociology). In 1975, I got a master's in education from William and Mary. After that, I taught emotionally disturbed adolescents in a Virginia hospital. When my husband and I moved to Florida, I taught gifted and homebound students.

(Received July 12, 1949)

Dear Naomi,

I count on your tolerance to forgive this delay in answering your letter. I have just been snowed under, and I didn't want to rush through your letter.

Sometimes I'm pretty certain that the fact that I am so busy means that I too am in science, but I refuse to face that possibility. I'll just say I haven't learned to say "no."

Your letter showed me that you've bounced back. By this time, you have made many friends, have trilled intimately with some, gazed longingly at others, and maybe fallen in love once or twice and vice-versa. All to the good.

---

[1] Editor's note: A year after Rothstein graduated from Forest Hills, she contacted Dr. Brandwein and, in the summer and fall of 1949, received two longhand responses from him. They are reproduced below. All of Brandwein's letters were handwritten unless he had a secretary, which in 1949, when he was still at Forest Hills, he did not. "I don't type," he once explained to me, "because I cannot type." A short, ineptly typed note proved him right.

Now there's no doubt that you have similar weaknesses—similar to those of other people. *All* people are influenced by many, many things—and when they are young—and searching for a road—they are influenced by all guideposts—if but for a moment. They ask themselves—Is this for me? Shall I go there? Here? Shall I be like him/her? Who am I? What am I? Where am I? etc., etc.

All this is very good. It leads to growth, eventual independence, happiness. Of course, if you were born into a family with a fortune going back—and you had no choice but to enswaddle yourself in a cocoon—an early death—you would live that warm fetal life devoid of the kind of poignant enjoyment you now have in life.

All those things that you want—you will have—if you work for them. A home over the cliffs is not hard to get—neither is being a psychologist or pediatrician improbable.

My thought is that you will make a fine teacher—and the human engineering lab won't help you decide. You'll have to try many things before you discover what you want.

If you have time to do any read[ing]—Read a biography of Albert Schweitzer. I think you will find something there that you are seeking. Also read *As I Knew Him* by Hans Zinsser.

Then there is Chaucer, who tells us—

... Strive not, thou earthen pot, to break the wall ...

To hoard brings hate, to climb brings giddiness[2]

It has always been the strangest thing to me to find that those who are the finest men and women I know are always castigating themselves—seeking perfection, not adequacy and security. Art, music, fresh air, sunshine, plants and animals, books, friends, family, the sea, a long walk, peaceful dreaming, pain and joy—closed to you? Do you really need to show that you can make your way in a predatory world by showing the "conspicuous consumption" that Veblen thinks is a degenerate activity and presages decay of civilization?

Come, come, dear one. I think much more of you than that. You have every capacity for enjoyment. Stay happy.

Sincerely,

P   B

(Received November 2, 1949)

Naomi, old worry wart

As long as you live, you will fluctuate along the normal curve—exaltation, contentment, depression—exaltation, contentment, depression. Don't you think it's time you got used to it? No?

Besides—no one is forcing you to go into science—or anything. Do what is going to make you *happy* (?????). Teaching?—Literature, Science, etc, etc?? Do what you

---

[2]Lines taken from two stanzas of Chaucer's *Ballade of Good Counsel.*

can *do best* (1) and what will *make you happy* (2) and what will help *you support yourself* (3). *Three criteria.*

Glad to share your burden. I've deposited it in my safe. Call for it when you come down next.

Sincerely,
Paul Fb

One Class Was Enough

Joanne Gallagher Corey

In the summer of 1956, I was fortunate to be able to take a course at Columbia Teachers College in New York City, taught by one Dr. Paul F. Brandwein, called "Teaching Science in Junior High School." After teaching general science to seventh and eighth graders in a small junior high school in Northfield, Vermont, for 1 year, I decided that I wanted to continue teaching science at those grade levels. I had majored in biology in college and had never taken any kind of education course prior to that summer in Manhattan. In the early 1950s, Vermont needed teachers badly and therefore hired me with a "provisional" certification, with the understanding that I would get the required education courses. At that time, I had no clue as to what a "behavioral objective" was or even how a lesson plan was composed.

The course I took from Dr. Brandwein met 3 days a week for 6 weeks. About 44 of us crowded into a small classroom. Because there were not enough seats for everybody, some people had to sit on a table in the back of the room or on the radiators. Nobody minded this inconvenience.

Dr. Brandwein was abrasive, sarcastic, and challenging. He was empathetic, perceptive, and sensitive. He was humorous and spellbinding. Even after 50 years, I have a vivid mental picture of him as he shared his philosophy of teaching with us. None of us wanted to miss a word of what he had to say, and no one ever came late to class.

As I recall, no one took a test for his course, but we all did a "project." He encouraged us to experiment—to choose unusual projects, ones that did not follow conventional accepted practices. I wrote a paper titled, "Using a Small Trucking Company to Teach Science in Junior High School." The family of a friend of mine had such a company, and I was able to visit their office, warehouse, and garage. I went on a trip with one of the truck drivers as he made deliveries in New York City. With Dr. Brandwein's encouragement, I got carried away but came to realize that I could not only use a trucking company to teach biology, chemistry, and physics but also could expand its pedagogical assistance to other disciplines. This project laid the groundwork for what would become my hands-on and interdisciplinary approach to teaching.

I got an *A* on that project, accompanied by a note to be careful, that I might get "burned out" because I was "too conscientious." I treasured that advice!

D.C. Fort, *One Legacy of Paul F. Brandwein*, Classics in Science Education 2, DOI 10.1007/978-90-481-2528-9_12, © Springer Science+Business Media B.V. 2010

The more I taught, the more I loved it and the more I wanted to learn how to be a better teacher. I gave many workshops to other teachers when I was teaching, and I would consider that mentoring. I was part of the New York State Middle School Mentor Program and part of the New York State Energy Education Project.

I went on to teach science at a middle school in upstate New York for 26 years. I continued to use Dr. Brandwein's *Concepts in Science* series as one of my classroom resources and stopped using a textbook, concentrating on hands-on, experiential teaching. We had fun.

My principal characterized my classes as "controlled chaos" but, generously, always gave me good evaluations.

In 1989, I was awarded the Presidential Award for Excellence in Science Teaching for New York State. I think Dr. Brandwein's spirit was sitting on my shoulder when I received the award. I donated its financial reward to establish a nature center at my school. A small brook runs through the school property, surrounded by a wetland and wooded area, a perfect setting for K–12 science studies.

Dr. Brandwein had a big impact on me. I followed his career through the years and attended a lecture he gave at a National Science Teachers Association conference. Dr. Brandwein was a remarkable man, and I felt honored to have taken that short course from him. It was a twist of fate that influenced my entire teaching career.

Science Education and Beyond: Colleagues

Paul F. Brandwein—A Personal Reflection

Sigmund Abeles

He was a stately gentleman. I remember seeing him walk down the street with his topcoat draped over his shoulders—much like a cape. Elegant and slender, he radiated an aura of self-assurance, competence, and warmth.

I had first learned of Paul F. Brandwein through his writings. When I started my career as a high school teacher of chemistry and physics, I was assigned some classes of gifted students and, of course, began reading about what others in the field had experienced. It was then that I picked up a copy of Paul's book *The Gifted Student as Future Scientist* (1955/1981).[1] In that book, he laid out many of the ideas he developed as a teacher at George Washington High School and Forest Hills High School in New York about how to work with gifted youngsters interested in science. His writing style was impressive, and the book's content was not only carefully researched but also filled with suggestions as to the means for identifying gifted students through their characteristics, behaviors, personality traits, and, yes, through tests as well. His favorite means, however, was "self-identification"—inviting promising youngsters into an environment that would allow them to choose courses and activities that would lead in the direction of a career in science. He also made a number of proposals for setting up school programs that would encourage these students to make the selections. It was a valuable book for me and was useful in helping me to introduce a new course in science for high-ability students in the school where I taught.

I was also aware of many of his other writings such as his *Teaching High School Science: A Sourcebook for the Physical Sciences*, which was crammed with laboratory activities and demonstrations—for me a really useful book (Joseph, Brandwein, Morholt, Pollack, & Castka, 1961). But still I had not yet met the man.

A number of years later, I took on a supervisory role with the New York State Education Department in Albany. During my work with the science bureau there, I was given the responsibility for the state's physics program and its Regents exam, as well as its elementary school science program. At this point, I became acutely aware of Paul's work in the field of elementary school science. One of his powerful pieces, "Elements in a Strategy for Teaching Science in the Elementary School"

---

[1] Reprinted in this volume.

D.C. Fort, *One Legacy of Paul F. Brandwein*, Classics in Science Education 2, DOI 10.1007/978-90-481-2528-9_13, © Springer Science+Business Media B.V. 2010

(1962), based on a Burton lecture that Paul delivered at Harvard University, laid the groundwork for what became one of the most successful—if not *the* most successful—elementary school (and eventually middle school) science series ever published, *Concepts in Science*[2] (Brandwein, Cooper, Blackwood, & Hone, 1966), with several revised editions, the last of which appeared in 1980.

Paul's philosophy was that it was not sufficient for young students to learn science by reading about it in textbooks. He believed that students had to *do* science—an approach we hear much about in the late 20th and early 21st centuries. This series of texts came with comprehensive laboratory kits to encourage teachers to use the materials and enable students not only to read about science but also actually to engage in investigation. A particularly salient point for me was Paul's statement that "In terms of developing science as an experience in search of meaning, *concept seeking and concept forming become the legitimate, indeed, the central, objectives of the science teacher*, even as they are the products of the scientist's processes" (1962, p. 7). I believe it was this approach that gave birth to the very influential *Concepts in Science* elementary school series. In addition, Paul, as author and/or coauthor of a number of extremely useful books for science teachers both at the elementary and secondary school levels, developed a path that America's presecondary science programs could choose to follow.

My experiences with Paul's publisher Harcourt, Brace, and World (later Harcourt Brace Jovanovich) were, at the beginning, independent of Paul's. I was called upon to help revise a physics test that Harcourt had been publishing for a number of years. As an outgrowth of that experience, I was asked to become a member of their science-testing advisory committee. Further along, Harcourt asked me to consult on several books as well as to write some and cowrite others.

After my stint in New York State, I came to the Connecticut State Department of Education in 1971 as their science consultant. At that time, environmental education was beginning to find a wider audience, and I led a number of workshops directed at environmental concerns. I knew of Paul's interest in this area and called to ask him if he would be willing to come and do some workshops for our elementary school teachers. As usual, he was generous with his time and effort. I arranged for the workshops, and Paul came to lead them, bringing only his powerful expression and his consummate knowledge of the content and pedagogy of his subject. He asked if a piano could be made available for the workshop. We were happy to comply, and Paul gave what I remember telling him was a tour de force. In current parlance, he "blew them away." He was funny, he was engaging, he used the piano not only as an instrument for entertainment but to correlate its sounds with a number of concepts he was discussing. The teachers were enthralled, and the large majority of their evaluations called the workshop an unforgettable experience.

Although I had known of Paul only through his writings until the early 1970s, he had given of himself in other science education activities in Connecticut much earlier. In the 1950s, for example, he was the keynote speaker at one of the Con-

---

[2]See Part IV, Appendix B, "Textbooks and Series," for a summary of these contributions.

necticut Science Teachers Association meetings. In addition, he gave a presentation for one of the *On the Shoulders of Giants* television programs organized by the then-president Donald P. LaSalle of the Talcott Mountain Science Center, in Avon. Paul also consulted with LaSalle in his preparation to open a school for gifted youngsters in science at the Center.

Paul's great interest in the education of gifted and talented youngsters brought him into a relationship—first in 1981 and then some years later—with the University of Connecticut's National Research Center on the Gifted and Talented, which was directed by Joseph S. Renzulli. Paul's last book, *Science Talent in the Young Expressed Within Ecologies of Achievement* (1995), was published by the National Research Center on the Gifted and Talented. Appearing just after Paul's death, in September of 1994,[3] it brought together many of his beliefs about education for children with talent. It focused on his belief that an "ecology of achievement" comprises three intereffective ecosystems—that of the family and school community, that of the culture, and that of the postsecondary systems. This overall theme, one to which Paul was deeply committed, appeared in a number of his writings in different forms.

In 1982, Paul came to visit me and my family to see if I might do some additional work for his publisher. My wife and I were pleased to welcome him to our home, where he was a gracious, warm, and engaging guest. Each time we met, I came away learning something about and from the man. Not only was he brilliant, but he was also caring, generous, and anxious to include many in his quest to provide a better education in science for young people.

I think our last meeting was in 1988, when Paul asked me to contribute to a volume on which he was working. I, of course, was pleased to do so. This resulted in a book, published by the National Science Teachers Association, titled *Gifted Young in Science: Potential Through Performance* (1989). It was coedited by Paul and A. Harry Passow, with Gerald Skoog as a contributing editor and Deborah C. Fort, the contributing editor of this volume, as association editor.

I did not see much of Paul after that, perhaps because of the illnesses that slowed (but did not stop) him in his last years.

Paul once called teaching a "mercy." He used that term in the inscription he wrote to me in his *Memorandum: On Renewing Schooling and Education* (1981), a book he gave to me that I value highly. I believe he sincerely felt that way and did everything he could to provide youngsters with the best possible experiences in science. In a sense, it was his gift not only to the "gifted and talented" but also to all youngsters. His approach to teaching science is fostered today in what we call science inquiry. In his words, "Why not then *confront* children at every stage of comprehension with learning situations in which their creativity in whatever form and of whatever excellence is brought to play? Whenever possible let them undertake *individual* investigations or inquiry" (1962, p. 14).

---

[3]Paul's colleagues Evelyn Morholt and Deborah Fort collaborated to put the finishing touches on his nearly complete work. The executive summary of that book is reprinted in this volume.

Paul F. Brandwein was one of those individuals who come along maybe once or twice in a century. His ideas and written creations were wide-ranging and prodigious. I was told that he usually wrote about a thousand words a day. His vision remains with us today. His impact on science education and on methods of teaching youngsters has been significant. His work is laced with a breadth of concept, a depth of research, an eloquent use of language, a generosity of attribution to those whose works he references, and an obvious concern for his audience. It was my pleasure to know him and to benefit from his knowledge, from his wisdom, from his warmth, and from his dedication to a field we all love.

Acknowledgments Grateful thanks to Drs. Donald P. LaSalle, chairman of the board, Talcott Mountain Science Center (Avon, Connecticut), and Frank R. Salamon, former director, Alternative Route to Certification, Connecticut Department of Higher Education (Hartford), for their careful reading of this essay and their helpful suggestions for its improvement.

References

Brandwein, Paul F. (1955/1981). *The gifted student as future scientist: The high school student and his commitment to science.* New York: Harcourt, Brace (Reprinted in 1981, retitled *The gifted student as future scientist* and with a new preface, as Vol. 3 of *A perspective through a retrospective*, by the National/State Leadership Training Institute on the Gifted and the Talented, Los Angeles, CA)

Brandwein, Paul F. (1962). Elements in a strategy for teaching science in the elementary school. In *The teaching of science* (pp. 107–144). Cambridge, MA: Harvard University Press. (Reprinted by Harcourt, Brace, and World, New York)

Brandwein, Paul F. (1981). *Memorandum: On renewing schooling and education.* New York: Harcourt Brace Jovanovich.

Brandwein, Paul F. (1995). *Science talent in the young expressed within ecologies of achievement* (Report No. EC 305208). Storrs, CT: National Research Center on the Gifted and Talented. (ERIC Document Reproduction Service No. ED402700)

Brandwein Paul F., Cooper, Elizabeth K., Blackwood, Paul E., and Hone, Elizabeth B. (1966). *Concepts in science series.* New York: Harcourt, Brace, and World.

Brandwein, Paul F., and Passow, A. Harry (Eds.), Skoog, Gerald (Contributing Ed.), and Fort, Deborah C. (Association Ed.). (1989). *Gifted young in science: Potential through performance.* Washington, DC: National Science Teachers Association.

Joseph, Alexander, Brandwein, Paul F., Morholt, Evelyn, Pollack, Harvey, and Castka, Joseph F. (1961). *Teaching high school science: A sourcebook for the physical sciences.* New York: Harcourt, Brace, and World.

Sigmund Abeles, BS (chemistry) and PhD (science education) from New York University, also did graduate work in chemistry and physics at Yale and Columbia universities. After working as a physical scientist for the U.S. Army and in industry, he taught in high school and college. He served in the Bureau of Science Education at the New York State Education Department, where he was responsible for the state's physics syllabus and the Regents exams in physics. Subsequently, he worked as the K–12 science consultant for the Connecticut State Department of Education. He has written or cowritten and edited or coedited articles for numerous journals and books for several publishers. Currently retired, he continues to evaluate science programs for the National Science Foundation.

Paul Brandwein and
the National Science Teachers Association

Marily DeWall

"Science. Teachers," said Paul Brandwein 20 years ago. "Those are sacred words." This statement expressed his feelings about the subject and its apostles. As a teacher for many years, he knew the extraordinary influence good science teachers could have on students, particularly at the elementary and middle school levels. Brandwein was not a joiner, unless he valued an organization and thought his endorsement could offer assistance in establishing and increasing its credibility and influence. Two education organizations occupied much of Brandwein's time and expertise, particularly during the 1960s and 1970s—the National Science Teachers Association (NSTA) and the Biological Sciences Curriculum Study (BSCS). In later years, he also was active in the Conservation Education Association, which later became part of the North American Association for Environmental Education, and the National Association for Gifted Children.

NSTA's historical documents indicate that it showcased Brandwein's unique teaching strategies in general sessions at its conventions, awarded him its highest honor (the Robert H. Carleton Award) in 1986, and published his papers in its journals and in a 1983 yearbook. In addition, in 1989, the NSTA published *Gifted Young in Science: Potential Through Performance*, a compilation coedited by Brandwein and A. Harry Passow, with the contributing editor of this volume, Deborah C. Fort, serving as association editor and NSTA past-president Gerald Skoog acting as contributing editor. Brandwein also occasionally funded NSTA projects, always with the stipulation that the gift be identified as coming from an anonymous donor.

Robert H. Carleton, NSTA's first executive director, mentions in his book *The NSTA Story: 1944–1974* that Brandwein was a member of the board of directors from 1946 to 1949, when the association was getting started. At the time, Brandwein was a teacher at Forest Hills High School, Forest Hills, New York, and president of the Federation of Science Teacher Associations of New York. He also served as department editor (biology) in the late 1940s for *The Science Teacher* magazine, the association's first journal.

NSTA's second executive director, Robert Silber, who served from 1975 to 1980, remembered that he met with Brandwein several times and found him "pretty independent." Silber said that Brandwein cooperated with the association when it was in his best interests, adding that Brandwein was a "great science educator." Brandwein

---

D.C. Fort, *One Legacy of Paul F. Brandwein*, Classics in Science Education 2, DOI 10.1007/978-90-481-2528-9_14, © Springer Science+Business Media B.V. 2010

protégé John "Jack" Padalino said, "Paul may have used NSTA to advance his agenda in the same ways he did at other organizations in which he was a part. At NSTA, however," Padalino continued, "because of its size and importance, the scale of Paul's involvement was greater."

Bill G. Aldridge, who served as the association's third executive director (1980–1995), had more contact with Paul Brandwein than either of his two predecessors. Aldridge first met Brandwein as a student at Harvard University, when Brandwein addressed his class as a guest lecturer. After Aldridge became executive director, Brandwein met with him on several occasions to discuss his experiences and use Aldridge as a sounding board for his ideas. When Brandwein came to Washington, D.C., he and Aldridge often went to dinner together. This was a treat for Aldridge, because "Paul knew the best restaurants and ordered the finest wines."

In the early 1980s, Brandwein provided Aldridge with donations to be used to advance the work of the association. According to Aldridge, when Brandwein dispensed the first check, he said, "Find something good to do for science education. Tell people the money is from an anonymous donor. I don't want anybody to know about it." Aldridge put the money in a discretionary account that was used for travel for special-task-force committee meetings.

In 1984, Brandwein gave Aldridge $40,000 to be used for an "innovative project." Aldridge's plan was to create a video, a new instructional medium at that time, demonstrating successful strategies for teaching science. A particular strategy he wished to illustrate was "wait time," a concept coined by noted science educator Mary Budd Rowe. Her research showed that the quality and quantity of student responses to teacher questions was directly proportional to the length of time the teacher waited after asking the question before moving on to another question or calling on another student. Aldridge commissioned Alice Moses, an elementary school teacher from Chicago and a past president of NSTA, who was working at the National Science Foundation, to teach a classroom of ethnically diverse third graders in an Arlington, Virginia, elementary school. Moses taught a lesson on magnetism, which was videotaped live with no student preparation. The resulting, edited version of the video was titled *Magnetic Moments* and offered as a teaching resource in the association's publications catalog.

Aldridge was disappointed that Brandwein did not fully appreciate the video when he viewed it. Instead, Brandwein put together a focus group to review the tape, and the conclusion of the group was that the teaching techniques were not sufficiently unique to be termed innovative. The students seemed too well organized (sitting at their desks, working in pairs rather than cooperative groups) and the lesson was too teacher-directed. Aldridge pointed out that the focus group members were not in the field of science education and did not understand the Socratic method of teaching. This experience with the making of and reacting to the video marked the end of Aldridge's and Brandwein's working relationship.

Interestingly, one of Brandwein's notes, written in 1991, stated, "The last time I saw Bill Aldridge, he asked me how to assert his power." "You don't tell them what to do," Brandwein replied, elaborating, "Authority is the discretionary use of power. Power is the discretionary use of authority."

"I don't understand you," Aldridge said.
They never met again.

During the 1960s through the 1980s, Brandwein often spoke at NSTA area and national conventions. Prior to accepting most invitations to speak, Brandwein often requested that a grand piano be placed on the dais. He used the piano as a prop and a metaphor in each of his lectures. He also made other requests. Alan McCormack, now a professor of science education at San Diego State University, remembers that, as a graduate student at New York University at New Paltz, he felt "intimidated" by Brandwein. McCormack, who was "awed" by Brandwein, had been appointed by his professor to get Brandwein what he needed. He had already secured the grand piano and was surprised when on meeting Brandwein, his idol commanded, "Hey, Boy, get me a bag of apples." Dutifully, McCormack purchased the apples and said that Brandwein used them as part of his presentation to demonstrate a black box activity. McCormack also remembered that Brandwein defined science as the explanation of the natural universe. He used Brandwein's *Sourcebook* as his "bible" through graduate school and in his teaching. Although McCormack's first impression of Brandwein was that he was arrogant, he later revised his thinking when he witnessed Brandwein's acceptance of the Robert H. Carleton Award in 1986. McCormack remembers that Brandwein said humbly, "I'm accepting this award on behalf of all science teachers."

NSTA past-president (1982–1983) Robert Yager recalls Brandwein's eloquence and appearance. "He was always the gentleman, always dressed in tailored suits and expensive ties. He was one of my heroes. I appointed him to the Horizons Committee (along with other such outstanding science educators as Paul DeHart Hurd, Fletcher Watson, and Willard Jacobson). He attended our 1983 board of directors meeting in Iowa City and spent much time sharing wisdom and ideas (including during visits to my home). I still regret that future presidents did not continue the Horizons Committee, building on the vision that some great minds in science education were willing and able to provide."

When Brandwein came to speak at the University of Iowa, Yager rented a grand piano (which was not easy to find). Brandwein spoke of his experiences, alternating between talking and piano playing. Yager remembers that Brandwein (perhaps somewhat facetiously) attributed his long life to his unusually slow heartbeat. Yager felt that Brandwein, who was such a creative teacher, "sold out" and succumbed to the textbook industry when he went to work for Harcourt Brace (later Harcourt Brace Jovanovich). Of course, Yager admitted, at Harcourt Brace, Brandwein also wrote the *Sourcebooks* for biology, physical sciences, and Earth science, which influenced and helped thousands of science teachers and their students.[1]

---

[1] The third edition of the *Sourcebook for the Biological Sciences*, written in 1986, still fetches a good price on the Internet. In 2006 and 2007, teachers participating in summer workshops at the Massachusetts Institute of Technology worked on revising it. If the revision process fails, it is hoped that the book's useful life can be extended by making the text available on the Internet.

Yager, along with Padalino, nominated Paul Brandwein for NSTA's prestigious Robert H. Carleton award for 1986. Below is Yager's letter of nomination.

COLLEGE OF EDUCATION
Science Education Center
450 Van Allen Hall
Iowa City, Iowa 52242-1478
319-335-1181 Fax 319-335-1188

July 1985

Awards Committee
National Science Teachers Association
1742 Connecticut Avenue, NW
Washington, DC 20009

Dear Committee Members:

It is with enthusiasm that I nominate Paul F. Brandwein for the Robert H. Carleton Award for 1986. Paul is one of our most productive, energetic, wise, and prolific member and supporter. His leadership has always advanced the field of science education and made it a true profession and a calling. His work on our Horizons Committee during my presidency made the Search for Excellence thrust a most significant theme – pointing to the kind of science designed to prepare teachers and their students for a better life and one where logic, thinking and learning, curriculum, and evidence collecting are all central rather than merely "dressings" and "something extra".

Paul Brandwein inspires people to achieve more and to participate actively in efforts to improve teaching, curricula, and quality of their personal involvement in dealing with the problems of our time. Paul's ideas and arguments concerning how generations and cultures change provide motivation and illustrate what is basic and vital to improve teaching and our whole profession.

Paul is not an idle philosopher; he is quick to share his ideas and to involve all around him in the debate and the process of science itself. His use of history, his piano abilities, and his innovative teaching experiences all unite to make him the kind of educational leader that exemplifies NSTA's mission. Paul Brandwein's selection as the 1986 Carleton Awardee would honor Bob Carleton and all that the organization professes.

I have asked other members of the Horizons Committee and other past award recipients to write letters of support for this nomination. Paul's joining the group of awardees will make it stronger and better able to work collectively in charting future events designed to make science teaching stronger and central to schools and in efforts to improve society.

I know of no one who is more deserving. Further, I feel his selection could encourage others to be more involved – making science education a discipline in its own right. I would hope that Paul's selection would set standards for future awardees with similar visions and leadership skills for advancing science education. Paul is a leader of leaders and will bring prestige and recognition to the award itself.

Sincerely,

Robert E. Yager
President, 1982-83

In response to a request for memories of Paul Brandwein, Marvin Druger, NSTA president 1985–1986, reported, "I went to a lecture by Paul Brandwein many years ago. The topic of the lecture was: 'Why You Should Not Lecture.'" There was a piano in the room. Paul played a boring, monotonous piece and said, "That's what you hear when the teacher is in the room." Then he played a creative, lively tune, asserting, "That's what you hear when the teacher is out of the room." He also demonstrated some of Piaget's tasks about conceptual understanding of volume. He pointed out that the learning results may vary if the teacher uses milk containers instead of Erlenmeyer flasks: "The students don't know about Erlenmeyer flasks, but they do know about milk cartons." Then he noted that learning results will be different if the students do the task themselves, rather than watching a demonstration. "The talk was brilliant, and I still remember it," confessed Druger.

"I went up to him at the end of the lecture, however, and said: 'That was a great lecture. But it was about why we shouldn't lecture.' He replied: 'At your stage of development, you can handle it.' He made me realize that teaching strategies are not unconditionally good or bad. A great talk by someone who knows how to lecture well may be a more effective teaching strategy than an inquiry lesson done poorly."

Gerald Skoog, who was president of NSTA from 1985 to 1986, reported, "I worked with Paul on the book (coedited by A. Harry Passow) *Gifted Young in Science: Potential Through Performance*. I don't think the book was a big seller for NSTA, but it made a contribution to the literature in many different ways. I worked on Part IV, which was titled 'Personal Reflections: From Gifts to Talents,' which included essays on what Paul called 'crystallizing moments' when people discover their professional choice. Contributors included Stephen Jay Gould, Joshua Lederberg, Lynn Margulis, and Glenn T. Seaborg. We printed a series of postcards and letters from Isaac Asimov to NSTA Executive Director Aldridge and to Association Editor Fort. The book contained many useful pieces written by various scientists and science educators. I received some very positive feedback from different educators about the book. Some of the feedback has been recent."

As told to Deborah Fort, Brandwein wanted *Gifted Young in Science: Potential Through Performance*, published by NSTA, to be paired with a book for those (the vast majority) whom he called "the science shy." A book with that title, he hoped, would be edited by Deborah Fort and Gerald Skoog. NSTA instead spearheaded a book on minorities in science.

Brandwein's love of the young extended across the spectrum. His concern for the gifted is well known, but that for slow learners was equally profound. "Have you noted," he once asked, "how close 'underserved' is to 'undeserved?'" He wanted "competence and constructivism in science, identifying science talent through an 'ecology of achievement'"[2] rather than through tests. He was deeply disappointed in NSTA's disinterest and noted, "To substitute a book on minorities in science for one on the science shy is a bad idea. The science shy are the majority. They get social studies and literature, but no science." In the future, jobs will require science, he pointed out, and to know no science is to be disadvantaged. Science shyness is the result of genetic, environmental, and cultural conditions.

According to Brandwein, the sociological factors holding up the entry of the science shy into science and into college included

- inadequate finances
- restrictive quotas as to race, religion, ethnicity
- limited family background
- poor academic preparation, including poor Scholastic Assessment Test (SAT) scores, inadequate achievement scores, and low grade point averages

Although Brandwein emphatically did not define the science shy in terms of minority participation in science, he hoped the book he proposed would take up the problem of the disadvantaged, including minorities. "Treating the disadvantaged as if they were equal," he said, "makes them less equal!" NSTA never picked up

---

[2]On Brandwein's use of this concept, so central to his philosophy, see his works appearing in Part III of this volume, particularly his last, posthumous contribution, *Science Talent in the Young Expressed Within Ecologies of Achievement* (1995).

publication of the science shy book, and Brandwein's project died on the vine. But Brandwein's commitment to NSTA and to teachers never wavered.

Gerald Skoog told another story about Brandwein's innovative presentations at NSTA conventions. "Paul was interested in showing examples of good teaching," Skoog continued. "In the fall of 1988, I chaired the NSTA/Science Teachers Association of Texas joint conference in San Antonio (4,800 participants). I worked with Paul to devise a plan and program whereby we'd broadcast a teaching episode live from an area classroom to a convention session. As the convention was approaching, the teacher involved backed out, and I think we had some technical problems with the broadcasting technology anyway. As a result, at the last minute, I recruited some students and had one of my colleagues teach a lesson, which Paul and I critiqued. Unfortunately, there were probably only 10 people in the audience— hardly a success.

"As a result of the aforementioned activities, I had a chance to meet and communicate with Paul from time to time. One day in Washington, D.C., I met with him in his suite at the elegant Madison Hotel, his hotel of choice for decades. During this visit, I learned about different aspects of his life, such as his military responsibilities during World War II. He also acknowledged that my research on the coverage of evolution in high school textbooks was accurate in relationship to the titles published by Harcourt. I appreciated that positive feedback.

"Paul gave me a copy of a book from his personal library by Thomas Henry Huxley. Published about 5 years after *The Origin of Species*, it was a collection of Huxley's 'sermons' (speeches). I have used this valuable book as a reference on many occasions, where it not only informs me but also serves as a frequent reminder of Paul's interests and generosity."

Marily DeWall, science education consultant and writer, is the former director of the JASON Academy at the JASON Foundation for Education, where she created and oversaw an extensive online professional development program designed for elementary and middle school science teachers. Prior to coming to the JASON Foundation, DeWall worked for more than 25 years for the National Science Teachers Association, serving in various capacities, including associate executive director for administration, editor of *Science Scope* magazine, and director of many corporately funded programs. DeWall is the author of numerous articles, is the editor of several publications, and is a frequent presenter at national and state conventions.

Dr. Paul F. Brandwein: Messages on Teaching and Learning for All Educators

E. Jean Gubbins

Dr. Paul F. Brandwein is noted for many achievements in the scientific and educational worlds. Some would describe him as a scientist; others would say that he was a writer; still others would say he was an educator. All of these roles carry incredible responsibilities because each one ultimately has an impact on the others. It is accurate to say that Dr. Brandwein melded successfully all of these roles into his highly productive life as a scholar who wanted to make a difference for the young. He often used this phrase—the young—when the more frequent word choice was "students" or "children." Reflecting on his choice of words leads to the realization that the use of "the young" implies a certain level of understanding. Think about the world of animals. The following phrase might come to mind: "caring for the young." The resulting inference is that all responsibilities for the offspring need attention or their survival is questionable. Perhaps, Dr. Brandwein chose to use "the young" to remind all of us that we need to view children in the same way. As adults, we must be responsible for caring and nurturing "the young." These are the children in our households, our schools, and our communities, who are dependent on adults for many more years than the offspring of most other animal species. Our children grow and develop in the ways that we nurture them. We all want this nurturance to be successful as parents educate their "young" and as teachers continue this process through "schooling."

Reflecting on the outstanding career of Dr. Brandwein creates a sense of awe because of the depth and breadth of his accomplishments and his ability to focus his messages about teaching, learning, and the young. I will attempt to share some of his messages with the current and future generations of teachers by reflecting on Dr. Brandwein's words from a select group of his published works. The small selection of publications is documented in the references for readers to search out and enjoy the journey of this polymath who made a difference in this changing world.[1]

---

[1] See Part IV, Appendix B, "Publications (Excluding Textbooks)," for Brandwein's non-textbook print contributions. The texts and series are listed in Part IV, Appendix B, "Textbooks and Series." Part IV, Appendix B, "Audiovisual Materials," lists many of his nonprint contributions.

D.C. Fort, *One Legacy of Paul F. Brandwein*, Classics in Science Education 2,
DOI 10.1007/978-90-481-2528-9_15, © Springer Science+Business Media B.V. 2010

On Teachers and Teaching

In many ways, Dr. Brandwein's melding of roles he so gracefully accomplished made him a philosopher. His writings are illustrative of how he thought deeply about teachers and teaching. He instructed others and was resolute in his beliefs about teaching and learning. He viewed the teaching profession as one of the highest goals a person could achieve. He lauded the talents of those who chose to work with "the young." His reference to "we" in his statement below about teachers helps us to understand his views of this profession of which he was a lifelong member:

> As teachers, we enter the minds of others; thus we live in eternity. We help others live better lives, thus teaching remains a mercy.
>
> Who else but those who have given their lives to producing changes in others can—or should—now take on the necessary change in their own personal behavior in the classroom? We stand not only on the shoulders of others, but, if we but look, we should realize—with effective surprise—that we stand on our own. (1983, p. 7)

Novices who are just starting to experience what it means to be a teacher may not greet his words with exaltation. The idea of entering the "minds of others" may elicit excitement because we want to make a difference, which is often cited as the reason why we became teachers. "Producing changes in others" ratchets up the responsibilities for educating "the young" to a higher level. Each child needs to be educated; we, as teachers, must welcome this responsibility. Standing on the "shoulders of others," as well as on our own, may be an overwhelming part of his message as we guide children through the learning process.

It takes time and experience to recognize the impact teachers had on one or more students. When a student describes how we made a difference, it is humbling. Sometimes the response is surprise because we have had no idea that we made a difference. Such divulgence from a student may happen months or years after we worked together. It is a precious moment that reinforces our decision to work with young minds. Dr. Brandwein's choice of language "with effective surprise" captures the realization that we, as teachers, do stand on our own shoulders.

As teachers, we spend years developing content knowledge, skills, and understandings. We are comfortable with our past and current knowledge that constantly expands and connects to prior learning. It may not seem that this large expanse of knowledge, skills, and understandings would be easy to share with all who are willing to listen. At times, telling students what we know may seem efficient and effective. Dr. Brandwein, however, asserted, "The teacher does not teach by telling (the lecturer is not always a teacher); he [the teacher] does not rob children of the right to discovery; he allows them the error which is prelude to truer understanding" (1971, p. 5).

The phrase "allows them the error which is prelude to truer understanding" is certainly poetic. It points to the dynamics involved in teaching and learning. Decades ago, the mind of a student was considered a *tabula rasa*. Many teachers accepted the metaphor of the mind as an empty vessel waiting to be filled. Dr. Brandwein rejected that notion with expedience. He believed that "Teachers, to be sure, have the skills of instructors in presenting subject matter, but teachers are more than instructors. The latter are no larger than their subject matter; teachers are as large as life" (1983, p. 7). What a great way to inspire teachers! By setting the standard for teachers high but attainable, Dr. Brandwein understood that authentic opportunities for learning would ensue. He believed in recognizing the qualities of successful teachers. In just three sentences, Dr. Brandwein (1983) presented a skill set integral to the dynamics of the teaching and learning process:

> Teachers insist on authentic activity, reflection, and reconstruction of concepts, values, and skills. They encourage the critical, solitary, even contrary thought, provided it is based on skillful grasp of information, careful analysis and synthesis of elements, and in the end—its applications to a revision of thought and action. In short, critical thinking. (p. 8)

Teaching is "the profession or practice of being a teacher" (*Encarta Dictionary*). Although the dictionary definition of teaching is sparse, one phrase—"practice of being a teacher"—makes a huge difference. The inclusion of the word "practice" signifies that teaching always requires work. On some levels, we think "teaching" just happens, as when a young child mimics another child's play behavior or when someone demonstrates a simple task such as how to use a ruler to measure a length of plastic tubing. Simple tasks; simple steps. Such simplicity may happen in various content areas. Are we "teaching" when a child connects the name of an object to a picture of the same object? How about showing a child how to label the parts of a three-dimensional model of the heart based on a diagram? What if we describe the attributes of geometric shapes and students must state the names of the shapes? There is some teaching, and there is some learning happening, but there is much more to be accomplished in the art of teaching. Dr. Brandwein (1983) described the characteristics of exemplary teaching:

> Where the art of teaching is a flow between students and teacher, where the student *expects* to respond and is prepared to respond, the atmosphere and effectiveness of the teaching act is entirely different from that in which the teacher is the main actor and the students are passive, not interactive and respondent. The heuristic act enters upon a discipline of responsible consent and dissent; it is not the gymnastic act of seeking the expected response nurtured by the lecture or the exercise in problem-doing, however well conceived. (p. 8)

This ebb and flow of ideas, concepts, skills, and the building of new knowledge and understanding is well beyond *sneaky telling*, in which the teaching process consists of asking questions and giving correct answers as soon as students offer some fragment of information. Teaching and learning promote "consent and dissent" as students and teachers test and retest their thinking and seek further knowledge to help them continue in the learning process.

Dr. Brandwein visited thousands of classrooms around the country and witnessed his philosophy in action. As a researcher with a quantitative orientation, he also carried the tools of the qualitative researcher—pens and journals—as he observed classrooms and recorded copious notes. His 1989 notes illustrate the art of teaching as flow between students and teacher:

Observations of a Combined Fourth and Fifth Grade Class (1989)

Aim: To study the concept of weight and lead to a concept of mass.

A boy brought up a problem one Friday: "I saw a boy balancing his father on a see-saw. The father was sitting near the hinge at the center; the boy at the end of the see-saw. How does this work?"

Several hands went up, but the class was ending, and the children and teacher agreed to take up the problem on Monday. By then, a girl had "invented" a model: A thin metal ruler on a pivot; four checkers on the ruler near the pivot; two at the end.

"If you know the length of the see-saw," she explained, "you can balance the weights. So W (weight of the body) x L on the other side." She drew a sketch of the apparatus on the board. "I checked it up in a high school textbook, but I thought up the checkers as weights and made the fulcrum using the edge of a box." She then answered questions, particularly about her "formula." (Brandwein, 1995, p. 44)

Do we teach because we see ourselves as a font of knowledge? Hopefully, that is an outdated view! Of course, we want to be knowledgeable. We do not want, however, to be the only source of knowledge, skills, and concepts. Maintaining sole guardianship of what students are exposed to places limits on their learning. Dr. Brandwein contended that "teaching is a personal invention. It is a result of many years of work completed with the interplay of factors of heredity, personality, and the larger forces of the environment—education, home, church, and community" (1955/1981, p. 63).

Those who choose to be involved in teaching realize that "the teacher is the guide to the archives, not their guardian" (Brandwein, 1983, p. 7). We strive to help students learn, and, we too, are always learners ourselves.

On Learners

Just as the learning process was not the same for all teachers as they developed their art and craft, learning is not the same for all students. Dr. Brandwein (1983) recognized the differences among learners:

Students are recognizably different not only in mode and manner of learning (so-called modalities), but also in background and ability. Accordingly, teachers are obligated not only to vary level of treatment, but also to select appropriate content, not to say methods and process that are designed to bring forth the *responses* which signal a student's participation in the lesson. (p. 6)

Imagine capturing the essence of teaching and learning in one paragraph. Dr. Brandwein's intellect and experiences led to these statements. Volumes have been written by other scholars; no one has codified the role of teachers and learners with such insight. We all know learners are different and we, as teachers, also learn differently. Learners' differences require attention and variation in treatment or educational options. To prepare learners, we have to ensure that they have "learned to learn, have learned how to learn, and have learned what, how, and to choose from whom to learn with a degree of independence required . . . ." (Brandwein, 1983, p. 6).

On Concepts

The "what" of teaching usually refers to curriculum. Curriculum is accessible from many formal and informal sources. It is not just a set of carefully selected books adopted by schools. Dr. Brandwein (1983) defined curriculum as the "reduction of complexity of the culture; the young are to have a roadmap of the culture so that their first years have a degree of surefootedness. Times, and the culture, will change soon enough" (p. 4).

As a "roadmap of the culture," curriculum cannot be just an accumulation of facts and events. Such details are important, but they need to be connected, rather than remain as disparate pieces of information. Facts are subject to scrutiny and change over the decades. As Dr. Brandwein (1971) claimed,

> Facts change with great rapidity, but the concept—a statement of relationships, a configuration or patterning of facts—remains somewhat stable over a few decades. Concepts offer us hope that we may have good—if temporary—moorings, or foundations, for building a science curriculum. (p. 7)

Dr. Brandwein (1971) guided teachers by developing conceptual schemes to be studied and analyzed. He established the following conceptual schemes for the elementary grades:

Conceptual Scheme A
When energy changes from one form to another, the total amount of energy remains unchanged.

Conceptual Scheme B
When matter changes from one form to another, the total amount of matter remains unchanged.

Conceptual Scheme C
The universe is in continuous change.

Conceptual Scheme D
Living things are interdependent with one another and with their environment.

Conceptual Scheme E
A living thing is the product of its heredity and environment.

Conceptual Scheme F
Living things are in continuous change. (p. 8)

Each of the conceptual schemes consists of seven concepts. The big ideas provide the working boundaries of the knowledge, skills, and understandings to be developed and applied. The concept levels are designed to be hierarchical, building on one another. As students develop their knowledge, they achieve the conceptual schemes in more depth. The interdependence of teachers, students, and their environments is evident in this conceptual approach to teaching and learning. "The scientific concept is the teacher's guide to planning, but it is not the road to discovery, never plainly viewed, for children" (Brandwein, 1971, p. 5). Consider these notes from Dr. Brandwein's classroom observation:

74   E.J. Gubbins

Observation of a Rural District of Fourth Graders (1964)

Aim: To illustrate concept formation, based on prior experience and leading to a construct.

In the introduction to the lesson, the teacher probed what his students knew, asking what kind of farms were in the area, what the crops were, what types of plants and animals they cared for, and so forth. He elicited all this information apparently not only to prepare the children's mind-set but also to set them at ease. Then, the teacher held up four hens' eggs—two brown, two white—and asked, "If these were hatched what would come of them?" The response, almost in chorus, "Chicks." One girl asked: "Are the eggs fertilized?" The teacher cracked one open; it was hard boiled. Laughter. "Nothing but lunch will come out of this one."

Asked the teacher, "Suppose they were fertilized—then hatched. What would happen in the next weeks or so?" The boys and girls described how a chick was brought to full development into a hen or a rooster. They discussed such matters as diet, for example. But the teacher noticed that one boy was silent, appearing inactive, and the teacher passed him an egg.

"Why not a duck, an ostrich?" the teacher queried. Softly, the boy said, "It doesn't have the DNA of these animals." With some encouragement, the boy was able to explain that DNA was in the cells of the growing chick. And, when asked—"What's DNA?"—he stood to answer, "deoxyribonucleic acid." He explained with some uneasiness that he learned about DNA first from a TV program; then, he went to an encyclopedia and to magazines; next, he consulted biology textbooks and had conversations with an older brother, then in high school. The construct developed before the end of the lesson: Living things inherit their traits from their parents. (Brandwein, 1995, pp. 41–42)

The fourth graders had different levels of the same concept, as documented in Dr. Brandwein's Table 1: Concept Levels for Conceptual Scheme E (A living thing is the product of its heredity and environment). The students' knowledge, skills, and understandings drove the aim of the lesson. While the teacher guided their thinking, students willingly shared what they knew. One child who remained silent during initial discussions moved the lesson to a more complex conceptual level, which he learned from TV. He pursued more knowledge independently using multiple resources and then initiated conversations with his high-school brother. This independent route to advanced knowledge documented, once again, the importance of Dr. Brandwein's comments about learning how to learn and what, how, and from whom to learn.

The multiple Concept Levels below (Beginning to Concept Level VI) also help us understand students' potential variability in knowledge and understandings of a big idea. If the goal is to have students reach Concept Level VI and understand that people are the product of their heredity and environment, then we have to determine their pre-existing knowledge. By posing questions, promoting the importance of inquiry, and seeking new knowledge, we can build and promote scientific concepts.

On Developing the Potential of the Young

Dr. Brandwein generated hypotheses about teaching and learning from perspectives as a scientist, teacher, researcher, and scholar. He experimented with instructional and curricular approaches and made adjustments as warranted. In the 1940s and 1950s, he created a talent-search learning environment for high school students whose potential in science was to be determined. He understood that science proneness was obvious as some students sought multiple opportunities to inquire, search, problem solve, and experiment. Questions as to how the nervous system functions, how chemicals are combined to make new products, or how weather patterns affect animals' behaviors provided intellectual sustenance. Other students showed preliminary signs of emergent science abilities illustrated by their level of engagement in science classes or investigations. Their scientific knowledge was in the process of development and they expressed a willingness to learn. Still other students' science abilities were latent. They lacked experiential background and scientific knowledge and needed exposure to the basics of scientific knowledge and understanding before any evidence of science proneness was indicated by their questions or comments. For the young whose potentials were latent, emergent, or obvious, he reflected on questions such as

- Whose curiosity is insatiable?
- Whose work is exemplary?
- Who goes beyond course requirements?
- Who has science-related hobbies?

Invited and self-nominated students were involved in laboratory work beyond their scheduled classes, such as preparing lab materials, assisting in experiments, maintaining a school museum, or participating in science clubs. Students continued to receive guidance and encouragement to pursue additional science opportunities. The increasing specialization required a considerable commitment to scholarly work. Students lived and worked as junior scientists, lending further research evidence to Dr. Brandwein's working hypothesis that the primacy of high-level ability in science was insufficient for a talented child to become a future scientist. As the breadth, depth, and complexity of the science work increased, Dr. Brandwein posed new questions and tested hypotheses about learning and teaching. He continually challenged his own thinking.

Table 1 Concept Levels for Conceptual Scheme E

Concept level	Conceptual Scheme E: a living thing is the product of its heredity and environment
Concept level VI	Man is the product of his heredity and environment.
Concept level V	The cell is the unit of structure and function in living things.
Concept level IV	A living thing reproduces itself and develops in a given environment.
Concept level III	Living things are related through possession of common structure.
Concept level II (Analogical)	Related living things reproduce in similar ways.
Concept level I (Analogical)	Living things reproduce their own kind.
Beginning concept level (Analogical)	Living things may differ in structure, but they have common needs and similar life activities.

(1971, p. 15)

On Curriculum

Dr. Brandwein translated theory into practice as he experimented with eyes-on, hands-on, brains-on, minds-on science techniques. Through the guidance and the talent of educators, the scientific minds of the young could be opened in many ways. Perhaps this belief was the impetus for one of his large-scale curriculum projects. Dr. Brandwein (1983) contended that "a curriculum . . . is not found in textbooks—or a sequence of textbooks—one for each grade" (p. 4). Dr. Brandwein's beliefs about teaching and learning were integral to creating the science series *Concepts in Science*, which was definitely more than grade-by-grade textbooks. It was not a stack of 25 books per classroom. This science series was more than just teachers' and students' editions for various grade levels—leaving teachers and students to navigate their way through the pages unassisted. The series may have been concurrent with or preceded other curriculum reform projects of the 1960s. I cannot trace the original release date for the *Concepts in Science*; however, the large closet-size cabinets and small tabletop compartments in green and purple for grade 5 students will never be forgotten because they held the tools and keys to experience the wonderment of science. *Concepts in Science* was a premier series with all the necessary conceptual schemes, instructions, materials, rocks, minerals, fossils, chemicals, beakers, plastic tubing, and measuring devices to turn traditional elementary classrooms into scientific laboratories. The laboratory atmosphere that Dr. Brandwein knew so well was available to all who accessed the well-designed, forward-thinking science series. (See Part IV, Appendix B, "Textbooks and Series" and "Audiovisual Materials," for further information on the *Concepts in Science* series and other texts, as well as teachers' manuals, tests, and additional printed matter, plus supporting audiovisual materials [filmstrips, audiocassettes, and the like].)

As a novice teacher working with fifth-grade students at the time of *Concepts in Science*, my science background was limited, at best. I was interested in many scientific topics, but believed that I was much more prepared to teach reading, writing, and math to young 10-year-old students who spent their early schooling mastering the "3R's" (reading, writing, arithmetic). To this day, I recall when each classroom in our elementary school received *Concepts in Science*. I remember vividly the book emblazoned with the author's name—Paul F. Brandwein. The series took on special meaning because it offered the novice teacher a hands-on investigative approach. Students, as well as their teachers, would think and act like professional scientists as they hypothesized and conducted investigations. Science went beyond words on paper—it was what it should be. Perhaps *Concepts in Science* was a beginning for many future scientists.

On Characteristics of Scientists

Dr. Brandwein reflected on "What Makes a Scientist?" (1955/1981). He probed the question by engaging in the following strategies:

Messages on Teaching and Learning for All Educators

- noting characteristics of scientists through observations
- checking the growing body of knowledge through discussions with colleagues, teachers, and supervisors
- preparing a booklet describing the high school program in which he worked
- asking for a critique of his findings and conclusions from 100 experts in the field of science teaching

Dr. Brandwein's study of research scientists noted the importance of genetic factors, such as high oral and written verbal ability and high mathematical ability. He believed that genetic factors

> appear[ed] to have a relationship to high intelligence and may have a primary basis in heredity. Naturally, Genetic Factors are altered by an environment. In fact, it is clearly understood here that... any individual is the product of his [her] heredity and his [her] environment. (1955/1981, p. 9)

Given the research scientists' central focus on their investigations, other factors related to commitment were evident. "Predisposing Factors" were characterized by persistence and questing. "Persistence" requires an extended time commitment to a research question that must be addressed despite failures and frustrations. "Questing" means "a notable dissatisfaction with present explanations and aspects of reality" (p. 10). These factors may be necessary, but they are not sufficient to explain the making of a scientist. Continued study revealed the importance of the "Activating Factor"—"opportunities for advanced training and contact with an inspirational teacher" (1955/1981, p. 11). As a researcher and scientist, Dr. Brandwein offered a working hypothesis:

> High-level ability in science is based on the interaction of several factors—Genetic, Predisposing, and Activating. All factors are generally necessary to the development of high-level ability in science; no one of the factors is sufficient in itself. (1955/1981, p. 12)

Dr. Brandwein applied his knowledge about the characteristics of scientists and his classroom observations from 1938 to 1986 to the creation of a list of characteristics that he offered as "teacher clues to the existence of nascent interest, even before it focuses on science proneness or definitive talent" (1995, p. 50). His list includes the following 10 inquiry-oriented, idea-enactive behaviors:

> A future in science may evolve in a child who

- participates readily in discussion after a science demonstration and in so doing defines his/her terms
- inaugurates an experimental (discovery) procedure in the mode of a hypothesis—for example, begins with "What if. . ."
- speculates by asking questions—(whether correctly or as guesses)
- invents equipment to solve a problem and/or shows ingenuity in devising experimental designs
- goes beyond the information known to the class, as evidence of individual initiative, interest, or reading
- prepares for the next day's work by self-initiated reading or investigation
- thinks conceptually or comes easily to an abstraction

- becomes absorbed in a subset of science—for example, life, matter, or energy
- acts spontaneously in uses of science vocabulary or uses imagery to call up pictures of scientific apparatus to solve a problem
- brings to class a project of his/her own, or is eager to enter a science fair and/or [become a] member of a science club (1995, p. 50)

These behaviors and the complete list (see Brandwein, 1995, pp. 50–51) are observable over time. Multiple observations and reflections would need to be conducted to assess "nascent interest" in science. Perhaps a subset of the list could be considered outcome measures of the extent to which the young had opportunities to experience science in their classrooms. In some ways, this list would determine: How are we encouraging the development of science talent in the young?

On Giftedness

Dr. Brandwein's research on the young, whose potentials were still being honed, and on adult research scientists, whose life works were progressing, influenced his views on giftedness. He recognized that the expression of giftedness in cognitive areas is not solely grounded in high IQ, "but is an interaction of personality traits (here called predisposing factors) and opportunities inherent in education and environment (here called activating factors)" (1955/1981, p. x). (See Dr. Brandwein's *The Gifted Student as Future Scientist,* reprinted in this volume, for further elaboration.)

He witnessed these interactions with his students and colleagues and provided sage advice related to the schooling and education of young, gifted children. His words are so meaningful that they serve as a rationale for identifying and serving gifted students. It is important to recognize

> that in one way or another we are obliged to ensure that gifted children preserve their integrity through childhood; (ii) that the behaviors of the gifted may be discerned early, at home and in school; and (iii) that providing these children with the special opportunities they seek and require is essential to the total environment we call schooling and education. The two, of course, require different strategies: *schooling* involves intended experience or "instructed learning" ([Jerome] Bruner's phrase) where the risk is minimized, while *education* per se may include unplanned, chance, and even risky experience. Education complementary to schooling, especially at home, is essential to a certain degree of success in school. (Morholt & Brandwein, 1986, p. 2)

We can learn from 50 years of Dr. Brandwein's research, teaching, and curricular guidance by reviewing the selected references and his more extensive publications. The dates of his publications are just a way to trace the release of his work. His cogent arguments for what we need in education and schooling are as relevant in the 21st century as in the 20th. The legacy of Dr. Paul F. Brandwein is codified forever in his words and actions to promote schooling and education. He understood how to identify and nurture the talents of the young and maintained that "culture provides a matrix for development" (Morholt & Brandwein, 1986, p. 8) of these abilities. We all need to attend to creating an ecology of achievement, which

Messages on Teaching and Learning for All Educators

demands "three interpenetrating ecosystems—that of the family, school, and community, that of the surrounding culture, and that of colleges and universities" (1995, p. 9). Dr. Brandwein's legacy on teachers and training, on learners, on scientific concepts, on curriculum, on characteristics of scientists, and on giftedness continues.

Personal Note

I was privileged to have several phone conversations with Dr. Brandwein as he prepared his research monograph for The National Research Center on the Gifted and Talented (1995). He talked about his work, his progress on the chapters, and his commitment to its completion. My comments about the brilliance of his work were always greeted with "you're so kind." This man of genius and scientific fame, who was also a master teacher, was profoundly humble. He was the one who was so kind in his unending commitment to science education. He definitely made science classrooms better places for children and teachers alike. Dr. Paul F. Brandwein was truly the kind person as a scientist, author, artist, master teacher, and humanitarian who contributed so much to the scientific and educational communities.

Dr. Brandwein's research monograph *Science Talent in the Young Expressed Within Ecologies of Achievement* (1995) for The National Research Center on the Gifted and Talented is a treasured resource. Unfortunately, he never saw the final published version because of his death in 1994. His esteemed colleagues, Evelyn L. Morholt and Deborah C. Fort, finalized his monograph. They collaborated with Dr. Brandwein on this and many other writing projects, and he acknowledged them with his unending gratitude. The research monograph is available online at http://www.eric.ed.gov/ERICDocs/data/ericdocs2sql/content_storage_01/0000019b/80/14/d6/2 f.pdf (Retrieved January 25, 2009)

—E. Jean Gubbins, 2007

References

Brandwein, Paul F. (1955/1981). *The gifted student as future scientist: The high school student and his commitment to science*. New York: Harcourt, Brace. (Reprinted in 1981, retitled *The gifted student as future scientist* and with a new preface, as Vol. 3 of *A perspective through a retrospective*, by the National/State Leadership Training Institute on the Gifted and the Talented, Los Angeles, CA)

Brandwein, Paul F. (1971). *Substance, structure, and style in the teaching of science* (Rev. ed.). New York: Harcourt Brace Jovanovich. (ERIC Document Reproduction Service No. ED108969)

Brandwein, Paul F. (1983). *Notes toward a renewal in the teaching of science*. New York: Coronado Publishers.

Brandwein, Paul F. (1995). *Science talent in the young expressed within ecologies of achievement* (Report No. EC 305208). Storrs, CT: National Research Center on the Gifted and Talented. (ERIC Document Reproduction Service No. ED402700)

Morholt, Evelyn, and Brandwein, Paul F. (1986). *Redefining the gifted: A new paradigm for teachers and mentors*. Los Angeles: National/State Leadership Training Institute on the Gifted and the Talented.

E. Jean Gubbins, PhD, is associate director of The National Research Center on the Gifted and Talented (NRC/GT) and associate professor of educational psychology at the University of Connecticut in Storrs. The NRC/GT, funded by the United States Department of Education, is the only federally funded center focusing on exploring the gifts and talents of young people and promoting the developing of research-based strategies to improve curricular options for students as well as their teachers. As an associate professor, her research interests and expertise include program design and development, gifted education pedagogy, professional development, and evaluation.

Some Reflections on Paul F. Brandwein's Impact on Science Education

Robert W. Howe

Paul F. Brandwein was a man with many talents and abilities, a variety of interests, and a high level of energy. He developed and refined his beliefs, philosophies, and visions for what he thought were desirable outcomes for science education. He also tried to work out conditions and ways to help people to achieve these outcomes.

Paul's work evolved from his experiences, observations, and insights. He was a teacher at all levels of education. He engaged in extensive observations and analyses of students, teachers, instructional materials, classroom environments, science education programs, school environments, and community environments. He was an avid and critical reader of materials from many disciplines and fields. He also worked with outstanding individuals and scholars from many disciplines and fields as a teacher, writer, editor, speaker, and presenter at conferences.

An exhaustive description of Paul's influences and impacts on science education would require a book-length publication. This essay limits itself to some of my contacts with Paul and his publications and highlights some of his ideas and works that I believe have influenced students, teachers, and others involved in science education, as well as scholars in other academic fields. Paul once stated, "Improving educational programs and practices is like developing a garden. In some cases, you can plant a seedling or grow a tree, and it can become an established plant in a relatively short time. In other cases, you plant a seed and hope that with the proper environment it will grow. In still other cases, you need a different plant in order to succeed." Paul experienced all of these situations as he worked to improve the science education experiences and learning of individuals young and old.

My first exposure to the ideas of Paul Brandwein was in 1956 when I read his book *The Gifted Student as Future Scientist* (1955/1981). This publication (reprinted in this volume) provides an introduction to Paul's ideas regarding able students; the science prone; the identification and development of scientific talent; enhancing variables (as well as inhibiting variables); and curricular and instructional approaches. I found the information useful for developing some science programs with special experiences and a course for science-prone students at the middle school level.

I was interested in his use of observations of teachers, students, programs, and schools to develop his theories, to distinguish important variables, and to develop approaches to instruction and curriculums. Early in my teaching career, I began to

D.C. Fort, *One Legacy of Paul F. Brandwein*, Classics in Science Education 2, 81
DOI 10.1007/978-90-481-2528-9_16, © Springer Science+Business Media B.V. 2010

collect similar types of observational data and to categorize and analyze the information related to students, teachers, programs, and schools.[1] I found these files and data useful for many purposes, including developing and teaching in secondary schools and undergraduate and graduate classes, designing curriculums and instructional techniques, presenting inservice activities, conducting research, consulting, and meaningfully integrating my research with that of others.

My next exposure to some of Paul's work and ideas was during the early 1960s when I was a graduate student at Oregon State University. One of the main resources for a biology practicum class was a version of *Teaching High School Science: A Sourcebook for the Biological Sciences* (Morholt, Brandwein, & Joseph, 1958).[2] Preservice teachers and their cooperating teachers found this book a valuable resource for class activities and projects. Students in my classes in the early 1960s frequently used the book to develop activities. Sometimes they had problems and sent comments and questions to the publisher and authors. They usually received responses from Evelyn Morholt and/or Paul.

During this same time period at Oregon State, I also read other materials incorporating Paul's contributions. These included *Teaching High School Science: A Book of Methods* (Brandwein, Watson, & Blackwood, 1958), which contained many of Paul's ideas about science education. A substantial portion considered his hypotheses about the science prone and their instruction and his ideas about developing scientific talent. It also contained sections on instruction and curriculum, offering a point of reference regarding his ideas on those subjects when the book was written. Some of Paul's later writings and comments reflect Philip Phenix's influence on his thinking, his concept of a quality curriculum, and the elements he would include in a discipline. Phenix's *Realms of Meaning* (1958) described six fundamental patterns of meaning—symbolics, synoptics, synoetics, empirics, ethics, and esthetics—that should be included in curriculums. The scope, content, and organization of material could be derived from considerations of human nature and knowledge and these six patterns of meaning for each discipline.

In 1964, when I was teaching and developing some courses at The Ohio State University,[3] I contacted staff of the Biological Sciences Curriculum Study (BSCS) for assistance and was referred to several people, including Paul, whom I encountered for the first time at a meeting. Later, discussions with him expanded to biology education, to instruction, to curriculum, to the science prone, and to a few educational research projects. He presented me with a copy of *Teaching High School Biology: A Guide to Working with Potential Biologists,* which he edited with Metzner,

---

[1]I would also be able to use some of these data gathered by me and my coauthors Stanley L. Helgeson and Patricia E. Blosser in a 1977 publication on the history of precollege science education, *The status of pre-college science, mathematics, and social sciences education: 1955–1975: Volume I. Science education.* Columbus, OH: ERIC Clearinghouse for Science, Mathematics, and Environmental Education (ERIC Document Reproduction Service No. ED153876).

[2]The third edition of this text (1986), written by Morholt and Paul, was being gradually revised in summer workshops at the Massachusetts Institute of Technology in 2006 and 2007. An unrevised edition may be digitized and made available on the Internet.

[3]Where I am now professor emeritus.

Morholt, Roe, and Rosen (1962) and which not only influenced my teaching and consulting but also provided a list of schools appropriate for visitations related to the science prone. I went to several of the schools listed to see their facilities, to observe classes and other activities, and to talk with students and teachers. I found substantial variation among the programs and discussed these with Paul. He felt most of the programs were making strong efforts to achieve their desired outcomes and thought that the variations were due to the culture of the settings and the ecologies of the schools and the communities. He also indicated that programs took time to develop and evolve and, without continuing effort, they would regress or discontinue. His comments at that time are still applicable to many school change efforts.

A few years later, I was serving on a BSCS committee to identify desirable elements and details for a biomethods course for secondary school teachers and to outline some prototype courses. Paul apparently saw a list of the people working on this committee and sent me a copy of *The Teaching of Science* (1962), which bound together his essay "Elements in a Strategy for Teaching Science in the Elementary School" with Joseph J. Schwab's "The Teaching of Science as Enquiry." In that book, Paul presented his ideas on elementary school science and the need for early and continuing experiences to develop science interests, understanding, and talent. Paul indicated that some of the ideas presented in this paper and a number of his other writings were influenced by Fletcher G. Watson (Harvard University) and Paul Blackwood (U.S. Office of Education), with whom he had worked in the past on a methods book and other publications.

In the late 1960s, I attended a conference at which Paul offered one of his many piano-assisted presentations honed to help convey his messages. Renowned curriculum theorist and evaluation expert Ralph Tyler, who usually, however, played for small groups, and Paul both used their superb piano skills to underscore their carefully crafted presentations.

The late 1960s, 1970s, and 1980s brought more contacts with Paul. After several colleagues and I started the ERIC Clearinghouse for Science Education at The Ohio State University in 1965, Paul contacted us to help identify materials we should include and to offer us some books.

ERIC—Development and Changes

In the late 1950s and early 1960s, the U.S. Office of Education established the Educational Resources Information Center (ERIC) to provide retrieval capability and access to research and other educational literature of central importance to administrators and researchers funded by the federal government. ERIC rapidly expanded in terms of users served, types and numbers of documents covered, and services provided.

An ERIC office with no money and no program was established in 1964. Information analysis, including trend studies, became an additional purpose of the system in 1964. Work in 1964 and 1965 led to the decision to maintain a decentralized system with a small office in Washington, D.C., and several subject matter/topical clearinghouses located in other parts of the country (generally at universities and associations).

In 1966, the ERIC system was officially established with (a) a central processing facility to coordinate the database development and to produce microfiche (sheet film) and paper copies of some documents submitted to ERIC; (b) a group of decentralized clearinghouses and other cooperating groups to collect, select, and analyze documents and journal articles; and (c) a small management group in Washington, D.C. The ERIC Clearinghouse for Science Education was one of the first clearinghouses; its scope of coverage was later expanded, and it became the ERIC Clearinghouse for Science, Mathematics, and Environmental Education.

Over time, the ERIC system was modified in several ways. The range of materials expanded from primarily research documents to include other items related to curriculums, instruction, evaluation, educational policies, and school operations. The system moved from print indexes and abstracts to computer and online retrieval. Information products produced by the system included digests, newsletters, trend analyses, and other materials requested by the users of the system. ERIC clearinghouses provided workshops on using the system and materials in the system, provided assistance to thousands of users each year in locating and using materials, and developed working relationships with many associations, state governments, school systems, and universities. An online information answering service, "AskERIC" (http://www.eric.ed.gov/), was developed to assist with answering questions. The number of ERIC clearinghouses and their scopes also varied up and down, finally shrinking to none.

ERIC was reviewed periodically to identify changes that might make the system more efficient and effective. Most of these reviews resulted in positive changes. The Education Sciences Reform Act of 2002, however, resulted in substantial changes to the system (some good) but eliminated the clearinghouses altogether. Guidelines for the new ERIC called for (a) central collection, review, and processing of materials to make processing more efficient and faster; (b) a comprehensive, easy-to-use, searchable, Internet-based, bibliographic, and full-text database; and (c) digitizing a substantial amount of existing microfiche to make it available online. Some live question-answering would continue. A steering committee and a Content Experts group were established to help guide the change and future directions.[*]

I am indebted to the work of Trester (1981) for elements of this short history of ERIC.

---

[*]According to its Web site summary, "ERIC provides free access to more than 1.2 million bibliographic records of journal articles and other education-related materials and, if available, includes links to full text."

Some Reflections on Paul F. Brandwein's Impact on Science Education   85

In the early 1970s, we hired Paul to develop two papers for us and to make some presentations related to science education. He and I had time during these events to discuss a variety of topics. Paul was interested in several surveys we were developing to analyze science instruction and materials used in schools. He was obviously interested in what our data would indicate regarding use of the *Concepts in Science* series,[4] which he first developed in the 1960s for elementary school science and which would be continually updated through the 1980s, with the assistance of others. We also talked about several of the usual topics including instruction, curriculum, educating teachers to help them fulfill their central roles, aiding able students, and making needed changes in science education.

When we added environmental education to our Clearinghouse in the early 1970s, Paul's interest sparked again. He offered additional discussions, questions, comments, and materials. He was concerned about how materials from several groups (nature study, outdoor education, conservation education, and environmental education) would be handled and wanted to be certain that members of our Clearinghouse understood the history of these movements. He supported our decision to add "environmental education" to the name of our Clearinghouse, was very positive, and provided assistance. We called him several times for his ideas and opinions and also when we were trying to maintain federal emphasis and support for environmental programs around the country.

Paul had contacts in and worked with several of the ERIC Clearinghouses that covered some of his other activities and interests. For example, he worked with the Clearinghouse on Rural Schools because of its support for personnel working in outdoor education. He worked with the Clearinghouse for Social Science and Social Studies Education not only because of its general interests but also because of its examination of social issues related to environmental education. He also worked with the Clearinghouse on the Gifted and Talented because of his work with able students and the development of their scientific talent. His involvement with these clearinghouses is additional evidence of his range of interests and level of involvement with various groups working on educational improvement, to say nothing of the power of his energy.

During the late 1970s, our Clearinghouse was involved in several major research projects, including reviews of the literature related to science education and some pilot surveys on instructional practices, materials, and resources used in the schools. Paul was interested not only in what we were learning but also in what some other investigators were finding. He was especially interested in data about the use of his *Concepts in Science* series; that our data and data from other studies (Weiss, 1978) indicated that his texts were the most widely used science textbooks for grades one through six pleased him.

My last personal contacts with Paul were through correspondence in the late 1980s and early 1990s. Our discussions focused on contemporary efforts to improve

---

[4]See Part IV, Appendix B, "Textbooks and Series," for a summary of these contributions.

science education, on attention being given (and not being given) to several persistent problems in science education, on current work related to creativity, and on ongoing research. He frequently stressed his continuing concern about many areas of science education, stating, "We know much that we should do and can do but are not doing."

The Influence of Some of Paul's Ideas

People have different opinions about Paul's influence based on their exposure to his ideas and work and their own experiences. I have selected some areas and some topics where I believe his ideas have been particularly powerful.

Identifying and Developing Scientific Talent

Many people endorse Paul's work and writings related to identifying and developing the science-prone student. Paul stressed several key ideas related to these processes, including (a) the importance of developing a positive ecology of achievement (home, school, community), (b) the need for early and continuing experiences to identify and nurture the science prone, (c) the use of self-identification of students for science-prone programs (not selection by tests), (d) the centrality of original research experiences for developing the science prone, (e) the need for a concept-based curriculum in science and other disciplines, (f) the importance of a broad range of investigative experiences to help students understand the many different methods for doing science, (g) the desirability of special programs (enrichment, grouping, and acceleration, in school and out of school), and (h) the essential experience where the student investigator presents research for critique by practicing scientists.

The Central Importance of the Teacher

Paul's observations and other experiences convinced him that a competent, interested, and motivated teacher could overcome many of the inhibiting or limiting variables in students' environments. He believed the most effective teachers were effective at mentoring not only their students but also their less experienced colleagues.

Paul thought that many programs designed to educate teachers did not place sufficient emphasis on helping teachers to provide the types of programs he advocated. He also believed that relatively few prospective teachers had school experiences that provided them with good models of programs for the science prone. He advocated preservice programs to educate teachers and inservice programs that would enable them to observe, study, and participate in exemplary programs. He also believed that schools and teachers should become involved in operating or developing programs for the science prone.

Resources for Students, Teachers, and Educational Leaders

Paul devoted much of his time and many years of his life to the production and dissemination of educational resources. He believed that, to bring about broad changes in education, it was necessary to put resources that would support desired goals in the hands of teachers and students. He also believed that effective materials were necessary to enable leaders to explain desired goals and how they could be attained. In support of this conviction, Paul authored and coauthored student textbooks, teacher guides, methods books, teacher-resource books, and supplementary materials. For this work, he is justly well known.

Paul's elementary school texts were the most widely used in the United States for many years. In addition, both his elementary school texts (and later ones appropriate for middle school) and high school methods books helped educate sometimes underprepared teachers for many years. The physical science and the biological science sourcebooks he wrote (sometimes with coauthors) also received wide use in college courses for teachers and as references for teachers at work. Few, if any, individuals have been the author or coauthor of as many widely used books for science education as was Paul. His books' impact on science curriculum and instruction is undeniable.

Research and Learning Using Disciplined Observation and Analysis

Paul was a strong advocate of doing research and learning through disciplined observations and analyses. His ideas were based on his extensive observations of schools, programs, teachers, students, materials, environments, and the interactions among them. He developed many of his theories and ideas from his observational research on (a) instruction, (b) teaching, (c) essential and less important variables in development and learning, and (d) characteristics of necessary experiences and materials. He believed that he learned more from his observational research than from strictly quantitative research and thought that teachers could do similar research and gain knowledge that could improve their own instruction and student learning. He also thought that his and similar data could explain some limitations in quantitative studies.

There has been an increase in recent years in the use of qualitative studies. Paul would probably recommend additional qualitative studies based in large samples. He felt he obtained more substantive meaning when he could identify patterns, and he observed that meaningful patterns usually do not emerge from a small number of observations.

Objectives for Elementary and Secondary School Science Education

Paul advocated broad goals for science education programs for all students. He argued for a curriculum that (a) was concept based, (b) was rooted in the disciplines,

(c) emphasized the scientist's multiple approaches for concept seeking, (d) defined and followed scientific attitudes, (e) stressed interactions of science and society, and (f) was activity based.

He advocated instructional approaches emphasizing the same elements. He stressed using instruction to (a) teach the concept-seeking ways of scientists, (b) help students gain meaning from understanding their world, (c) increase interest in science, (d) become curious, (e) develop investigative and reasoning skills, (f) gain skills in using information from investigations for future research and knowledge-building, and (g) nurture the science prone. Paul also called for instruction that involved a variety of resources and settings. Paul stressed that there are many ways to investigate and many places where instruction and investigation should take place, including classrooms; school libraries; school, industrial, and college laboratories; outdoor areas; homes; and other sites.

In short, Paul's influence on science education was profound. His ideas about curriculum and instruction gained wide attention. His publications served as change agents in many schools. His interactions with his students, colleagues, and others influenced their thinking and their work.

Paul's Last Book

Paul's last book (Brandwein, 1995) was published after his death. Deborah C. Fort, contributing editor of this book and association editor of *Gifted Young in Science: Potential Through Performance* (coedited by Paul, Passow, and contributing editor Skoog, 1989), and Paul's trusted colleague, the late Evelyn Morholt, assisted him in completing it. The executive summary of that work is reprinted in this volume. I encourage readers to read the entire monograph to learn more about his most mature theories and ideas and how they relate to the work of others, changing conditions, and the future.

References

Brandwein, Paul F. (1955/1981). *The gifted student as future scientist: The high school student and his commitment to science*. New York: Harcourt, Brace. (Reprinted in 1981, retitled *The gifted student as future scientist* and with a new preface, as Vol. 3 of *A perspective through a retrospective*, by the National/State Leadership Training Institute on the Gifted and the Talented, Los Angeles, CA)

Brandwein, Paul F. (1962). Elements in a strategy for teaching science in the elementary school. In *The teaching of science* (pp. 107–144). Cambridge, MA: Harvard University Press. (Reprinted by Harcourt, Brace, and World, New York)

Brandwein, Paul F. (1995). *Science talent in the young expressed within ecologies of achievement* (Report No. EC 305208). Storrs, CT: National Research Center on the Gifted and Talented. (ERIC Document Reproduction Service No. ED402700)

Brandwein, Paul F., Metzner, Jerome, Morholt, Evelyn, Roe, Anne, and Rosen, Walter G. (Eds.). (1962). *Teaching high school biology: A guide to working with potential biologists* [Biological Sciences Curriculum Study Bulletin No. 2] (Report No. SE0088). Washington, DC: American Institute of Biological Sciences. (ERIC Document Reproduction Services No. ED011002)

Brandwein Paul F., and Passow, A. Harry (Eds.), Skoog, Gerald (Contributing Ed.), and Fort, Deborah C. (Association Ed.). (1989). *Gifted young in science: Potential through performance*. Washington, DC: National Science Teachers Association.

Brandwein, Paul F., Watson, Fletcher G., and Blackwood, Paul E. (1958). *Teaching high school science: A book of methods*. New York: Harcourt, Brace.

Joseph, Alexander, Brandwein, Paul F., Morholt, Evelyn, Pollack, Harvey, and Castka, Joseph F. (1961). *Teaching high school science: A sourcebook for the physical sciences*. New York: Harcourt, Brace, and World.

Morholt, Evelyn, and Brandwein, Paul F. (1986). *A sourcebook for the biological sciences* (3rd ed.). San Diego: Harcourt Brace Jovanovich.

Morholt, Evelyn, Brandwein, Paul F., and Joseph, Alexander (1958). *A sourcebook for the biological sciences*. New York: Harcourt, Brace.

Phenix, Philip (1958). *Realms of meaning*. New York: Holt, Rinehart, and Winston.

Trester, Delmer J. (1981, September). *ERIC—The first fifteen years, 1964–1979: A history of the Educational Resources Center*. Columbus, OH: the Ohio State University Science Mathematics Environmental Analysis Information Reference Center.

Weiss, Iris R. (1978). *Report to the 1977 National Survey of Science, Mathematics, and Social Studies Education*. Washington, DC: U.S. Government Printing Office.

Paul F. Brandwein and Conservation Education

Charles E. Roth

I have chosen to follow a path less traveled than those taken by many of my friends and colleagues. After earning my bachelor's degree at the University of Connecticut in 1956, I taught environmental education at all levels, from elementary school through graduate courses.

I did get a master's degree in conservation education in 1960 at Cornell University. Then I took a fork in the path that led me to work in conservation and science education while keeping one foot in academia. I wanted to face the realities of the workaday world of teaching children and adults while exploring the academic world for the unifying ideas it might provide. It has been a fascinating journey, one stimulated by Paul F. Brandwein, who believed that "the value of a person's advice about teaching is inversely proportional to the square of the distance he or she is away from the classroom" (quoted in Fort, 1998, p. 10).

Although PFB is generally associated with science education in its broadest sense, he had a particularly strong commitment to conservation and to education that would lead to its fulfillment. Over the years, ideas about the nature of conservation have gone through several cycles and even some name changes, but the need remains to find ways to educate *everyone* about the ongoing relationships between people and the dynamics of the planet we share with a wide variety of organisms.

During the first decades of the 20th century, Americans were having a vast impact on the seemingly unlimited resources of this great land. Space does not permit discussion of this interaction here except with a very broad brush. In general, the early years of the century saw vast deforestation across the country, increasing erosion of soil (particularly in the 1930s), and extinction or near-extinction of many species of wildlife, from passenger pigeons to bison. America's population in the early 1900s centered largely in rural areas or small towns.

Youngsters growing up in such environments roamed freely around their homes and communities and became acquainted with the world of nature. This contact with the natural world led many of them to seek education and employment in resource management work. They became the backbone of many of the early investigations of our natural resources and eventually gravitated to the new local, state, and federal agencies inaugurated under President Theodore Roosevelt, who served from 1901 to 1909, and expanded by later administrations.

D.C. Fort, *One Legacy of Paul F. Brandwein*, Classics in Science Education 2, DOI 10.1007/978-90-481-2528-9_17, © Springer Science+Business Media B.V. 2010

92   C.E. Roth

Among the early progeny of the conservation movement Theodore Roosevelt led were people like

- Gifford Pinchot (1865–1946), often called the father of American conservation because of his concern for the protection of the American forests and his service as the founder and first chief of the U.S. Department of Agriculture Forestry Service. Pinchot was also twice governor of Pennsylvania.
- Hugh Hammond Bennett (1881–1960), often called the father of soil conservation, who led this movement in the United States in the 1920s and 1930s, urging the nation to address the "national menace" of soil erosion. Bennett also created a new federal agency—the Soil Conservation Service, now the Natural Resources Conservation Service in the U.S. Department of Agriculture—and served as its first chief.
- Aldo Leopold (1887–1948), the Wisconsin-based father of wildlife ecology, a renowned scientist and scholar, an exceptional teacher, a philosopher, and a gifted writer.

Men like these, with Roosevelt's enthusiastic support, realized that, were they to accomplish their goals, the general public would have to be educated. So conservation education became a program, one that the federal resource agencies promoted strongly. These early leaders brought pressure on textbook publishers, who eventually often added a chapter on conservation at the end of their science textbooks. Unfortunately, many teachers never got to that chapter during a normal school year.

PFB, as a textbook author, saw this flaw and tried to encourage some changes in how textbooks dealt with the concept of conservation. In the 1960s, he and Matthew Brennan (with PFB codirector of Grey Towers, in Milford, Pennsylvania, which housed the Pinchot Institute for Conservation Studies) convened several meetings of people from the science education community, the resource agencies, and other constituencies at Gifford Pinchot's Grey Towers estate.[1] William B. ("Bill") Stapp[2] and I were among the young professionals invited to participate in periodic 2- to 3-day conferences with senior participants at Grey Towers. The Grey Towers discussions had powerful effects on both of us and set us both on new paths. Bill and I would remain friends and colleagues throughout our careers.

Discussions at Grey Towers were often mind-blowing.

PFB had a much broader view of what should be included under the rubric of "conservation" than did many of the other participants. He saw a broad role for social science as well as natural science in conservation education. He often talked of the need to create "sanative" environments for all living things, by which term he meant surroundings that were healing, curative, and healthy. He realized that

---

[1]Grey Towers was donated in 1963 to the federal agency Gifford Pinchot had founded by his son, Gifford Bryce Pinchot, and had become a conservation education center.

[2]See his essay in this volume. Bill, at that time, was teaching in the Ann Arbor School System in Michigan and working on a doctorate in conservation education; I was director of education for the Massachusetts Audubon Society, with a focus on teacher training in conservation education.

our settlement patterns had changed over the years away from rural and small town living to urban living. He knew that urban settings were often unhealthy and that we had to put people into conservation of the entire environment, not just of natural resources. Bill and I were in full agreement and focused on this principle in our teaching.

We saw cities as natural environments of a distinctive kind. They were the creations of human organisms just as coral reefs were the work of fragile coral polyps, paper-wasp nests the constructions of the wasps, and termite mounds the results of excavations by millions of tiny insects. We felt it was necessary to include this broader view into the amalgam of conservation education, nature study, outdoor education, and geography that was beginning to shape a new, broader kind of conservation education.

Bill and I decided to call what we were doing "environmental education" to reflect the broader base of our new thinking. The term did not catch on immediately. It took the rise of environmentalism and the concept of Earth Day, first celebrated on March 21, 1970, to set the stage. Environmentalism flourished during the 1970s, 1980s, and 1990s, and environmental education became the dominant means for working toward PFB's sanative environment.

But language usage changes over time, and, as the country grew more conservative, environmentalists were often accused of political activism, which some saw (disapprovingly) as "liberal," even perhaps dangerous. Although environmental education is noncoercive, the work of people like Harold R. Hungerford, who in the 1970s introduced decision-making skills, examining and choosing among alternatives, and acting on individual choices, appeared threatening to archconservatives. In the next decades, he published many of his findings. See, for example, Hungerford and Volk (1991) and Hungerford, Litherland, Peyton, Ramsey, and Volk

(1992). Environmental education began to develop political enemies. During the presidency of George W. Bush, industry was often allowed to thrive at the expense of the environment. Vice President Al Gore's alarm about global warming, eloquently expressed in his 2006 film *An Inconvenient Truth*, warned the public of its imminent, fast-growing, and profoundly disturbing presence.

While Gore's film focused on the danger to the natural world, the need for conservation education in its broadest sense remains as well. Where do we go from here? How do we regain public concern for having children understand the importance of sanative relationships between them and the world around them? Heraclitus remarked that no one can step into the same stream twice because the world is in constant flux. This is as true of history as it is of the flow of energy, materials, and time. A new label solves nothing: What we need is a clear-cut understanding of what concepts people need to know about their relationship with the natural world in order to give them the skills to put those concepts to work in our universe. Change occurs both regularly and abruptly, and today's appropriate answers or solutions may become tomorrow's old wives' tales.

Much of what we have taught as conservation or environmental education has been future-oriented. This is not bad. In the 21st-century, "me-first" world, where many are content to let the devil take the hindmost and to sacrifice conservation to apparent immediate reward, however, we must demonstrate how conservation benefits people *today*, not just how it will serve future generations. Current political realities suggest that it is important to incorporate the language of religion into the science language of environmental understanding, such as promoting stewardship of God's Earth as a main goal of such education.

Educators need to understand that until their basic needs for food, shelter, health, and companionship are met emotionally and physically, people will give little consideration to broader ecological issues. People need to see that they are *part of*, not *apart from*, the natural world. People need to attend to their inner feelings and fears as well as to external positions in ecosystems. They need to understand that societies are ecosystems, ecosystems similar to those of other species. They need to perceive that economics and ecology are basically the same in terms of energy and materials movements that comprise these two sciences. And they need to understand that all actions have consequences and learn to think about those consequences before they take action.

While most teachers know and emphasize the importance of the liberal arts, much education today focuses on developing skills for jobs. Professional education is undeniably important, but just as important is developing skills for everyday life, citizenship, parenthood, relationship building, and the like. We have to realize that education is much more than schooling. Education occurs in many venues—in homes, in families, on playgrounds, in youth groups, in schools, and in workplaces. We have tended to focus our efforts to introduce environmental education over the years in schools because they provide a captive audience. But pretending that most learning goes on in schools is increasingly invalid today, thanks to computer learning, home schooling, television, movies, museums, and human interactions.

Indeed, PFB always emphasized this dichotomy between schooling and education. Over a quarter of a century ago, he wrote,

> It happens that one, and only one, of the institutions society has sensibly invented to press forward the individual's claim to the good life, for the fulfillment of self-expectation, is schooling. The argument—as old as Plato, and surely older—runs like this: The good life is inconceivable without the good society; the good society is surely impossible without the kind of education that will sustain and maintain a particular society; the aims and ends of a good society are surely envisioned in its educational policies and practices; and, conversely, educational policies and practices are implicitly if not explicitly expositions of the good life. ... Plato's view of education gave only passing attention to schools; he did not equate schooling with education. Indeed, for him, and for scholars in the centuries following, schooling was a useful part, but certainly not all, of education. (1981, pp. 9–10)

Much more effort needs to be placed on creating tools that help people in all educational venues contribute to conservation understanding.

For PFB, the idea was to teach students key concepts and organizing principles about the need for conservation. This goal he attempted in part to achieve through his enormously influential series of elementary school textbooks, *Concepts in Science*.[3] I believe that those organizing concepts are mostly still viable, that they can make sense of random bits of knowledge and experiences. This useful approach, which became the backbone of most science courses, worked when students shared many similar experiences in and knowledge about the world of nature, but because today's much more diverse and much more urban population rarely shares daily contact with natural rhythms and cycles, the concepts PFB taught are less apt to enable students to organize information. Worse (and this misuse would profoundly dismay and sadden PFB), the *Concepts* series sometimes became material to be memorized at the behest of some authority figure. Since this approach is generally used only in school, it can become an isolated activity unimportant for daily living, for *education*.

One way to deal with this problem is to make sure that students are exposed to common experiences. The information we take in through our senses becomes our experiences. In schools, this exposure to nature can happen through schoolyard exploration and outdoor education activities that provide common experience.

In a project involving what I called "remedial experiencing," several teachers and I tried to teach the water cycle to middle school students in Lowell, Massachusetts. We established a survey instrument to see what experiences the students had already had. The items included bouncing a ball off a wall, watching a puddle evaporate, and making dams in gutters to slow down water's force. (Some phenomena, like the bouncing ball, of course, had nothing to do with water. The ability to recognize irrelevance was also significant.) Once the survey was processed, we selected the students who lacked the key experiences linked to the water cycle and gave them outdoor activities to supply those basic experiences. When the whole class took up the water cycle, our students learned much more deeply and quickly about the water cycle than had prior classes. Remedial experiencing worked.

---

[3] See Part IV, Appendix B, "Textbooks and Series," for a summary of these contributions.

Kurt Lewin has written that little substantive learning takes place without involving experience and reflection (1951). Learning, he continues, involves feeling things about the concepts (emotions) and doing something (action). David A. Kolb describes learning as a four-step process. He identifies the steps as (1) watching and (2) thinking (mind), (3) feeling (emotion) and (4) doing (1984). Jean Piaget described intelligence as the result of the interaction of the person and the environment.

In the world outside classrooms, nonschool organizations such as Girl and Boy Scouts, 4H, and YMCA and YWCA programs, as well as parents and other mentors, should be encouraged to provide activities that will bring students to school with some common experiences of the natural world. Having experience in an area makes studying the concepts underlying it more valuable. With experience, the student has something to organize through the concept and can begin to see its importance. When this process continues lifelong and is encouraged by programs fostering "experiential education," people will continue to have a rich understanding of their environment, as John Dewey knew well.

One of the criticisms aimed at conservation and environmental education is that the information presented in such courses is sometimes outdated, and some of the conclusions reached may have been already discredited. This difficulty is not surprising, since it takes almost 10 years from the time a textbook is written until it has been cross-checked and gone through the publishing process to reach the marketplace. In today's world, new data are being uncovered by science much faster than textbooks report them. We are trying to teach final solutions, when they do not exist.

As I began to develop the concept of environmental literacy (Roth, 1992, 1996), as I tried to discover its basic concepts and facts, I again recognized that Heraclitus' view of a world in constant flux was essential. *The key lay in the questions we ask, not in the solutions we find.* Science teaches a process for asking questions and seeking reliable, verifiable answers; thus, science education should focus on helping learners ask and frame suitable questions and develop tools for seeking answers to these questions. Unfortunately, instead, much science education has worked to provide answers to questions the students have never thought of asking. While the kinds of questions and answers sought must be age-appropriate, children can and should begin these processes early in science generally and in conservation and/or environmental education[4] in particular.

Based on these thoughts I developed a question-based *Framework* for shaping environmental literacy (in cooperation with Earthlore Associates and The Center for Environmental Education of Antioch New England Institute, 2002). It is built on the premise that although conservation issues and concepts change with time and new technologies, the questions that underlie them remain the same. My *Framework* is a curriculum-building device based on the premise that questions lead to and provide the *context* for basic concepts about the environment and our relationships with it. Such concepts underlie the conservation issues that many programs use as the

---

[4](I use these terms interchangeably.)

starting point of their efforts in environmental education. In my view, environmental issues are inappropriate starting points. Instead, conservation education that leads to true environmental literacy should begin with the context-setting questions.

I further believe that although the answers to the questions may change with increased knowledge, experience, and technologies, the questions remain essentially the same over long periods of time, and many should be revisited a number of times over one's lifetime.

Our task as educators—be we teachers, parents, youth leaders, mentors, or adult instructors—is to help learners formulate and frame good questions to explore. To become environmentally literate is to ask basic questions and to revisit them periodically in order to discover, based on continuing advances in knowledge and technology, what solutions are currently optimal. Our knowledge about the environment changes rapidly; our ignorance grows equally fast, often more rapidly. This is because for every question answered or problem solved, several new ones are generated.

The questioning *Framework* is basically *integrative*. One might describe it as interdisciplinary, or even adisciplinary: The questions explored often require information from more than one traditional discipline to arrive at viable answers. In the day-to-day world that makes up our environment, we seldom work only within a single academic discipline. Rather, we blend information from several disciplines to create new information or identify new alternatives—or at least we should.

The *Framework* is based on the understanding that our environment, which has at least three interconnected, interacting components, surrounds us totally. Those three elements include the bio-geo-physical (nonhuman); the social; and the mind/body (psycho-physiological, or inner). Environmental literacy results from our response to the questions we learn to ask about our world and our relationships with it, the ways we seek and find answers to these questions, and the ways we use the answers we have found. Thus, conservation literacy demands the understanding, skills, attitudes, and habits of mind that empower individuals to take day-to-day and long-term actions to maintain or restore sustainable relationships with other people and the environment.

A major goal of the *Framework* is that by the completion of 12 years of formal and informal schooling, the average high school graduate should have attained at least a functional level of conservation literacy. The questions in the *Framework* explore four major environmental arenas:

- nonhuman surroundings
- society
- self
- connections

These arenas define a structure that aims to assure that students will develop the knowledge and skills needed to think, feel, and act in ways that preserve the integrity of the environment. This structure is intended not only to meet the immediate personal, societal, physical, social, and emotional needs of today's students, but also to serve the needs of generations yet to come.

A major potential new player in conservation education is place-based learning, which is actually a new term for some activities practiced earlier, particularly in Bill Stapp's Global Rivers Environmental Education Network (GREEN). (See his essay in this volume.) In addition, in Maine in the 1960s and early 1970s, Dean B. Bennett[5] did some significant work for which I was then a consultant.

Two stories from that work bear retelling for the light they shed on the power of place-based learning. One group of first graders was engaged in our "Ugly/Beautiful" activity and was exploring their school grounds to determine what they thought was ugly and what was beautiful. They then looked over their lists to determine those things that everyone agreed was ugly. The hands-down winner was the swing set, which had lacked maintenance for years.

Their next exploration was to determine who or what was responsible for this ugliness. Their conclusion was that it was the school custodian's neglect, and he was invited in to explain why things had gotten to their current condition. His response was that the school was generally so messy after each day that he didn't have time to work on things like the swing set. Not happy with that response, the students invited in the principal to discuss the issue with them. He agreed with the custodian, leaving the students without a solution. Then one of the students remarked that his dad had a lot of old paint cans in the garage. Perhaps the students could use them to paint the uprights in bright colors. Others chimed in that they too could get some paint. Some had access to brushes. The upshot was that they brought in the paint and brushes and painted rings around uprights as far as they could reach. They thought the swing set now looked much more beautiful. While adults still thought that the swing set was pretty unattractive, the fact is that these young students perceived a problem, investigated it, and proposed and implemented a solution in an age-appropriate fashion.

A second opportunity arrived with a sixth grade class that had just completed a health unit. The teacher felt that the kids had generally been bored with it and thought he hadn't done a particularly good job of presenting the material. Then one student brought in an article from the nearby Portland, Maine, newspaper telling how rats were moving out of the city into the nearby suburbs. The youngsters remembered the pictures they had seen in class of carts full of dead bodies being hauled away in Europe after a plague, which had been attributed to rats and their fleas. The student wanted to know if their town had rats. The other students decided to check around to see if they could find any evidence of rats.

The next day, one child came in, all excited. He had seen a rat near the local grocery store and had discovered its burrow. A sense of panic went through the class, and the teacher came to us as consultants to find out how he should deal with this controversial issue. Bennett gave him extensive advice on how to protect himself from miscommunications and then recommended that he have the kids study

---

[5]Now professor emeritus at the University of Maine at Farmington.

rats and determine how many rats and what conditions would constitute a true rat problem.

After this explanation, the kids surveyed the community for rats. They determined that the one rat seen near the grocery store did not constitute a significant rat problem, but they did find one area of town with a large thriving rat population—the town dump. They explored the dump to find out what was supporting the rats and discovered that unrinsed cans, particularly those formerly containing pet food, provide an ample food supply.

They then agreed that in most of their homes they were the ones responsible for trash collection. If they didn't check to see that the cans were rinsed, they were the ones feeding the rats. To paraphrase Walt Kelly's Pogo—they had met the enemy, and it was they. They decided they needed to work harder at home to prevent further rat problems in town. In this case, the original class had bombarded the kids with facts for which they had no experience to build on. It took the local rat scare to provide the experiences they needed to reflect on the class work.

Although current place-based learning can be traced back to such early work, the focus now is quite different, arising in part from thinking about bioregionalism and knowing intimately the area in which we live. Where does your drinking water come from? Where do your community wastes go? What are the natural and human resources of the area? What are the good things about the place and what things need to be improved? To me, these questions (and their answers) are important not only for youth but also for adults. Each time we move to a new location, we need to explore it in the same way and compare the differences between the old and new locations. This chore can make use of the questioning framework and encourage the revisiting of many of the questions at different times in our lives.

While place-based learning is no panacea for developing conservation literacy, it offers a good starting point.

References

Brandwein, Paul F. (1981). *Memorandum: On renewing schooling and education*. New York: Harcourt Brace Jovanovich.

Fort, Deborah C. (1998). *Science in an ecology of achievement: The first Paul F-Brandwein symposium*. Dingmans Ferry, PA: Paul F-Brandwein Institute.

Hungerford, Harold R., Litherland, Ralph A., Peyton, R Ben., Ramsey, John M., and Volk, Trudi L. (1992). *Investigating and evaluating environmental issues and actions: Skill development modules*. Champaign, IL: Stipes. (ERIC Document Reproduction Service No. ED368557)

Hungerford, Harold R., and Volk, Trudi L. (1991). Changing learner behavior through environmental education. *Journal of Environmental Education, 21*(3), 8–21.

Kolb, David A. (1984). *Experiential learning: Experience as the source of learning and development*. Englewood Cliffs, NJ: Prentice Hall.

Lewin, Kurt (1951). *Field theory in social science: Selected theoretical papers* (Dorwin Cartwright, Ed.). New York: Harper.

Roth, Charles E., Earthlore Associates, and The Center for Environmental Education of Antioch New England Institute (2002). A questioning *Framework* for shaping environmental literacy. Retrieved December 13, 2007, from http://www.anei.org/download/82_questioning.pdf

Roth, Charles E. (1992). *Environmental literacy: Its roots, evolution, and directions in the 1990s.* Columbus, OH: ERIC/CSMEE Publications, The Ohio State University. (ERIC Document Reproduction Service No. 348235)

Roth, Charles E. (1996). *Benchmarks on the way to environmental literacy.* Boston, MA: Executive Office of Environmental Affairs.

Environmental Education and Paul F. Brandwein's *Ekistics*

Rudolph J. H. "Rudy" Schafer

In December 1967 I went to work for the California State Department of Education as their first full-time environmental education staff person. My background included serving as a classroom teacher, public information officer, and conservation education specialist with the Los Angeles Unified School District. In addition, I worked 11 summers as a National Park Service Ranger, a major factor in developing my personal environmental commitment.

I had little or nothing to build on in my new position. One departmental staff member was allocated 10 percent of his staff time for conservation education, and he was only months from retirement. So, in effect, I started from zero. Also, the department, having been pushed by the legislature and the resource agency people into approving my position, was fairly unenthusiastic about the program.

I did, however, have a lot of outside support. The 1960s were the "ecology now" days, when nearly everyone wanted the nation to turn from our environmentally profligate ways and adopt a new stewardship toward planet Earth and its resources. The State Resources Agency, the legislature, the public, and the media were demanding action at all levels, particularly in the schools. So here I was. While I had to create a new program, I could see that I had friends and potential resources with which to work. Norman Livermore, then state resources secretary (a member of the governor's cabinet) was, in particular, most supportive.

In all such situations, first, one tries to see what is already going on, where, and how successful these efforts are. One also reviews the literature, not easy in those pre-Internet days. I soon learned that the various state agencies having to do with natural resources were doing quite a bit, but most of their approaches were in the form of classroom or auditorium Smokey Bear talks, coloring books, field trips, and the like, promoting the work and views of their particular agency. In checking with neighboring states, I found that this pattern was also typical elsewhere.

It became evident that the natural resources people were spending lots of money on largely ineffective educational programs reaching a limited audience. Let us face it: How many dog and pony shows could an agency deliver statewide to three million school kids, and how much good would they do? But it also became evident that the leadership wanted to update their school programs and that an important task ahead of me would be bringing the education and the resource-management people closer together.

In talking with resource and education people in neighboring states, I found that the California picture was replicated in various degrees throughout the Western states. It occurred to me that something worthwhile might be done if I and others could work on a regional basis. Accordingly, I applied for and obtained a federal grant to bring together state-level education and resource-management people from 13 western states.

Founded in 1970, the Western Regional Environmental Education Council (WREEC) produced two quality education programs, Project Learning Tree and Project WILD, and partnered in a third, Project WET. Almost 40 years later, all three programs are still in existence and are still reaching teachers and students throughout the United States and several other countries.

Early in my career with the state, Dr. Paul Brandwein, a recognized name in science education and writer of many textbooks, contacted me and offered to produce a curriculum outline and supporting materials for a very modest price. We discussed the matter at some length, and I got permission from the department to spend $10,000 on the project.

In my meetings with Brandwein, we spent much time discussing terms we would use to describe our work. The term *conservation education* had been widely used in years past and was defined as helping learners gain knowledge and skills along with a commitment to use natural resources wisely. The term did not, however, address such matters as population, land use, or environmental pollution. The term *environmental education*, which was coming into wide use at the time, *did* address these issues, but the term did not carry with it a commitment for constructive action as did the term conservation education. The latter implied that you learned about the environment—period.

Brandwein proposed the term *Ekistics*, a Greek term which relates to the care of the home—our earthly home. The term was coined by Constantinos A. Doxiadis (1913–1975), a city planner based in Athens. Well, Brandwein was a persuasive man, so after much discussion we agreed that the title of the proposed publication *would* be *Ekistics*, but I insisted on a subtitle, *A Guide for the Development of an Environmental Education Curriculum.*

The publication began to take shape. It had three major components, which included

- a study of San Francisco Bay, with its many and complex natural and human components and their interrelatedness, to show that a study of this or nearly any environment involves the interaction of many scientific, social, and cultural factors that cannot be understood piecemeal
- a conceptual K–12 framework that addressed such complex actions and interactions, organized under three major "pathways": (a) Humans are interdependent with their natural and physical environment (science and technology); (b) the social behavior of humans is basic to maintaining, altering, adapting, or destroying the environment (social studies); and (c) humans use symbolic and oral traditions to maintain or alter the environment (humanities)

Environmental Education and Paul F. Brandwein's *Ekistics*

- a chapter on instruction, which reiterated Brandwein's strongly held views about helping learners gain conceptual understandings and critical-thinking skills that would help them to process and evaluate information rather than the rote learning of "facts"

When Brandwein and his associates delivered their final draft, I found that I had a most innovative and valuable piece of work. But there was a problem. Erudite and scholarly as it was, I felt it was too long and would be a bit over the heads of most of my potential readers. I decided that some editing was in order, and set to work. I felt that if I could produce a volume that *I* could readily understand and put to use, any educator could do likewise. I spent many days writing and rewriting and sent my version on to the departmental publications office for their further editing. I worked with an editor, Ted Smith, who was a hard taskmaster, but he made one's work shine. Then there were illustrations, layout, and other factors to be selected and implemented. At last I got my final sign-off, and *Ekistics* was on its way to the state printer.

Not quite. Superintendent of Education Max Rafferty had failed in his bid for re-election and one of his top staff members, Wilson Riles, had been elected in his place. The new superintendent and his staff began taking over and implementing new programs and policies. A directive came down from on high that any and all publications awaiting printing produced under the Rafferty regime were to be discarded. Period.

After all our work!

I tried to appeal through channels, but to no avail. The order became departmental policy, and there was no appeal.

Taking my bureaucratic life in my hands, I went to see Resources Secretary Norman Livermore and laid out the problem. He was aware, through my advisory committee, that the publication was in the works, and had been most supportive. "What do you suggest?" he asked. "I would like to draft a letter for your consideration," I replied. "If you approve it, you can send it over your signature to the new superintendent." He smiled and told me to go ahead.

In my draft of the letter to the new superintendent I, speaking for the secretary, emphasized the interest of the Resources Agency in conservation education and the need for effective school programs in this field. I also noted that the secretary was interested in working closely with the new superintendent toward this end and hoped to meet him one day soon.

The letter went on to say that the secretary was aware that a new publication, soon to be published by the department, would aid educators in developing suitable classroom programs. "I am certainly looking forward to receiving a copy of this publication," the letter concluded. Two days later I got a panic call from a deputy superintendent. *Ekistics* rose from the dead.

Ekistics proved to be a valuable contribution to the environmental education field, which was to become most useful to curriculum writers for some time to come. It went through several printings and finally suffered a death only possible in bureaucracies. Ted Smith, now director of publications, requested permission to reprint *Ekistics* at no cost to our branch, because it was a good seller. I completed the necessary paperwork and sent it up the line for what I assumed would be routine approval.

Unbelievably, my request came back disapproved. I went to see the person responsible for the disapproval and tried to turn the thing around. "How will this action benefit *me*?" I was asked.

"Well, we would be continuing to make available a publication many people find valuable, and it won't take anything out of your budget," I replied.

"Well, I still don't see of what benefit this would be to *me*, so my disapproval stands." End of story.

I saw Paul off and on over the years. He once favored our multistate organization (WREEC) with one of his premier performances and was always supportive of my work. I owe him much for his generous help when I really needed it. He contributed an important intellectual and prestigious building block for the program, which I was able to build over my 20 years of state service.

References

Brandwein, Paul F. (1971). *Ekistics: A handbook for curriculum development in conservation and education* (Report No. SO 002497). Sacramento, CA: California State Department of Education Bureau of Elementary and Secondary Education. (ERIC Document Reproduction Service No. 064196)

Brandwein, Paul F. (1973). *Ekistics: A guide for the development of an interdisciplinary environmental education curriculum.* Sacramento, CA: California State Department of Education.

Rudolph J. H. "Rudy" Schafer (1927–2007) was a pioneering conservationist and educator whose innovative programs helped many people of all ages learn about the importance of respecting and preserving the environment. When he became the first environmental education director for the California Department of Education in 1967, he brought together educators, conservationists, and representatives of government and industry to create environmental learning programs for schools. In 1973, he founded the Western Regional Education Council, comprising representatives from 13 states. The group founded Project Learning Tree, which shows schools, zoos, aquariums, and other institutions how to increase environmental literacy and activism, K–12. The regional group grew nationwide to become the Council on Environmental Education, which has trained hundreds of thousands of teachers, who in turn reached millions of students about the importance of sustaining the environment. The recipient of many awards for his work in conservation education, Schafer also collected and sold antiques.

Watershed Education for Sustainable Development

William B. Stapp[1]

I first met Paul F. Brandwein 35 years ago, when he served as director of the Pinchot Institute for Conservation Studies, in Milford, Pennsylvania. He convened a meeting of formal and nonformal educators to identify ways to integrate conservation into the mainstream of American education.

This contribution in many ways builds on the educational philosophy, cognitive framework, instructional techniques, and interdisciplinary curriculum strategies Brandwein developed (1955/1981, 1966, 1973), following in the footsteps of John Dewey (1933, 1963). Brandwein strongly believed that if students were to contribute to society, they needed to be

- grounded in all areas of the curriculum
- linked to real-life experiences
- committed to school and community interaction
- experienced in individual and group investigations
- problem-solvers, not problem-doers
- persistent in seeking explanations
- allowed the time to think and seek solutions
- mentored to work toward responses
- informed at the local and global level

Brandwein was an advocate of models of instruction that view the teacher not just as a conveyor of knowledge but also as a facilitator of learning. This learning takes full advantage of community resources and the abilities of the students, recognizing that

---

[1] Editor's Note: Keith A. Wheeler, first chief executive officer of the Global Rivers Environmental Education Network (GREEN) and president of the Paul F. Brandwein Institute, and I have crafted a shortened and revised version of William B. Stapp's essay, a published version of the first talk in the Paul F. Brandwein lecture series, that originally appeared in 2000 under the same title in the *Journal of Science Education and Technology, 9*(3), 183–197 © 2000 Plenum Publishing Corporation, excerpted with the kind permission of Springer Science and Business Media. William B. Stapp, professor emeritus of resource planning and conservation at the University of Michigan School of Natural Resources and Environment, died May 21, 2001.

experience feeds knowledge, and knowledge feeds experience; the known feeds the unknown, and the unknown catalyzes knowing. The teacher is a guide, not the guardian of the archives. A child can and should learn in all ways—by voice, by book, by machine, by investigation, but above all, by example. A teacher cannot be replaced; a lecturer can. (Brandwein, 1973)

Paul Brandwein was instrumental in helping me to explore the educational field for an instructional model for improving the teaching and learning climate in education. The field of environmental education cannot be seen as isolated from other emerging fields that focus on human rights issues, development issues, conflict issues, and peace. Environmental issues, for instance, involve ethical questions regarding the sharing of the world's natural resources. As we look to the future, we need to take a closer look at issues of human rights, development, conflict, and peace (Wals, Stapp, & Cromwell, 1993).

This essay focuses on the past, present, and future of the Global Rivers Environmental Education Network (GREEN), an innovative, action-oriented approach to education based on an original interdisciplinary watershed education model that reflects many of Brandwein's ideas. GREEN's mission is to improve education through a global network that supports local efforts in watershed education and sustainability. As a nonprofit organization linking teachers, students, administrators, and professionals in watershed education programs in all 50 states and in over 135 nations, GREEN serves as a resource to schools, youth organizations, and communities that wish to study their watersheds and rivers and work to improve the quality of life.

The Educational Setting

Our world is intertwined by massive environmental issues that transcend national and international boundaries—issues that can be addressed only through an unprecedented degree of global cooperation and action (Stapp, Wals, Moss, & Goodwin, 1996).

One major challenge that will increasingly confront environmental educators is to develop curriculums and instructional strategies that emphasize the global component of local environmental issues but neither overwhelm students nor make them lose hope (Stapp, Wals, Moss, & Goodwin, 1996). How can we educate and empower students to take action on local issues while simultaneously developing within them an international cross-cultural perspective? How can we best encourage this first generation of truly planetary citizens to assume responsibility for their shared home?

Rivers were selected as the central focus of this global, experiential, interdisciplinary, action-taking approach because they are a reliable and informative index of environmental quality. Rivers also naturally link chemistry to biology and the physical sciences to the social sciences and humanities. They bind together the natural and human environment from the mountains to the sea and from farmland to the inner city. Rivers were the sites of early civilization because they provided fresh water, fish and other food, and a means of transportation. Floodplains provided rich areas to grow crops. Rivers also contain a historical perspective of cultural diversity and can be a vehicle for cross-cultural dialogue. For these reasons, the study of rivers forms a coherent curricular framework for study of a wide range of environmental studies and issues (Stapp & Mitchell, 1996).

Environmental educators have long recognized the educational potential of water resources. Many activities and lessons and much fieldwork in environmental education revolve around water-quality monitoring and protection. Such programs often involve schools and their surrounding communities; responsible water resource management can maintain a high quality of life for both human and nonhuman species. Some programs focus primarily on local watersheds, whereas others also explore the global dimension of local issues by linking communities worldwide. Many groups tie into GREEN, which for more than 10 years has actively promoted not only interactive, community-based water-quality monitoring but also international networks.

Rivers

Water at one point seemed an unlimited resource. The use and abuse of water resources by an increasing human population and an ever-expanding industrialized society, however, have begun to affect the very nature of water and the water cycle itself. Increasing global urbanization places new demands on both surface and underground water resources. Throughout history, successful joint use of river basins has depended on cooperation among riparian states. Failure to reconcile the competing interests of upstream and downstream users can generate considerable political conflict.

For example, in the states on the Arabian Peninsula south of the borders of Iraq and Jordan, including the Kingdom of Saudi Arabia, Kuwait, the United Arab Emirates, Bahrain, Qatar, Oman, and Yemen, not a single stream flows year-round. The over 23 million people living in this area depend completely on underground aquifers, which do not exist in isolation from the hydrologic cycle but must be recharged by precipitation elsewhere in the watershed. If water is managed irresponsibly in the upper reaches of the watershed, millions of people downriver will suffer. Citizens of the Arabian Peninsula must practice effective water management and work in cooperation with nations in the headwater regions to guarantee their survival (Stapp & Mitchell, 1996).

Mostafa K. Tolba, former executive director of the United Nations Environment Program and current president of the Center for Environment and Development, noted that international competition for water and land could endanger political ties. Tolba stated that food security depends on how well countries manage their land. He connects ecological stability with political stability.

Degraded water quality in the developing world will need to be tackled before sustainable development can be accomplished. In the Third World, over five million children under the age of five die each year from the water they drink. Eighty percent of all sicknesses result from unsafe water corrupted by land drainage and poor sanitation. One of four hospital beds is occupied by people with waterborne diseases (the World Conservation Union, 1991).

Most citizens in industrialized countries have access to safe drinking water, but water quality still needs monitoring. While point source pollution has been decreasing in many countries, nonpoint sources continue to cause problems, especially with groundwater supplies. The water that flows over land and into rivers also percolates through the soil to recharge deep groundwater aquifers. Land use practices polluting streams and rivers can also pollute groundwater sources. The industrialized countries of Europe and North America face increasing contamination of groundwater reserves. In the United States, up to 2 percent of water in our deep aquifers may be unsafe to drink. Pollution from nonpoint sources such as fertilizers and pesticide residues in farm runoff, deicing salts from city streets and highways, leaking underground sewer lines, and inadequate disposal of chemical and other hazardous wastes are mostly to blame. In addition, many arid regions are mining their deep aquifers. Because aquifers generally recharge slowly, continuous groundwater mining can lead to immense future water shortages (Stapp & Mitchell, 1996).

Many international watersheds are threatened by activities, such as agricultural chemicals washing into surface waters, causing cultural eutrophication and lower species diversity; industrial waste released by food factories and chemical plants increasing biochemical oxygen demand; accumulation of toxic substances, like heavy metals and phenols in the sediments near industrial outfalls, poisoning benthic organisms and shellfish; drainage of thousands of small settlements established on the shores of river systems that lack wastewater treatment facilities and contribute to hepatitis outbreaks and other diseases; and poorly designed irrigation schemes creating conditions leading to oxygen loss, water logging, salinization, infertile soil, and lower food production.

The need for a national and international framework to integrate development and conservation is clear. All nations need a foundation of knowledge and information exchange, a framework of law and institutions, and consistent economic and social policies if they are to advance rationally. National and international programs for achieving sustainability must be adaptive and continually redirected in response to new experiences and to new needs.

GREEN's Roots

GREEN started in 1984, when a biology class at Huron High School in Ann Arbor, Michigan, became concerned about the water quality of the river (the Huron) that flowed by their school. The students knew that the Ann Arbor Parks and Recreation Department leased a windsurfing concession at a city park bordering their school. Several windsurfers, who had sailed on the river, had contracted hepatitis A. The class realized that water quality formed a potential health threat, and their concerns fueled a 3-week investigation of the Huron River. With the cooperation of university resource people, the biology class tested the water for the nine parameters designated by the National Sanitation Foundation as critical indicators of water quality.

The investigations revealed that after heavy rains, the fecal coliform count reached levels high enough to categorize the river water unsafe for body contact. The question quickly became, "Is windsurfing a body contact sport?" Findings and opinions were sent to the County Health Department, the City Parks and Recreation Department, city council members, and published in the school newspaper and in the "Letters to the Editor" section of the local newspaper. In response to these data, the concerned city council funded the University of Michigan's School of Public Health to test the river's water quality at regular intervals. Approximately 70 businesses and homes in the city had sanitary waste lines illegally connected to storm water lines that drained directly into the river. The raw sewage accumulated in these lines flushed into the river after rainfall, posing a health hazard to people using the river for recreation. These findings triggered the development of a formula that could help predict when the river should not be used for windsurfing.

The students then collaborated with the city on the design of a sign posted on the riverbank to warn windsurfers. The sign stated,

DUE TO STORM RUNOFF FROM RECENT RAINS, BACTERIA LEVELS
IN THE RIVER EXCEED STATE WATER QUALITY STANDARDS.
FULL BODY CONTACT IS NOT RECOMMENDED

Later, the city decided not to renew the lease of the windsurfing concession.

The students' and teacher's interest in this high-school water-quality monitoring project, along with the collaboration from the University of Michigan, led to other high-school classes' involvement along the Huron River. These high schools came together to share their data and to form a collective picture of the water quality and to recommend taking actions along the river.

In 1987, the Huron River project caught the attention of Friends of the Rouge, a Detroit nonprofit community organization that arose when citizens realized that the Rouge River in Southeastern Michigan had been abused and neglected for too long. The Friends of the Rouge funded students and educators from the University of Michigan's School of Natural Resources and Environment to establish a pilot project involving 16 schools distributed throughout the Rouge River Watershed. With the assistance of the pilot schools' teachers and Friends of the Rouge, the university team developed an interactive water monitoring program for schools. The program is now widely known in environmental education circles as the "Rouge River Model." Within 3 years, over 100 elementary, middle, and high schools participated in the Rouge River Watershed program (Wals, Monroe, & Stapp, 1990).

Programs using this interdisciplinary cooperative model expanded into over 200 Midwestern and 2 Canadian schools. A key element in these programs was the ability to share ideas and questions among students, teachers, community groups, and agencies throughout the watershed via computers linked through a telecommunications network to one central host that acted as a clearinghouse for storing and forwarding data.

GREEN's Educational Components

Watersheds form the central focus of investigation in GREEN-type programs, which transcend physical, political, and social boundaries. Most rivers originate in rural areas and flow through suburban and urban areas, collecting water and draining in larger watersheds and/or the ocean. Such linkages provide the opportunity to bring diverse socioeconomic and racial populations together to learn more about each other and work jointly to help resolve issues that affect us all. The initial 6 years of program development led to the following model used in most GREEN programs (Mitchell & Stapp, 1998). Through experiential learning, participants investigate

- history and culture
- land use practices
- sensory evidence
- benthic organism studies
- laws and regulations
- cross-cultural exchanges

They make use of

- nine water-quality tests
- computers, networking, and group cooperation
- planning for the future

They

- take action
- are evaluated

An ongoing *evaluation procedure* determines what changes are occurring at the student, teacher, institution, and/or community levels. In carrying out the *watershed educational model*, core educational approaches help interested students, teachers, schools, and communities in GREEN-type programs form an excellent medium for teaching systematic analysis and integration of information and knowledge. Our land use practices have a strong influence on the quality and quantity of water throughout the watershed in which we live. Whether we remove vegetation along the banks of a stream or leak toxic chemicals into a river from a factory, our actions affect the river system. Aerial images, maps, excursions, interviews, and field studies note changes in land use practices and the impact of these changes on river water quality (Stapp, Wals, & Stankorb, 1996).

The *experiential approach* toward education brings students out of the classroom and provides opportunities for them to study and help resolve real-world issues. People are characteristically most interested in what affects them directly. When people's interest is captured on a personal level, the result can be genuine learning that leads to action motivating change on both the individual and the community level. Developing deep understanding of local issues and learning about the actions of others can empower participants to make a difference in their world. Further, students come to experience the value of science, mathematics, and technological knowledge as they engage in their practical application when monitoring and analyzing the watershed. It is critical that the students, not just the teacher, plan, collect, analyze, and take responsible action (Wals et al., 1993).

Becoming a student of the world's rivers requires an *interdisciplinary approach* that goes beyond any one way of thinking. Studying local water systems can link science, humanities, social studies, and geography classes and enhance the development of more scientifically and environmentally literate students capable of understanding and addressing specific issues. A watershed study can meld water ecology field studies with historical and sociological investigations of the local water and land use systems. Through this approach, students build understanding of the complex cause-and-effect links that create local problems and realize that these problems are often local manifestations of global issues. By addressing water-quality issues, students enter the cross-section of three important domains of contemporary education—science, technology, and society. Resolving these issues requires knowledge of all three domains and an ability to integrate this knowledge into one's own decision-making process (Bybee, 1987; Yager & Penick, 1990).

Integrated problem solving enables students and teachers to resolve an issue that learners have identified. This process differs from more traditional studies, which more narrowly focus on goals and objectives presented by outside experts. Rich immersion into the local environment can empower students to become effective problem solvers. They learn to recognize and define problems from a variety of perspectives; they collect, organize, and analyze information; they identify, study, and select actions that may lead to improvement; they develop and carry out a specific plan of action; and they evaluate the outcome of the entire process (Brody, 1982; Wals, Beringer, & Stapp, 1990). During the problem-solving process, students become engaged in group work, joint decision making, and other skills needed to function in a democratic society (Wals et al., 1993).

It is important not only to test the river to determine its health, but also to prescribe "treatment," or *action-taking strategies*. The network (computers, newsletters, cross-cultural partners' exchanges, and workshops) facilitates the exchange of different methods of problem solving and action taking. Students can share successful and problematic experiences as they develop strategies on what to do in their own community. A focus on water-quality studies and watershed land uses encourages students to look beyond their books for education and to try to improve their local environment. Students can find creative ways to educate their communities about water quality. Students involved in a GREEN water-quality monitoring program took some of these culminating action-taking steps: They

- established a communitywide hazardous waste collection and disposal day
- promoted laws and posted areas to prohibit powerboats or reduce their speed limits in an effort to save the endangered manatee
- stenciled messages on street curbs to alert people to avoid dumping wastes into storm drains directed to the river
- prepared and delivered weekly radio announcements on local river history, river monitoring results, and student water-quality concerns
- developed a school policy concerning use of lawn chemicals on school property
- worked with their city council to post signs along the river against water-based activities until water quality could be improved
- appeared on national television to call attention to untreated sewage effluent being discharged into an ocean-based recreational area
- performed environmental audits for local businesses and provided advice on soil runoff and erosion

If carefully guided and properly focused, the problem-solving and action-taking components of the program can be an empowering experience for both teachers and students (Wals et al., 1993; Mitchell & Stapp, 1998).

Through involvement in a river network, students can form *support networks* to share information, technologies, and actions taken on local rivers. They can also learn that their investigations are valued by their peers. Acknowledging the challenges of interdisciplinary and experiential education, GREEN tries to facilitate the formation of collaborative support systems of the schools and the teachers involved in a watershed monitoring program. These systems consist of workshops, newsletters, computer telecommunications conferences, and multimedia forums to share expertise and to provide internal support. Within each program, GREEN encourages the formation of links between the school and the community and the creation of a community of learners (including parents and representatives from neighborhood groups, businesses, industry, and government) willing and able to support each other in creating a stimulating environment in which students can explore their watershed. GREEN also encourages classrooms to consider a cross-cultural component to their watershed education program. Cross-cultural partners further the idea of global citizenship by linking students from different regions of the world and allowing participants to share technical, scientific, social, and cultural information on their respective river systems.

GREEN has developed an ongoing partnership network for carrying out its educational, scientific, and cultural mission. GREEN's partners include businesses, industries, and corporations; local, state, and national governments; international organizations; foundations and consortiums; natural resource commissions; universities and colleges; and international agencies and programs, among others. These partners have helped to conceive, develop, promote, and finance GREEN's multiple goals and objectives. More specifically, these partners have helped GREEN to develop low-cost water-quality monitoring kits, formulate cross-cultural partner programs, develop software and computer technology, access remote sensing maps, prepare instructional materials, establish global communication connections, sponsor specially designed workshops, and improve global networks to promote watershed sustainability. GREEN also uses the 50 or more country coordinators to identify and adapt programs for unique cultural settings. This partnership network has been essential in helping both to create and to expand GREEN's effort in watershed education and sustainability (Stapp & Wals, 1994; Stapp, Pennock, & Donahue, 1996; Wals et al., 1993).

GREEN's International Development

There are over 200 international river basins (shared by two or more nations) in the world, representing over 47 percent of the global land surface and 60 percent of the combined land area on the continents of Africa, Asia, and South America (Frederick, 1996). Our initial international watershed educational model was developed in the Great Lakes region between Canada and the United States and on the Rio Grande (the *Rio Bravo*) bordering the United States and Mexico.

Between 1984 and 1989, various communities in different regions of the world contacted the University of Michigan wanting to share and obtain more information on watershed education. Based on this expanding international interest, 26 university students and faculty with background in water quality and environmental education responded by organizing an advanced environmental education course in January 1989. The GREEN students and faculty laid the foundation for this international network. Students then traveled to and helped organize 22 workshops in 18 countries during the summer of 1989. Each country workshop brought together educators, administrators, citizens, resource specialists, students, and representatives from governmental and nongovernmental organizations to exchange thoughts and ideas on watershed education. Each workshop aimed to discuss approaches to experiential field trips and cross-discipline teaching and to explore how watershed education programs might support the educational and instructional goals of the nations involved.

GREEN's First International Congress, in Sydney, Australia, in 1995, formed international regional teams of country coordinators in Africa, the Middle East, Latin America, Asia, Europe, Australia, and Pacific Oceania. The conferees developed a full matrix of short- and long-term goals and activities. Their work built the foundation for the planning, operations, and evaluation of GREEN's international network over the next 5 years and provided a blueprint for GREEN's activities globally.

GREEN has kept a strong commitment to both national and international river basins since its founding. In 1996, my colleagues and I prepared a publication, *International Case Studies on Watershed Education* (Stapp, Walls, & Moss, 1996), which highlighted GREEN activities, programs, and data on over 40 international rivers in the 7 regions. From these international studies we obtained environmental information ranging from the pristine to the polluted. For instance, nitrogen levels in some Swedish estuaries have increased by 200 percent since the 1950s. In Japan, 600 tons of inorganic mercury has been dumped into Minamata Bay since 1953; the Tansui River in Taiwan contained at midday 0.0 mg/L of dissolved oxygen, 138 ppm of biochemical oxygen demand (BOD 5-Day), and high concentrations of heavy metals and toxic organic compounds.

Almost eight million Third World people die each year from the air they breathe and the water they drink. In many parts of the world, no pipes distribute clean water to residents, and wastewater treatment facilities to reduce pathogens from entering waterways are rare.

We also obtained useful educational information that allowed us to further adapt GREEN to local circumstances. For example,

- Third World nations need low-technology solutions and access to low-cost water-quality monitoring kits.
- Many geography teachers in Africa now can sit on a riverbank and interpret certain phenomena: rocks covered with algae may be a sign of nutrient enrichment; meandering rivers may cut into banks and deposit the eroded soil on a downriver flood plain; certain macrophytes indicate water temperature and pH levels; riffle areas contribute dissolved oxygen to the aquatic ecosystem; and small concentric rings on surface water may indicate the site of emerging insects.

Observations are often consistent with water-quality data obtained with scientific water monitoring kits and systematic macroinvertebrate assessments. Aerial and satellite images are useful in monitoring watersheds to determine current and changing land use practices and the resulting impact on water quality. Cross-cultural partner watershed programs spark interest among involved schools worldwide.

Furthermore, there is an international need to develop a framework in which upstream nations understand implicitly the natural rights of downstream nations for a stable and unpolluted water supply. This need calls for watershed action plans, which should include a basin-wide network system on monitoring of land and water systems for environmental quality; development of a network for monitoring and assessment of resources for environmental quality; instillation of environmental quality to sustain health and well-being of the watershed; and provision of national planners with critical information necessary for making environmentally sound decisions to increase sustainable development.

In a world where rivers no longer conform to political boundaries, respect for diverse cultural viewpoints created through effective cross-cultural communication forms the basis for better international cooperation and action. As an international network that connects people from around the globe, GREEN provides students and teachers with opportunities to share their perspectives on teaching and learning about water quality and quantity concerns (Wals et al., 1993).

An important component of GREEN is the Cross-Cultural Watershed Partners Program, developed to provide a framework for sharing observations, knowledge, ideas, and solutions with people from diverse backgrounds. Typically, each participating school in this exchange program begins with steps developed in a GREEN publication: contacting and formalizing a cross-cultural relationship with the assistance of GREEN; preparing students for a cross-cultural exchange; developing pen-pal relationships; researching the historical background of one's river; monitoring the benthic macroinvertebrates and physical-chemical quality of the water; visualizing the desired condition of the local river at a time in the future; identifying the laws, policies, and responsibilities of relevant regulatory agencies; developing and carrying out an action plan; and evaluating the program. GREEN partnerships include schools from over 20 nations (Stapp, Walls, & Moss, 1996).

Many environmental problems are caused by human activities, and education should be part of the process of changing attitudes into a new system of thinking, new values, and new policies directed at improving the quality of human life. Education can be the instrument to change development into a sustainable future. International participants have helped to identify and clarify the direction of the GREEN program.

GREEN's mission is to improve education through a global network that promotes watershed sustainability. It serves as a resource to schools, communities, and businesses that wish to study their watershed and work to improve their quality of life. Its programs are designed to foster environmental ethics, cross-cultural sensitivity, and respect, thus contributing to a more trusting, caring, peaceful, and sustainable world.

GREEN's strategies aim to

- promote watershed programs based on sound scientific analysis, values, and action
- support community-based education through local and regional partnership initiatives
- develop model educational programs aimed at enhancing teaching and learning at the local level
- create an educational environment in which empowerment for responsible environmental behavior can occur
- provide leadership in the application of interdisciplinary, system-based watershed education
- enhance cooperation through a global network that promotes ties within and between communities
- invent mechanisms for greater accumulation of, and access, to baseline environmental information on a national and global scale

To help these aims succeed, GREEN distributes its quarterly newsletter to individuals and country coordinators in over 135 nations and promotes an international computer network making use of telecommunications, especially e-mail and the Internet, to allow students worldwide an easy way to share watershed information and to participate in a global community. GREEN's Web site—retrieved January 28, 2009, from http://earthforce.org—provides links to watershed education resources elsewhere on the Internet.

Designed to improve access to technology and to increase the opportunities for people engaged in watershed education projects to interact directly with one another in ways they find valuable, GREEN is developing a Web-based program for online exchange of standardized water-quality data sets and off-line data analysis. In addition to recording raw data, the program allows participants to add field notes and information on monitoring methods and equipment.

GREEN has also made available a series of books (published by Kendall/Hunt, in Dubuque, Iowa) on subjects relevant to cooperative watershed education. In addition, in collaboration with the LaMotte Chemical Company, GREEN has put together an innovative program containing two safe, inexpensive kits to gather information on local environmental quality to help people on all continents become more familiar with their local watersheds. GREEN will be forging new partnerships with environmentally conscious corporations that want to involve youth in improving watersheds.

GREEN sponsored the first National Student Watershed Congress. For 3 days this congress united students, teachers, and community partners who had worked together to ensure the well-being of local watersheds. School teams from across the country held seminars on how they had adapted the GREEN model to watershed education to work locally and explained the research and actions they had conducted on behalf of their watersheds. The congress gave participants the opportunity to highlight qualities of their local programs, while recognizing the successes, concerns, and challenges they share.

GREEN has also trained a cadre of workshop facilitators who work worldwide and has established special branches in Australia and Greece.

Final Reflections

GREEN believes that teachers and students should play a major role in shaping their own education. Teachers' personal classroom experiences and insights are not used enough in designing good education. At the same time, students are often an untapped source of renewable energy and creativity with ideas and concerns of their own (Robottom, 1985).

The pedagogical model with which GREEN has been working is called "Action Research and Community Problem Solving" (Wals, Monroe, et al., 1990; Stapp, Wals, & Stankorb, 1996). Kurt Lewin named and refined the "action research" methodology (1946). He believed strongly in democratic decision making and equitable distribution of power, and he believed that practical problems were a never-failing source of ideas and knowledge. Rather than asking for outside experts to resolve disputes, Lewin involved the affected groups in articulating, discussing, and eventually acting on particular problems. Through analysis, conceptualization, fact-finding, planning, implementation, and evaluation—and then a repetition of this spiral of activities—participants became engaged in a cyclical process of task resolution, marked by critical reflection and action (Kemmis, 1985).

Community problem solving focuses on resolving or improving local issues through a problem-solving process. This process originated in the grassroots community's organized efforts to help groups concerned about local problems and conditions become effective in accomplishing their goals. The process of community problem solving recognizes a difficulty and collects, organizes, and analyzes information about it; then defines the issue from a variety of perspectives; identifies, considers, and selects alternative actions; develops and carries out a particular action; and, finally, evaluates the outcome and the entire process.

Action research and community problem solving is a process that enables students and teachers to participate fully in the planning, implementation, and evaluation of educational activities aimed at resolving an issue that the learners have identified. During this process, students become the practitioners of their own education, taking on the role of explorers, researchers, theorists, planners, and actors. Through the process, they come to assume responsibility for their own learning. The teacher becomes the guide and facilitator in this process and shares the role of learner when reflecting upon his/her own teaching practices and when learning about the issue in which the students immerse themselves during the action.

Several key assumptions underlie the action research and community problem-solving process. First, it is crucial for society to solve critical issues with the full participation of its young members. Second, students need to know that they can be forces of constructive change and that their involvement is essential. In other words, education should be geared toward substituting for feelings of apathy and powerlessness the belief that one, be it as an individual or as a group, indeed can make a difference. Giving students a chance to investigate and act upon problems of their choice will increase their motivation to learn. Lastly, the school and its community contain an untapped abundance of rich material for making education more meaningful to the students (Stapp, Walsh, & Stankorb, 1996; Bardwell, Monroe, & Tudoe, 1994).

It is important in education to involve people, including students, in the challenges of our time. Nobody knows the right ethical lifestyle, but we all have to be responsible for seeking a world that is built upon equity and sustainable sharing of natural resources, not only between members of the Western world, but the world as a whole.

GREEN is designed to bring individuals closer together and to encourage them to develop a sense of responsibility for their communities and planet Earth simultaneously. Through direct learning experiences, students are more likely to recognize the relevance of science for improving their own lives, to adapt better to an increasingly technological world, and to contribute to resolving science–technology–society issues responsibly. With GREEN, the legacy of Paul F. Brandwein continues in the new millennium.

Acknowledgments Grateful thanks to Arjen E. J. Wals, a former staff member of GREEN and currently associate professor, Communication and Innovation Studies Department, Wageningen University, The Netherlands, for his substantive contributions to this article; and to past and present staff of GREEN/Earth Force for their support and leadership.

References

Bardwell, Lisa, Monroe, Martha, and Tudoe, M. (1994). *Environmental problem solving: Theory, practice and possibilities in environmental education.* Troy, OH: North American Association for Environmental Education.

Brandwein, Paul F. (1955/1981). *The gifted student as future scientist: The high school student and his commitment to science.* New York: Harcourt, Brace. (Reprinted in 1981, retitled *The gifted student as future scientist* and with a new preface, as Vol. 3 of *A perspective through a retrospective*, by the National/State Leadership Training Institute on the Gifted and the Talented, Los Angeles, CA).

Brandwein, Paul F. (1966). *Notes towards a general theory of teaching.* New York: Harcourt, Brace, and World.

Brandwein, Paul F. (1973). *Ekistics: A guide for the development of an interdisciplinary environmental education curriculum.* Sacramento, CA: California State Department of Education.

Brody, Richard (1982). *Problem-solving: Concepts and method for community organizations.* New York: Human Sciences.

Bybee, Rodger W. (1987). Science education and the science-technology-society theme. *Science Education, 7,* 667–683.

Dewey, John (1933). *How we think.* Boston, DC: Heath.

Dewey, John (1963). *Experience and education.* New York: Collier Books.

Frederick, K. D. (1996, Spring). Water as a source of international conflict. *Resources, 123,* 9–12.

Global Rivers Environmental Education Network. (1994). *Walpole Island first national water quality monitoring and environmental education handbook.* Ann Arbor, MI: Author.

Kemmis, S. (1985). Action research. In T. Husen and T. Posthlethwaite (Eds.), *International encyclopedia of education: Research and studies* (Vol. 1). Oxford, UK: Pergamon.

Lewin, Kurt (1946). Action research in minority problems. *Journal of Social Issues, 266,* 3–26.

Mitchell, Mark K., and Stapp, William B. (1998). *Field manual for water quality monitoring.* Dubuque, IA: Kendall/Hunt.

Robottom, Ian (1985). School-based environmental education: An action research report. *Journal of Environmental Education and Information 4* (1), 29–44.

Stapp, William B., and Mitchell, Mark K. (1996). *Global low-cost water quality monitoring.* Dubuque, IA: Kendall/Hunt.

Stapp, William B., Pennock, T., and Donahue, T. (1996). *Cross cultural watershed partners.* Dubuque, IA: Kendall/Hunt.

Stapp, William B., and Wals, Arjen E. J. (1994). An action research approach to environmental problem solving. In Lisa V. Bardwell, Martha C. Monroe, and M. T. Tudor (Eds.), *Environmental problem solving: Theory, practice and possibilities in environmental education* (pp. 49–66). Troy, OH: North American Association for Environmental Education.

Stapp, William B., Wals, Arjen E. J., Moss, M., and Goodwin, J. (1996). *International case studies on watershed education.* Dubuque, IA: Kendall/Hunt.

Stapp, William B., Wals, Arjen E. J., and Stankorb, S. (1996). *Environmental education for empowerment.* Dubuque, IA: Kendall/Hunt.

Wals, Arjen E. J., et al. (1993). Proposal to national science foundation (pp. 1–23). Ann Arbor, MI: GREEN.

Wals, Arjen E. J., Beringer, A., and Stapp, William B. (1990). Education in action: A community problem-solving program for schools. *The Journal of Environmental Education, 21,* 13–19.

Wals, Arjen E. J., Monroe, Martha, and Stapp, William B. (1990). Computers: Bridging troubled waters. *Australian Journal of Environmental Education, 8,* 45–58.

Wals, Arjen E. J., Stapp, William B., and Cromwell, M. (1993). *GREEN program characteristics: National science foundation proposal.* Ann Arbor, MI: Global Rivers Environmental Education Network.

World Conservation Union (1991). *Caring for the Earth: A strategy for sustainable living.* Gland, Switzerland: International Union for Conservation of Nature.

Yager, Robert E., and Penick, John (1990). Science teacher education. In W. R. Houston (Ed.), *Handbook on research in science education* (pp. 657–673). New York: Macmillan.

The Paul F-Brandwein Institute: Continuing a Legacy in Conservation Education

Keith A. Wheeler, John "Jack" Padalino, and Marily DeWall

During his lifetime as educator, author, lecturer, editor, scientist, conservationist, leader, friend, mentor, and humanitarian, Paul F. Brandwein entered the minds and hearts of many who had the opportunity to work and learn with him as well as many others who knew and learned from him through his published works and presentations. With his wife and partner, Mary, he transformed an historic farmhouse and the surrounding worked lands into a homestead. They became stewards of the pastoral and forested lands around them. Throughout his life, Paul Brandwein helped people become better environmental citizens by providing them with the tools for literacy needed to understand and act to solve environmental problems.

On March 1, 1994, Paul and Mary, together with their friend (and Paul's coauthor on many projects) Evelyn Morholt, created the Rutgers Creek Wildlife Conservancy of the Brandwein-Morholt Trust, as a means to sustain and restore a small but important area of that historic region of New York State in perpetuity. The decades of hard work and loving care Paul and Mary Brandwein devoted to the land and community resulted in a living nature laboratory, a special place—The Rutgers Creek Wildlife Conservancy—for learners of every age and background to investigate, to discover, to learn, and to enhance their sense of wonder.

Years 1993–1999

For many years, Paul and Mary surveyed their property and planned its use as a conservation and learning center. At Paul's request, in the summer of 1993, John "Jack" Padalino, then president of the Pocono Environmental Education Center (PEEC), followed up on Paul's early initiatives. He brought together a panel of leaders in nature center planning from across the nation to meet with the Brandweins, to examine the 63-acre site, and to consider the possibilities for its development and operations as a nature center. The consulting team, drawn from the leadership of the Association of Nature Center Administrators, included Steve Coleman, director of Manitoga, Inc. (New York); Ken Finch, director of the Glen Helen Center (Ohio); Tracy Kay, director of the Rye Nature Center (New York); Corky McReynolds,

D.C. Fort, *One Legacy of Paul F. Brandwein*, Classics in Science Education 2, DOI 10.1007/978-90-481-2528-9_20, © Springer Science+Business Media B.V. 2010

director of Treehaven (Wisconsin); Mike Riska, director of the Delaware Nature Society; and Pat Welch, director of the Pine Jog Environmental Education Center (Florida); as well as scientists James Montgomery, Bill Olson, Alan Sexton, and others. The group made site visits, reviewed the land, and suggested ways that it might be managed and monitored.

After Paul's death on September 15, 1994, conversations began among Mary Brandwein, the Brandwein-Morholt trustees, and Padalino with the purpose of creating an organization to perpetuate Paul's legacies—pedagogical, scientific, humanistic, and environmental. They decided to create an institute named for Paul, which would reflect his wisdom and vision and be dedicated to educating teachers and students about their responsibility for sustaining the environment.

On October 23, 1995, a meeting took place at the Brandwein home with Mary Brandwein, Padalino, Bill F. Hammond, Keith A. Wheeler, and Alan R. Sandler to determine what programs could be focused on lands at the Rutgers Creek Wildlife Conservancy. These efforts were managed by PEEC and funded by the Brandwein-Morholt Trust. In addition, the formation of the Paul F-Brandwein Institute was suggested to serve as a keystone organizational structure that would be jointly supervised by the Brandwein-Morholt Trust and PEEC.

Shortly after the 1995 meeting, the PEEC Board of Trustees authorized the development of a collaborative partnership between PEEC and the Brandwein-Morholt Trust to use the trust's lands for scientific investigations, monitoring and restoration, service, and special projects within guidelines provided by the trust. In 1995, the Paul F-Brandwein Institute was established as a "Center of Excellence" at PEEC with support from Mary Brandwein and the Brandwein-Morholt Trust. Brandwein-Morholt Trustees Henry Burger and William D. Bavoso appointed an advisory board to guide the institute. The board included Dean B. Bennett, professor (now emeritus) at the University of Maine; Mary Brandwein, for the Brandwein-Morholt Trust; Marily DeWall, then associate executive director of the National Science Teachers Association; Hammond, president of Natural Context; John "Jack" Padalino, president (now emeritus) of PEEC; Alan R. Sandler, executive director of the Architectural Foundation; and Wheeler, executive director of the Global Rivers Environmental Education Network (GREEN).

A Place-Based Program with Local Impact

During its first 5 years, the institute's programs focused on the gifted and talented at all educational levels. Basing its approach on Paul's teaching philosophy, PEEC initiated the Junior Natural Scientist program, involving students in informal scientific study of nature. The program helped young people acquire the skills, concepts, and values of the sciences and the humanities that are the basis for environmental decision making in the context of global citizenship.

The institute, while housed at PEEC, also focused on continuing education for teachers and future leaders to foster the skills, concepts, and values of the sciences and humanities. A primary focus of the Brandwein Institute at PEEC was to encourage the development of mentorship programs that linked teachers, teachers of teachers, and scientists and engaged them in teaching and learning about the environment. This focus informs today's organizational principles of the institute of sharing Paul Brandwein's values and beliefs with the next generation of conservation-minded educators.

The Brandwein Institute advisory board met often in the early years to determine the context in which the institute would do its work. The board members recognized that real-estate pressures were consuming and transforming the farmlands of the Rutgers Creek region into suburban housing developments and weekend retreats for haggard urbanites. The Rutgers Creek Wildlife Conservancy and the Morholt home, both trust properties, would provide the ideal opportunity for demonstrating scientific monitoring and ecological succession of farmland back to native forest. While active farming continued on the adjacent lands, the property of Mary Brandwein could become an exemplary demonstration site for the process of "permaculture" (a term that describes a design system that encompasses both permanent agriculture and appropriate legal and financial strategies). Engaging students, neighbors, leading environmental scientists, land-restoration specialists, and educators from the region and across the nation with this mini-conservancy makes it possible for the legacy of Paul Brandwein's Human Habitat Study to evolve and grow. (The Human Habitat Study, a nonprofit organization, dedicated itself to the maintenance of a healthful and healing global environment.)

In 1996, the institute hired David Foord as assistant director to manage its day-to-day work. In the spring of 1996, the first PEEC survey teams of staff and Junior Natural Scientists from the American Museum of Natural History visited the Rutgers Creek site and developed a plan for monitoring and establishing long-term surveys of the land and its ecosystem. A comprehensive monitoring, survey, and land-restoration program plan was initiated with connections to the White House Global Learning and Observations to Benefit the Environment (GLOBE) program and the GREEN program. Monitoring and survey work continue to track ecosystem changes today. A web site for the institute (www.brandwein.org) was designed to provide information about the ongoing projects at the Rutgers Creek property. Field botanist Bill Olson developed herbaria to document existing plant species and began a vascular plant collection that resulted in over 350 species being housed in a permanent herbarium at the institute. Breeding bird surveys were initiated in 1996, as well as automated weather monitoring. Members of the New Jersey Mycological Association initiated "fungal forays" on the conservancy grounds. Since 2002, they have identified over 500 species of mushrooms on this property.

From Local to National and Global Impact

The Paul F. Brandwein Endowed Lecture. In 1995, Mary Brandwein, the Brandwein-Morholt Trust, and the Human Habitat Study endowed an annual lecture series in Paul F. Brandwein's memory. The institute directors select speakers who have made significant contributions to science and conservation through education and research that reflect the work and contributions of Paul F. Brandwein. The lectures are featured events at annual National Science Teachers Association's conferences on science education. The *Journal of Science Education and Technology* publishes each lecture.

Past lecturers include

1996—William B. Stapp, University of Michigan
1997—William F. Hammond, Florida Gulf Coast University
1998—Cheryl Charles, Hawksong Associates
1999—Joseph S. Renzulli, University of Connecticut, Storrs
2000—Lynn Margulis, University of Massachusetts
2001—Robert Tinker, Concord Consortium
2002—Barbara Barnes, Educating Future Generations
2003—Rodger Bybee, Biological Sciences Curriculum Study
2004—Dean B. Bennett and Sheila K. Bennett, University of Maine
2005—F. James Rutherford, formerly of the American Association for the Advancement of Science
2006—Charles E. Roth, Massachusetts Audubon Society
2007—Richard Louv, author and futurist
2008—Rodger Bybee, formerly of the Biological Sciences Curriculum Study
2009—Cheryl Charles, Children and Nature Network

The 1997 Brandwein Fellows Symposium. The Brandwein Institute assembled leading science teachers and ecological scientists for a 3-day retreat at the Pinchot Institute for Conservation Studies, in Milford, Pennsylvania, in November. This core group made up the first class of Brandwein Fellows, individuals who are recognized for their contributions to conservation education and the institute. The symposium focused on defining the key characteristics for the development of field-based science research projects. The fellows considered the problems and issues that frame field-based science research, as well as the strategies employed and the roles played by different individuals within the environmental community. The working sessions were structured to allow teachers and scientists working as teams to explore ecosystems using state-of-the-art field-testing equipment and then to reconvene to share their findings. The teams provided the institute with recommendations on what and how programs should be conducted. One of the key recommendations was to formalize the Brandwein Fellows Program. Biographies of the first group of fellows appear in the symposium monograph *Science in an Ecology of Achievement* (Fort, 1998). Today, there are over 100 Fellows from across the nation comprising both educators and scientists.

A New Organization. At the end of the first 5 years of the Paul F-Brandwein Institute's existence, it became evident that significant opportunity existed for the institute to grow and become an independent organization. In 1999, it was incorporated as a not-for-profit 501(c)(3) organization, known as the Paul F-Brandwein Institute, Inc. Bylaws were written and approved, and the board of directors was elected, with Mary Brandwein serving as chairwoman.

Mary Brandwein

Padalino was selected as the institute's first president. This series of activities set the stage for the growth that was to follow. The founding board of directors included Mary Brandwein (chairwoman), Padalino (president), Bavoso (vice president), Burger (treasurer), DeWall (secretary), Hammond, Sandler, and Wheeler.

Years 2000–2003

The Summer Leadership Programs

The Paul F-Brandwein Institute, in partnership with PEEC, was awarded a 3-year Toyota USA Foundation grant to initiate Summer Teacher Leadership Institutes beginning in 2000. The program was also supported by the Brandwein-Morholt Trust.

According to Paul Brandwein, the best way to encourage the young in science is to help them early to do original work, and the best way to make that happen is through mentoring. Mentoring relationships can be scientist–teacher mentoring, teacher–student mentoring, teacher–teacher mentoring, or scientist–student mentoring. The teacher, the key to the success of these relationships, must be supported.

The goal of the Leadership Institutes was to develop teacher and scientist mentors who would share their expertise with other teachers and students nationwide.

For each of 3 years, 20 outstanding teachers were selected to attend a Leadership Institute. The program focused on three critical topics related to environmental science education: "Establishing Guidelines and Protocols for Field-Based Research"; "Integrating Technology into Field-Based Inquiry Studies"; and "Creating Instruments to Assess Field-Based Learning." At the Leadership Institutes, the teachers established guidelines and protocols to help students to collect and share environmental data. Teachers gained experience in using the latest technology and learning to integrate it into programs where students are working with and learning from scientists. The teachers also developed ways to measure how practicing science outdoors can enhance learning. Strategies for enlisting community support and securing grant monies to sustain and grow outdoor learning programs were also emphasized.

The first Leadership Institute, held in August 2000, focused on establishing guidelines and protocols for field-based research. The program assisted participants in identifying opportunities for studying air, water, soil, land, plants, and animals in the context of a changing environment. Participants formulated essential questions to inspire field-based research studies and modeled an atmosphere in which problem-based learning is the central paradigm for learning science. The goal of the first workshop was to help participants create an ecology of learning that encourages classrooms, local and regional scientists, and the community (both at large and engaging, in particular, other educators) to come together to engage in authentic inquiry about environmental research studies and change in a local context. The monograph *Ecology in Action: Biodiversity Field Studies* (Rapp, 2001) resulting from the first Leadership Institute provided strategies that enable teachers to establish outdoor learning laboratories. It documented field study activities that can be replicated on school and community sites and described how mentoring partnerships can be formed to conduct field-based research.

The second Leadership Institute, held in July 2001, emphasized the integration of technology to support authentic inquiry in field-based studies. The technology used at the workshop ranged from Internet searches to leading-edge technology tools for monitoring the local environment to communication tools for communication with other teachers, scientists, and students. Participants employed digital mapping technologies, such as Global Information Systems and other interfaces, including computer-assisted probes that monitor water quality, soil pH, and meteorological phenomena to develop land use management and habitat assessment at research field sites. Some of the institute program's time was spent on working with the latest software for analyzing, mapping, displaying, and communicating results. The monograph resulting from the second Leadership Institute, *Ecology in Action: Biodiversity Field Studies, Vol. II* (Paul F-Brandwein Institute, 2002), defined prerequisite technology needed to perform effective field-based study as well as to report and communicate the results of study.

The third Leadership Institute, held in July 2002, focused on designing and implementing assessment instruments to evaluate field-based instruction and learning. Participants devised methods to measure success with problem-based science field study. They created instruments to measure field-based learning and evaluated alternative assessments and performance-based examinations. Participants found ways to measure not only what students learned but also whether their learning had an impact on the students themselves, on their society, and/or on the environment. Teachers and scientists reviewed different models and metrics to enable them to demonstrate effectively the success of the inquiry approach to field investigations. The monograph resulting from the third Leadership Institute, *Ecology in Action: Biodiversity Field Studies, Vol. III* (Paul F-Brandwein Institute, 2003), provided a sample of problem-solving performance activities that teachers can implement to measure the success of their program.

The Paul F-Brandwein Summer Leadership Institutes trained a core of science teachers in environmental science education nationwide. Teachers at the summer institutes acquired skills to foster mentoring partnerships with students and with other teachers and scientists. Following the institutes, the teachers conducted outreach activities at workshops and meetings at local, regional, and national gatherings to share their skills in building outdoor programs and mentoring relationships with their peers.

The Sanibel Retreat

In June of 2002, a small group of Brandwein directors and key fellows met on Sanibel Island, Florida, to determine where the Paul F-Brandwein Institute should be focusing its efforts over the next 5-year period. A consensus emerged that the institute should capitalize on its exemplary local work and begin to focus on creating a national impact. In his life and work, Paul Brandwein catalyzed new trends in science and conservation education, convened the nation's brightest individuals to address critical issues, and communicated findings to key decision makers to insure the change necessary for the transformation that he sought would actually happen. This process of convening, catalyzing, and communicating became the key element that would guide the institute's future activities.

Years 2004–2007

The Conservation Learning Summit

Paul F. Brandwein and his colleagues played a key role in defining and shaping conservation education in the United States through the Pinchot Institute for Conservation Studies conferences in 1965 and 1966. These conferences brought together a community of leading thinkers and practitioners of conservation education. The participants defined their goals in terms of assuring a citizenry that understood and supported the value of scientific and rational planning for the efficient use of natural resources. These conferences had a long-lasting impact on many of their participants who, in turn, exerted influence on the various informal and formal conservation education communities of which they were members. The conference participants moved forward in response to this challenge from Paul F. Brandwein:

> We must develop new structures, new strategies, and new techniques of teaching. We must test and revise until we have developed a culture which recognizes man's interdependence with his environment and all of life, and his responsibility for maintaining that environment in a condition fit for life and fit for living. (1966, p. 13)

The Paul F-Brandwein Institute directors and fellows believed there was a need to rekindle the power of the 1965 and 1966 conservation education conferences to bring together education and conservation leaders old and new, to look carefully and creatively at our current conservation situation, and to identify how to address our challenges in conservation education. The institute assembled an advisory committee comprising members from many organizations to collaborate in the conference planning.

The Paul F-Brandwein Institute and its partners launched a national Conservation Learning Summit in 2005 and a series of supporting projects to focus the national leadership of the conservation education community and key stakeholders on identifying the root causes of the human resource and intellectual capital drain in the conservation and wildlife management sciences. One of the major drivers for the summit was the fact that a projected 60 percent of the nation's senior leadership in the government work force will be eligible to retire in 2007. Many of those retirees are natural resource scientists and conservation workers; their withdrawal will create a serious need for conservation workers to take their places.

In a letter that appeared in the summit conference proceedings, Stewart L. Udall, honorary chairman of the summit and former secretary of the U.S. Department of the Interior, wrote,

> Every generation has a rendezvous with the land. Nothing is more important than the legacy we leave for the future. It will take whole communities and whole nations, working together, to solve the problems we face. We are better off when the community is more important than the individual—when people are judged by how much they contribute to their community. That is why we created this national Conservation Learning Summit. Now I look forward to seeing tangible results and a renewed conservation ethic. (2006, p. 4)

Eighty representatives of federal agencies, nongovernmental organizations, philanthropy, academia, and business attended the invitational national Conservation Learning Summit, held during November 4–6, 2005, at the National Conservation Training Center in Shepherdstown, West Virginia. They were drawn together by a shared concern about an imminent loss of expertise to ensure the future health of the environment, including the nation's heritage of abundant natural resources. Among the many national leaders who participated was Richard Louv, author of the best-selling book *Last Child in the Woods: Saving Our Children from Nature-Deficit Disorder* (2005); Deputy Secretary, U.S. Department of Interior Lynn Scarlett; and Congressman Tom Udall of New Mexico. After lively debate, recommendations for specific programs of action emerged. The proceedings are available in PDF format on the Brandwein web site (www.brandwein.org).

An Action Agenda

The Conservation Learning Summit led to a national agenda for action and commitments from key stakeholders to implement it. Following the summit, representatives from a diverse group of organizations and public agencies came together at the National Press Club in Washington, D.C., to announce an ambitious agenda to reconnect youth with nature and prepare a future work force to take care of the nation's natural resources. The Paul F-Brandwein Institute convened the event, and Louv delivered the keynote address. A videotape taken at the summit was shown, and copies of the proceedings, *Conservation Learning Summit: A Re-Commitment to the Future* (DeWall, 2006), were distributed.

Louv's inspiring book matched his call to participants:

> Western society is sending an unintended message to children: nature is the past, electronics are the future and the bogeyman lives in the woods. This script is delivered in schools, families, even organizations devoted to the outdoors and codified into the legal and regulatory structures of many of our communities. Healing the broken bond between our young and nature is in our self-interest, not only because aesthetics or justice demand it, but also because our mental, physical, and spiritual health depend upon it. So does the health of the Earth. Conservation-oriented groups are beginning to realize that a generation that has had little or no personal connection to nature is unlikely to produce passionate stewards of the Earth.

In response, the institute announced the formation of a new campaign to "Leave No Child Inside."

The long-term health of everyone, from children to the planet itself, will benefit by reconnecting children with nature. Along the way, the Brandwein Institute and organizations working with it will prepare the next generation of informed and committed resource professionals. Over 40 years ago, Paul Brandwein identified the strong linkages between the interactions of the very young to inquiry and the natural world and showed how critical this process was as a step in the path of self-selection that leads young adults to pursue careers in the sciences. Leading representatives of the nation's largest conservation organizations, senior officials in U.S. federal agencies, deans of colleges and universities, business leaders, and government officials have joined the Paul F-Brandwein Institute to embrace the purpose and vision of this new campaign.

The Brandwein Medal

As part of the commitments made at the Conservation Learning Summit, the Brandwein board of directors voted to sponsor a national conservation educator award called the Brandwein Medal. The first award was presented to an outstanding elementary teacher on March 31, 2007, in conjunction with the Brandwein Lecture at the National Science Teachers Association's National Conference on Science Education. The recipient of the award was David Brown, a fifth-grade teacher from Quincy, Illinois. Brandwein Medal recipients receive national recognition, a bronze medallion, and a cash award.

Changes in Leadership

In January 2006, Padalino retired to the role of president emeritus and the Brandwein Institute board of directors elected Wheeler to serve as president of the board. Mary Brandwein, the driving force behind the Brandwein Institute, died at 94 on September 4, 2006, after a brief illness. The Paul F-Brandwein Institute will continue to serve a role as convener, catalyzer, and communicator, carrying on the legacy that Mary defined and that the institute has excelled at over the past decade.

Steven C. Hulbert and Cheryl Charles were elected as new members to the board of directors at the board meeting in December 2006. Burger was elected as the new chair, and Sandler, the new treasurer, of the institute.

Years 2007–2012

The institute is continuing its mission, while looking to the future.

The Carmel Retreat

The Paul F-Brandwein Institute board of directors met in June 2007, in Carmel, California, to develop a strategic plan for programs over the next 5 years. The board's consensus was that the institute should capitalize on what has been done in the past and should carry on the Conservation Learning Summit legacy to sustain the environment with children and adults. Each year, board members will decide on an annual theme, and then the institute will partner with an organization that also studies and supports that theme. The theme will also be reflected in the Brandwein Lecture and the criteria for the Brandwein Medal awardee. Possible themes include working with the "built environment" (architecture) and connecting children with nature.

Institute Expansion

The Brandwein Institute plans to relocate to a new administrative office in the Callahan House, an historical building owned and operated by the National Park Service, in Milford, Pennsylvania. The institute headquarters and collections will remain at the Morholt home on the Rutgers Creek Wildlife Conservancy in Greenville, New York.

Paul F-Brandwein Institute: Milestones

1995: An initial meeting among Mary Brandwein (Brandwein-Morholt Trust), John "Jack" Padalino (PEEC), Bill F. Hammond (Natural Context), Keith A. Wheeler (GREEN), and Alan R. Sandler (then of the American Institute of Architects) provided core ideas for a collaborative relationship between PEEC and the Brandwein-Morholt Trust to form the Paul F-Brandwein Institute. The Evelyn Morholt home on the Rutgers Creek Wildlife Conservancy was designated as the site for the institute's administrative offices and laboratories.

1996: The Paul F. Brandwein Lecture series was established as an ongoing session at the National Science Teachers Association Annual Meeting. The late William B. Stapp, professor emeritus at the School of Natural Resources (University of Michigan), delivered the first lecture.

The institute initiated several field studies and baseline data collections on the Rutgers Creek property. Field botanist Bill Olson began a vascular plant collection that found over 350 existing species, which are housed in a permanent herbarium at the institute.

1997: Educators and scientists attended a 3-day symposium in November to identify issues to be addressed by the Brandwein Institute.

1999: The institute was incorporated as a not-for-profit 501(c)(3) organization, known as the Paul F-Brandwein Institute, Inc. Bylaws were written and approved, and the board of directors was elected, with Mary Brandwein serving as chairwoman.

Toyota USA Foundation awarded the institute a grant to plan and run 3 years of Paul F-Brandwein Summer Teacher Leadership Institutes at the Rutgers Creek Wildlife Conservancy.

2000: The first Summer Teacher Leadership Institute met from July 28 to August 6, serving 20 teachers and numerous resource people. Teachers learned to implement long-term ecological techniques for use with students.

2001: The second Summer Teacher Leadership Institute met from July 19 to 29, again serving 20 teachers and research scientists, implementing long-term ecological research, integrating field-based inquiry with technology, and exploring assessment strategies.

2002: The third Summer Teacher Leadership Institute met from July 18 to 28, with 20 teachers and numerous resource people, focused on exploring assessment strategies. Participant Kelly Nolan wrote a report about aquatic and terrestrial studies done at the Rutgers Creek Wildlife Conservancy.

Members of the New Jersey Mycological Association initiated "fungal forays" on conservancy grounds. Since 2002, they have identified over 500 species of mushrooms at the conservancy.

In June, a small group of Brandwein directors and key fellows met on Sanibel Island, Florida, to develop a 5-year plan for the institute.

2004: The institute's board of directors voted to host a conference on conservation education to commemorate the conferences led by Paul F-Brandwein and his colleagues at the Pinchot Institute in the mid-1960s. An advisory committee of members from many organizations collaborated in the conference planning.

2005: The Conservation Learning Summit, with representatives from 75 entities, including government agencies, academia, nongovernmental organizations, philanthropies, and businesses, met to discuss critical issues in conservation education and the conservation work force.

2006: Proceedings from the Conservation Learning Summit (featured speaker Richard Louv) were printed, distributed, and discussed at a press briefing in May.

The institute board of directors approved the Brandwein Medal, to be awarded each year to an outstanding educator.

On September 4, Brandwein Institute founder, inspiration, and chairwoman Mary Brandwein died after a brief illness.

As a follow-up to the Conservation Learning Summit, the Brandwein Institute hosted a breakfast on October 14 for the Conservation Education Commission delegates and life members of the North American Association for Environmental Education. Steven C. Hulbert screened a summit report via DVD and presented recommendations from the summit, and Cheryl Charles announced the formation of the Children and Nature Network.

Hulbert and Charles were elected to the board of directors of the Paul F-Brandwein Institute at the annual board meeting in December. Henry Burger was elected as the new chair, and Alan R. Sandler the new treasurer.

2007: The first Brandwein Medal was awarded at the Brandwein Lecture, whose audience exceeded 500, to David Brown on March 31 at the National Science Teachers Association annual meeting in St. Louis, Missouri.

In June, the Brandwein Institute board of directors met in Carmel, California, to create a 5-year plan of activities for the years 2007–2012.

References

Brandwein, Paul F. (1966). Techniques of teaching conservation. In *Conference on Techniques of Teaching Conservation, October 10–12, 1966*. Milford, PA: Pinchot Institute for Conservation Studies.

DeWall, Marily (Ed.). (2006). *Conservation Learning Summit: A re-commitment to the future*. Unionville, NY: Paul F-Brandwein Institute. Retrieved February 10, 2008, from http://www.brandwein.org/modules&name=PagEd&file=index&topic id=0&pa ge id=32

Fort, Deborah C. (1998). *Science in an ecology of achievement: The first Paul F-Brandwein symposium*. Dingmans Ferry, PA: Paul F-Brandwein Institute.

Louv, Richard (2005). *Last child in the woods: Saving our children from nature-deficit disorder*. Chapel Hill, NC: Algonquin.

Paul F-Brandwein Institute (2002). *Ecology in action: Biodiversity field studies* (Vol. II). Unionville NY: Author.

Paul F-Brandwein Institute (2003). *Ecology in action: Biodiversity field studies* (Vol. III). Unionville, NY: Author.

Rapp, Kathleen A. (2001). *Ecology in action: Biodiversity field studies*. Unionville, NY: Paul F-Brandwein Institute.

Udall, Stewart L. (2006). Letter opening C*onservation learning summit: A re-commitment to the future*. Unionville, NY: Paul F-Brandwein Institute.

Keith A. Wheeler, president of the board of directors of the Paul F-Brandwein Institute, also heads the Foundation for Our Future and the International Union for the Conservation of Nature and Natural Resource's Commission on Education and Communication.

Wheeler served on President Clinton's Council for Sustainable Development's Education and Communication Task Force and cochaired the White House Initiative Education for Sustainability. He was founding executive director of GREEN and soil scientist for the U.S. Department of Agriculture and for Cornell University.

He speaks to conferences on sustainable development, conservation, and education issues and has edited *Education for Sustainability: A Paradigm of Hope for the 21st Century* (New York: Kluwer/Plenum, 2000).

John "Jack" Padalino, president emeritus of the Paul F-Brandwein Institute, was president (also now emeritus) of PEEC for three decades. There, he initiated numerous conservation programs in national parks and elsewhere. Padalino also led the National Science Education Leadership Association, the American Nature Study Society, the John Burroughs Association, and the Alliance for Environmental Education. As an American Association for the Advancement of Science fellow, he received numerous awards recognizing his leadership in science and conservation education. Padalino was principal investigator for two National Science Foundation projects and summer institutes funded by Toyota and a private trust. For 14 years, he chaired the National Science Teachers Association's Tapestry environmental science panel. He is a member of the International Union for the Conservation of Nature and Natural Resource's Commission on Education and Communication.

Marily DeWall, science education consultant and writer, is the former director of the JASON Academy at the JASON Foundation for Education, where she created and oversaw an extensive online professional development program designed for elementary and middle school science teachers. Prior to coming to the JASON Foundation, DeWall worked for more than 25 years for the National Science Teachers Association, serving in various capacities, including associate executive director for administration, editor of *Science Scope* magazine, and director of many corporately funded programs. DeWall is the author of numerous articles, is the editor of several publications, and is a frequent presenter at national and state conventions.

Part II
Paul F. Brandwein in His Own Words—Reprints 1955–1995

The Gifted Student as Future Scientist

Paul F. Brandwein

A Perspective Through a Retrospective

It is possible, in going on an important journey, to fix our eyes firmly on a distant goal and become so intent upon the pursuit of it that we take little stock of where we are and lose sight altogether of where we have been. In doing so, we may be spending our time rediscovering what is already known and repeating past mistakes. For this reason, it is wise to pause once in a while to reflect, to look back over the paths traveled yesterday and gain perspective on the challenges we face today.

Interest in gifted education is not new. In fact, a wealth of information on the subject has been collected over the past few decades. Unfortunately, much of the literature in which these theories, research studies, and practices are reported is now not readily accessible in convenient form. The purpose of this series of publications entitled *A Perspective Through a Retrospective* is to make available once again the concepts, principles, and applications which are a part of our heritage and, thus, to encourage a better understanding of the present in gifted/talented education through a reexamination of the past.

Irving S. Sato[*]

Preface: A Quarter of a Century Later—In Retrospect

Irving Sato and his colleagues on the board of National/State Leadership Training Institute on the Gifted and Talented have given me the opportunity to write a second preface to this small book which they have graciously undertaken to reprint.

Copyright holder of *The Gifted Student as Future Scientist* Deborah C. Fort gives permission and authorizes its reproduction here.

[*]Irving S. Sato was director (now emeritus) of the National/State Leadership Training Institute on the Gifted and the Talented. His organization in 1981 reprinted *The Gifted Student as Future Scientist* (1955 Harcourt, Brace and World) without its subtitle *The High School Student and His Commitment to Science*.

Some thirty-five years ago, my colleagues and I began a study of gifted students. Twenty-five years ago, I wrote this preliminary report entitled *The Gifted Student as Future Scientist*. Since that time, I have been privileged to follow the course of a number of students who furnished the basis for the study and to collect additional data which might assist us in verifying certain of our initial findings, certain assumptions, and, if you will, hypotheses on giftedness in general and on giftedness in science, particularly. During these twenty-five years, I have also had the privilege of consulting with certain state departments of education, cities, and school districts, as well as with some forty individual schools which had undertaken programs designed to give special opportunity to students who were deemed to be gifted or talented. I have drawn on certain additional experience with the administrators, teachers, and students in these states, cities, school districts, and schools in support of the following general statements by way of a preface to this reprint.

First, it is significant that there is a revival of interest in school programs for the "gifted and talented." Depending on an observer's point of reference, this interest has waned and then lagged for some fifteen to twenty years. Other movements in schooling have suffered a similar fate. It is important for those who wish to lay factual foundations for the study of education to note that pendulation of movements and innovations in schooling is, indeed, an observable phenomenon. It may be even a phenomenon worthy of careful study.

Second, with regard to giftedness in science, it may well be that the foreword to this book, "These Critical Years," could be rewritten as applying to the next quarter century without much emendation. In the opinion of a decent number of students of our society (particularly Daniel Bell), we are about to enter—or have already entered—a post-industrial era in which competence in science and mathematics will again be of utmost importance. However, in a post-industrial society—taking its impulse from the lives of its citizens—a grounding in the social sciences as well as in the humanities will surely be requisite, particularly as part of the equipment of gifted students who become our scientists and mathematicians.

Third, the hypothesis stated on page 150 (a triad model, if you will) finds general support in my observations over this quarter of a century, as well as in my examinations of the literature. It is generally accepted that the expression of giftedness in cognitive areas is not solely (and not mainly) grounded in high IQ, but is an interaction of personality traits (here called predisposing factors) and opportunities inherent in education and environment (here called activating factors). Factors such as high verbal and mathematical ability, use of fantasy (imagination), and high physiologic vigor are not, in themselves, sufficient as generators of highly productive work in science or scientific endeavor, namely, original researches which add new knowledge. Character-rooted traits, factors of personality and environment are, from this view, essential. (On this point, see also Joseph Renzulli's triad model.)

Fourth a mode of selection or identification of the gifted suggested as useful in Chapter 2 of this book deserves reexamination. It appears that, once a rich curriculum in science is developed and instructional modes are invented which aid in fulfilling the powers of the young in the pursuit of their special idiosyncratic excellence, then the young select themselves for the program. They identify themselves

The Gifted Student as Future Scientist   137

by their inquiries, by their original work and researches, if you will. One useful mode of identification of the gifted is, then, "self-identification." *By their work we come to know them.* (On this point, see page 165, "An Inference for the Present."

Fifth, our observations cause us to wonder whether we can or should identify the gifted and/or talented young in the schools on the basis of IQ or test scores. We may argue that, in the absence of judgments based on original contributions by the gifted, we are able, in truth, to select only the *potentially* gifted or talented. It is also clear to us that too often the gifted and/or talented who do, indeed, contribute original work are not necessarily to be found in the groups with IQ levels of 130 or above. (On this point, see also the work of Torrance and Getzels.) Far too often, the gifted individual is lost in the arms of the larger statistic. The gifted individual is one of a kind. For this reason, it is perhaps feasible to consider the kind of "operational approach" to the self-identification of the gifted suggested in this book.

Sixth, whatever model is appropriate to the identification of the young who become future contributors to science (or perhaps any other area of work), a prominent factor in the development of a scientist-to-be is a *key figure*—in the form of a "teacher" inside or outside the school or university. None of the potential or working scientists or mathematicians we observed and interviewed were willing to suggest that they could have made their journey without the intervention of a key figure who gave them opportunity and guidance—and, perhaps, destination.

Seventh, we have had occasion to speculate why the work in investigation (an "original" piece of work, to be sure) done by the young is considered by us to be a surer way to identify a contributing scientist than tests that either predicted high mental ability or demonstrated high achievement in a given subject matter. In following the course of work and contribution of the young who were part of the program, as well as their personal development, we have come to know of the *conflicts* they have faced in investigating the "unknown." To accept the task of solving a scientific problem—an unknown—means, in a sense, being alone. It means facing uncertainty and ambiguity, the possibility of failure, the highs and lows in emotion—from exultation and joy to depression and despondency—and eventually, the coalescence of dissimilar units in a "solution" (a "sudden live image," a "flash of insight") which leads, of course, to new problems. This constant alternation of problems with solutions, of conflict and coalescence, of exploration and illumination requires courage and persistence—in a word, character.

This is not to say that Wallas's description of mental process is valid. Aruti's "The Magic Synthesis" suggests another model. We are, nevertheless, obliged to consider that the making of a scientist or mathematician or artist or writer or any contributor of new works (of additions to the culture) depends not only on an *innate ability* and on *opportunity*, but also on the *maturing of personality* as well. *Actual work in problem solving*, not in the *problem doing* of laboratory exercises whose solution is foreordained, gives evidence of an ability to face the alternating conflicts and coalescence characteristic of original work. The school should be a place where original work is possible—at least, for the scientist-to-be.

138 P.F. Brandwein

Eighth, we are now assured that equal educational opportunity of achievement is to be made available to *all* young. But herein lies a paradox. The gifted in science and mathematics—even as the gifted in athletics—require special opportunity to develop their gifts. Eventually, we shall come to know and practice what is becoming obvious: given the fact of great variety in ability of the young, *special* opportunities as well as *equal* opportunities are to be afforded to all. This may mean that, in curriculum and instruction, nothing is so unequal as the equal treatment of unequals. This may mean that, once equality of educational opportunity is safeguarded for all, the young can be trusted to fulfill their special powers in the pursuit of excellence.

Thus, both difference and likeness will become precious, as they should.

When they do, we shall outwit time.

Paul F-Brandwein

Foreword: **These Critical Years**

These are critical years for our country. The tremendous advances in science during the last quarter-century have resulted in a demand for more scientists, more engineers, more technicians, more trained workers. These are years that pose grave problems for industry, for government, and for education. These are years, also, of challenge and of opportunity for our young people.

Henry Chauncey, president of the Educational Testing Service at Princeton, states the problem this way:

> Manifestly, if we are to maintain our technological superiority throughout the next quarter- or half-century, our educational system must produce a large number of very able scientists and engineers. Yet the future of America depends also on the quality of leadership available in government, business, and the professions. If we are to produce more scientists, as we must, and if we are to do so without making undue inroads upon other important specialized fields, we must make fuller use of our raw intellectual resources than we have made in the past.

This is tantamount to saying that the issue will ultimately be decided by the quality of the education we make available to the youngsters now in school and those who will shortly be entering.

General Omar Bradley points up the challenge in another way:

> Our scientific lead depends on many factors ... on the ability of our educational system, from junior high school upward, to find and encourage scientific talent.

President Eisenhower, recognizing the gravity of our responsibility to provide better-trained people in science, has turned the problem over to a special committee on the training of scientists and engineers. It consists not only of the directors of the offices of Defense Mobilization, the National Science. Foundation, and the Atomic Energy Commission, but also the Secretaries of Labor, of Commerce, and of Health, Education and Welfare.

Beyond these statements is the fact, constantly stated in articles in the newspapers and magazines, that underdeveloped areas of the world are drawing upon trained

The Gifted Student as Future Scientist 139

American scientists for technical assistance in improving their living standards. Our skills in science are helping to place these countries on the side of the free world.

Scientists are trained, not born. Therein lies the challenge—and the opportunity. It is a challenge for all citizens—a challenge to find an answer to the problem of providing boys and girls with the opportunities to recognize their full intellectual potential.

Benjamin Fine, writing in the New York *Times* of October 10, 1953, states the general nature of the problem in these words:

> The question of what to do with students of superior ability has troubled school and college officials for many years.... Generally the student on the upper intellectual scale is left pretty much to shift for himself.

America's future—indeed, the future of western civilization—lies with our young people, many of them gifted in special ways. And it is a growing responsibility of our schools and colleges to discover methods of preparing them for the highest use of their individual gifts. All our future scientists go through our secondary schools. It is there where they may be identified and trained.

Who among our young people becomes a scientist? What starts a boy or girl on the road of becoming a scientist? Who teaches the future scientist? How do we get more scientists? Some guides, based on many years of search for a practical method of developing the abilities of young people, are to be found in this book.

We count it a privilege to place its message before American educators, businessman, and parents. It has implications worth weighing by all who are interested in young people as the future scientists in the critical years that lie ahead.

Harcourt Brace Jovanovich[*]

Preview: **Intent and Content**

The purpose of this little book is sixfold.

First is the formulation of a conceptual scheme, or large working hypothesis, which may be useful in determining the nature of giftedness or high ability in science. Such a conceptual scheme, based on first-hand observation of working scientists and youngsters who have high promise of contributing to science, is presented here for consideration. Three factors are considered as being significant in the development of future scientists: a Genetic Factor, with a primary base in heredity (general intelligence, numerical ability, and verbal ability); a Predisposing Factor, with a primary base in functions which are psychological in nature; an Activating Factor, with a primary base in the opportunities offered in school and in the special skills of the teacher. High intelligence alone does not make a youngster a scientist.

Second is the description in detail of one type of program which has been somewhat successful in stimulating the development of youngsters with high ability in

---

[*] Harcourt, Brace and Company was the original publisher of the book, in 1955. By 1981, Paul Brandwein worked for its successor, Harcourt Brace Jovanovich.

science. The assumption is that even without adequate tests to identify such students, a program with adequate opportunities for youngsters with the level of Genetic and Predisposing Factors described here will cause such youngsters to come forward and identify themselves.

Third is a description of available tests, and tests now being developed, which seem to have promise in identifying these youngsters. Particularly are attempts made to present "operations" which busy teachers may develop for their own particular situation—e.g., Man-to-Man Rating Scales, and a Rating Scale of Predisposing Factors. The performance of youngsters of high ability in science on existing tests and the promise of a Test on Developed Ability in Science in identifying such youngsters are reported.

Fourth is an attempt to describe the behavior of these youngsters at work. Theirs is a picture of high "drive" and a tendency to introversion. An attempt is made to analyze their ways of approaching a problem. This approach appears to involve discernible periods of "Incubation" and "Illumination" which seem to be characteristic of the way these young people work.

Fifth is an attempt to describe the kind of teacher who seems to be successful in working with these youngsters. The indication is that aside from his superior training and wide interests, such a teacher also serves as a "father" or "mother" image.

Sixth are some proposals which may be useful. They are concerned with certain specific problems on the local level and with suggestions for action on the local, state, and national levels.

Finally, in adopting any program, any testing device, any proposal, it is worth remembering that there is no one way of doing what is worth doing. Teaching is a personal invention.

The boys and girls of this country are national resources upon which we must draw to solve the pressing problem of our dangerous shortage of scientists. All the boys and girls who will be our scientists pass through our secondary schools. It is there where efforts to increase the number of youngsters who enter science professionally will bear the greatest fruit.

The teacher remains the key to the solution of this problem, which is claiming national attention. And so our teachers have also become an important national resource.

Caution: To the Reader

Although this account is based on observations, even research, done over some twenty years, it is aimed at the teacher and the school administrator—not at the researcher per se. Our prime purpose is a report to teachers who, day after day, bloody their cerebral noses in the attempt to unite the gifts of their students with the opportunities at hand. The reader should be aware, therefore, of the purposeful brevity of detail and the emphasis on useful practices which may be adapted for use in teaching the gifted.

The Gifted Student as Future Scientist

Anyone who travels about the country, visiting schools and speaking to teachers on the tactics and strategy of modern science teaching, finds wide acceptance of the notion that gifted boys and girls can be left to their own educational devices, more or less. It is as if to say that the vast educational appetitie of a gifted boy or girl can feed itself. Somehow equality of opportunity for all has been equated with identity of opportunity for all.

Nothing could be more contradictory to one of the major concepts of our American educational and political philosophy.

The development of each and every individual to his fullest is at once the basis of our effort as well as our goal.[1]

Jean de La Bruyère, writing in 1688 in his *Caractères*, said "There are certain things in which mediocrity is insupportable—poetry, music, painting, public speaking."

Now we must add—mediocrity in science, and possibly science instruction, is insupportable as well. Urgent, and of the first order of priority, is our need to look to one of our critical national resources—youngsters of high level ability in science.

True enough, boys and girls of high level ability in science are of no greater importance *as human beings* than are students without particular gifts in science—whether they be of average or low level ability. First, the student with average or low ability in science may be a contributor of first rank in other work. Second, while it is true that students with high level ability contribute more—in original work and discovery—to science, high level work in modern science cannot go on without the support and understanding of the majority of our citizens. In fact, one of the major assumptions of this booklet—and some evidence is available on this point—is this: Successful work with students of high level ability on the school level does not usually go on unless successful work with the average student and the so-called slow learner goes on as well. Nevertheless, individuals are not interchangeable; their contributions are unique; each deserves the opportunity to realize his own potential.

The observations detailed here have one major flaw, among others, which the reader will readily detect. The teacher on the job, surrounded by youngsters and helping them grow, often loses the objectivity necessary to impartial observation. Even worse—he hasn't the time or opportunity to sit back and examine and re-examine his findings, possibly to plan improvements of his experimental design—a major need of the researcher. This work reflects such a need. However, the research worker in education often hasn't the opportunity of the teacher who is daily at work with the raw materials of his observations—youngsters at work, youngsters acting and reacting, youngsters growing.

---

[1] Benjamin Fine, writing in the New York *Times* of October 18, 1953, states the general nature of this problem as follows: "The question of what to do with students of superior ability has troubled school and college officials for many years. For the most part the educational curriculum on virtually every level, from elementary through university, is geared to the average student. Frequently the below normal or mentally retarded boy or girl receives special attention. But generally the student on the upper intellectual scale is left pretty much to shift for himself."

The observations described here actually began in the spring of 1929 when, fresh out of high school, I had the good fortune of securing employment in the Littauer Pneumonia Research Laboratory under the patient guidance of Dr. Isidor Greenwald. From 1929 through 1944, I had the good fortune to be associated with working scientists in different kinds of situations, in different kinds of laboratories, working on different kinds of problems. As I undertook graduate work (for the doctorate) I came into closer contact with scientists in various fields, particularly the large areas of Biology and Chemistry. It seemed even then that scientists were different from others. But was this true?

What makes a scientist? Or as the question became transmuted in time and thought: Is there such a charactersitic as "science talent"? This question served as the underlying motivation for an investigation which fluctuated in intensity over the period 1931–1953. Many notebooks were filled with observations and case studies (two abbreviated case studies, 1 and 2, of graduate students are to be found in Appendix E).

How often have we wished for a formula like

$$1 = \frac{E}{R} \quad \text{or} \quad E = mc^2$$

(where E = educatability, m = materials of instruction, c = creativity) or some similar scheme which would enable us to say with dignity and certainty that such is such, and that is that. We can't at present. Hence the observations presented here are subject to all the weaknesses inherent in a *social science* (albeit with scientific overtones), which lacks a body of data upon which the organization of conceptual schemes depends.

Let us use the phrase "body of observations" with the greatest care. When I took my doctorate work in Biology, I could rely on a body of observations, carefully confirmed by patient observers throughout the world. First came the observations, to repeat, carefully confirmed. No such body of observation, except in the areas closely related to psychology and measurement, exists as yet in education. The observations are being gathered. In education per se there is not, as yet, a universal desire to substitute painstaking, slow, varifiable observation for intellectual schema which appeal primarily to the imagination. In the area of the study of what makes high level ability in science we have few confirmed observations. But we are beginning to get observations—very slowly. And these must be winnowed from opinion and from urgent exhorations to increase the number of trained persons with science talent.[2]

We want here to add, if we can, to the taxonomic descriptions of *Homo sapiens* variety *high ability in science* so that further steps can be taken toward recognizing him and helping him attain fruitful maturity. Execusions into statistical evaluation are deferred mainly because this book is aimed at the teacher and administrator;

---

[2]This is not to say that there is a lack of studies on the "gifted." The research of pioneers like Cox, Hollingworth, Terman, and associates has given us some understanding of the characteristics of highly intelligent children.

The Gifted Student as Future Scientist   143

statistical analysis would have been made, however, if deemed necessary to add meaning. Perhaps the most important function of this effort will be to show the state of our ignorance in the field of developing students with high ability in science.

Whether or not this be so, our nation needs scientists; our schools must supply them. One practical method is described here.

Chapter 1: **Who They Are–*A Working Hypothesis***

Consider how much easier the task of making rich opportunities available to gifted students in science would be if we had a fairly reliable way of identifying these specially gifted people. And so the first task seems to be to determine whether in fact there are gifts, or syndromes of gifts, or special operations which characterize the individual who is specially able in science. There are several ways of going about this; one approach is described here. This approach, time and special circumstances developed into a fruitful design.

Almost at the outset, there was felt a need for a framework in which to view the matter of "giftedness" in science. Those of us in science have learned the value of some grand idea or scheme which James B. Conant has called a "conceptual scheme"[3] —a large idea which furnishes the framework for an experimental design. In 1951, when the opportunity came to work with Dr. Conant at Harvard, it became almost a necessity to seek some framework, a conceptual scheme, in which work directed at understanding what was and is called "science talent" could proceed with meaning. Within this conceptual scheme, several working hypotheses might be fashioned.

Definition

The literature is full of the term "science talent." We have preference for the expression "high level ability in science," rather than "talented" to describe the student we have in mind. It is worth while going into the reasons very briefly. The word "talented" has been associated in the popular mind with rather specific talents—with music and painting. Furthermore, there exists the notion that the core of these talents is to some extent hereditary. It may be that high level ability in science is in the nature of a "talent" similar to these: that is, that it is partly hereditary in nature. In any case, it seems best not to use a term which has so many connotations.

The phrase "high level ability in science" as used here shall be taken to refer to past high school students now in college who fall into the categories determined as follows:

---

[3]Conant, James B., *On Understanding Science*, Yale University Press, 1946; reprinted by Mentor Books, 1951. See also Dr. Conant's *Modern Science and Modern Man*, Columbia University Press, New York, 1952.

1. Contribution to science by means of original work in science, as evidenced by publication of a research paper or acknowledged assistance in the prosecution of such research.
2. An advanced degree in science or apparent successful prosecution of work toward the degree.
3. Reasonable evidence of commitment to research in science, such as having been selected for responsible assistance in laboratories or research work during the undergraduate years.
4. Reasonable evidence of successful commitment (on the college level) and competence in fields preparatory to work in science (whether research or applied, engineering, medicine, etc.).[4]

The division into categories may seem to be, at first glance, an exercise in snobbishness. Such is not the intention; indeed the purpose here is to shed further light on a disastrous situation. The situation is this: many students who should go to college simply do not get there.

Wolfle[5] and Wolfle and Oxtoby[6] give us good reason to doubt our educational efficiency. They supply figures which substantiate the generalization that there is a sufficiently large supply of students with high ability in this country to double, or more than double, the number of college graduates which now constitutes the upper half of college graduating classes. In short, rather than emphasize our personnel shortage, we should emphasize our waste of human resources.

In the publication cited,[7] it is reasonably established that a score of 120 on the Army General Classification Test is "certainly high enough to justify using it as a minimum for estimating the number of individuals capable of college work ..." "Yet *half of the persons with an A.G.C.T. of 120 or higher do not enter college and only about one-third of them graduate.*"[8] In the writer's opinion, an A.G.C.T. of 120 is a severe standard; this is in agreement with the belief of those responsible for the statement cited.[6] Hollinshead, in his *Who Should Go to College* (Columbia University Press), lends further support to the general estimates contained here.

Further statements substantiating the notion that we are wasting students with high level ability are to be found in *A Policy for Scientific and Professional Man-*

---

[4](An A to B+ average in two years of science on the college level is taken to mean competence in science. This was accepted as a useful criterion for those who had been graduated *only* recently from high school. The inadequacy of this criterion is recognized, since grading systems are not absolute.)

[5]Wolfle, Dae, "Future Supply of Science and Mathematics Students," *The Science Teacher*, XX:159, 210, September, 1953.

[6]Wolfle, Dae, and Oxtoby, Toby, "Distributions of Ability of Students Specializing in Different Fields," *Science*, LXVI:311-314, September 26, 1952. See also Dael Wolfle's *America's Resources of Specialized Talent*, Harper & Bros., New York, 1954.

[7]*A Policy for Scientific and Professional Manpower*, National Manpower Council, Columbia University Press, New York, 1953.

[8]*Ibid.,* pp. 79 and 82. Italics mine.

The Gifted Student as Future Scientist

power. The categories indicated on page 144 refer, therefore, only to those who get to college.

Will the tragic waste of our human resources be stopped? Not so long as we worry more about the destination of our soil than of our young people. Both need unsparing attention.

Further, although there may be the fullest intention of commitment by high level students to science (Category 4), they may be forced by many circumstances, financial and otherwise, into an area other than the one originally planned. It may be, however, that, although the opportunity to go to college may be denied students with high level ability, their commitment to science is of such a high order that they will choose to accept work as technicians in order to remain in science. That this is indeed so in a number of instances has been gleaned from the case histories with which the writer is familiar.

Aside from this, there is the full realization that science has become a team effort. A top-flight researcher cannot operate without a competent technician. Observations of the operations at "Manhattan District," Oak Ridge, The Rockefeller Foundation, Sloan-Kettering, hopelessly confuse one who would try to determine whose contribution is most important—investigator, engineer, doctor, inventor, etc., etc. The categories listed on page 144, therefore, are included only in an attempt to introduce definitions determined operationally.

Raw Material for the Study

When Forest Hills High School was first organized in 1941 as a four year high school, the writer was appointed Chairman of its Science Department and a science program was developed by a devoted group of teachers with the fullest support of the administrative officers of the school.[9]

The faculty of Forest Hills High School is committed to the fullest of its resources to the education of *all* boys and girls; these vary as widely as it is possible for a heterogeneous population to vary in gifts and opportunities Within the organization of a modern public school, within the goals of a modern American community with its divergent attitudes and ways of life, attempts are made to meet the needs of all the students.

It soon became apparent that we had to meet the varied needs of a heterogeneous population—among them the needs of a small group intensely interested in science and committed to it as a way of life even on the ninth grade level. As we set up the program it became clear that very little was known about meeting the needs and interests of high level students in science. It was accordingly desirable and feasible to develop both a program of work for these students and a program of research to determine the nature of high level ability in science.

---

[9]Forest Hills High School, New York City, at this writing has a student population of approximately 3,300 students and a faculty of 150. See Acknowledgments for the people who have aided the work described here.

146   P.F. Brandwein

The observations detailed here, however, are based not only on a study of high school students but also on others, as can be seen from the following:

1. 31 working scientists, from graduate students to those engaged in research on the highest levels. These were observed during the period 1931-1941.
2. 431 boys and girls who at one time or another indicated an interest in making science their vocation in one form or another. These have been studied over a period of thirteen years and will hereafter be designated as the *Major Group*. These students furnish the base for our observations and conclusions. These were observed during the period 1943-1953.
3. 263 students of the freshman classes (1944, 1945) who were involved in a study of the stability of science interests.
4. 201 students who were involved in a testing program to shed light on the question of whether high level ability could be predicted by tests.[10]
5. 82 teachers whose success in stimulating students of high level ability was well known (e.g., their "reputation" was clearly established) were also studied. These observations furnish some of the data to be adduced in Chapter 5.

A Working Hypothesis

If there is any lesson to be gleaned from a study of modern genetics, it is that the organism is a product of its heredity and its environment. Studies in sociology and anthropology lend support to this. Indeed one of the major conceptual schemes of our century is that the individual and his environment react upon each other; he is the product of his heredity and environment. Even if we were to assume that high level ability was primarily an inherited factor and was identifiable as such, we could not assume that it was fixed; in other words, we should need to assume that high level ability would be *expressed most favorably in the most favorable environment*. Further, very early in this work, it became crystal-clear that there were two major areas in the selection and identification of students with high level ability in which we could operate with profit. One was in the development of testing procedures which would identify students of high level ability and predict their success. The second area was in the development of a program which could reasonably be expected to appeal to such students so that they might *identify themselves*. Then we might use our observation of this work during their total training period in school as a base for predicting success.

This second area of operation was in a sense forced upon us at the beginning because there were no established procedures for testing students with high ability in science to which we could turn. Second, we could develop testing procedures as we went along. Third, suitable working hypotheses could emerge. Mainly, however,

---

[10](In addition, with the special help of two teachers, Evelyn Morholt and Sylvia Neivert, tests were conducted not only at Forest Hills, but other high schools. These add to our observations.)

The Gifted Student as Future Scientist

there was no body of observations on youngsters with high ability in science to which we could turn with confidence.[11]

It seemed reasonable, therefore, to assume that whether or not high ability in science could be identified as a hereditary factor, whether or not it could be identified by testing devices, it would still be expressed at its highest level in the most favorable environment.

This approach to a study of high level ability in science determined by the need to furnish the most favorable environment for its development is called here "The Operational Approach." It is in a sense one that is used in our graduate schools. Briefly, in our graduate schools a student commits himself to advanced training. He is given a program of courses and a program of research. How well does he stand up? The "experts," the members of the faculty, evaluate his progress in training and predict his success in contributing to science. That in essence is the broad plan of "The Operational Approach" to be described here—with two important major differences.

1. We did not select those who had been successful in previous science courses. The program was available to all who would take part in it, regardless of previous record, I.Q., or reading scores. In short, the boys and girls selected themselves.
2. We did our best, no holds (except unethical ones) barred; we tried whatever methods seemed practicable within our situation to give the youngsters every opportunity for development.

It seems clear even now that an "Operational Approach" of one type or another can be adopted for all schools, large or small, and that in itself it can furnish a useful means for identifying youngsters with high level ability in science and for predicting their future success. However, the prediction of success in science of the Major Group on which most of our observations are made is based on an Operational Approach and on a Testing Approach as well. Nevertheless, to point ahead, it is clearly desirable, practical, and useful to let youngsters identify themselves. They do so simply by coming forward to take advantage of opportunities which are available to them.

Early in 1931 when the writer first began to ask himself what made a scientist, and whether there were characteristics peculiar to them and not to others, he began to note characteristics of individual scientists as well as of graduate students in science. These scientists were of different ranks and abilities. Slowly there accumulated a mass of descriptive detail which was sifted as time permitted.[12] The mass of observations dealt with the observable behavior of these scientists. The observations were used first in an attempt to sketch a conceptual scheme of the scientist-in-embryo.

---

[11] In the past decade, however, there have been increasing reports of such observations (see bibliography footnoted throughout this study).

[12] To save space and the reader's time, the mass of observations will not be detailed but the hard kernels of these observations will be described.

Eventually they formed the basis for a *working hypothesis* on ability in science; this is stated in the summary of this chapter. Finally, they formed the basis for a conceptual scheme on ability in science; this is stated in the summary of this little book. (See "Intent and Content," page 139.)

Initial observations were made of the operations of scientists during 1931-1941 in two universities, three research laboratories, a botanic garden, and a certain combined vacation ground and work place of scientists (mainly biologists). In addition, other observations were made in initial school work at Benjamin Franklin and George Washington High Schools in New York City.

Finally, the notions presented in this book were checked against the opinions of selected teachers who had given the matter of the identification of students of high level ability much thought. First, the notions were checked in the summer of 1952 in discussions with a nationally representative group of teachers selected by President James B. Conant, Professor Fletcher G. Watson, and the writer, for a summer course dealing with the "Methods of the Scientist." In the succeeding summer these notions were checked against the opinions of a nationally representative group of science supervisors gathered at Harvard to deal with the problems of science teaching, under the leadership of Fletcher Watson, assisted by the writer. In the fall of 1953, the writer was enabled (as a Ford Fellow) again to put these notions to the test of other teachers' and supervisors' opinions.

In 1953, in response to continued requests for a description of our science program at Forest Hills, this booklet was put in a very rough draft form and sent out to those who had requested information, as well as to some 100 "experts" in the field of science teaching. Revisions were made in the light of criticism received. However, the notions expressed here—and the method of their expression, aside from the kind editing this has received—are those of the author.

From the very beginning of the writer's early observation of research scientists at work, it seemed clear that high verbal ability (oral or written) and high mathematical ability generally characterized those who remained in scientific research. The operations of the modern scientist indeed require that he report his findings to his colleagues orally and in writing (in acceptable English, and often acceptable mathematics).

These the writer has called Genetic Factors. They appear to have a relationship to high intelligence and *may* have a *primary* basis in heredity. Naturally, Genetic Factors are altered by an environment. In fact, it is clearly understood here that, generally speaking, any individual is the product of his heredity and his environment.

It soon became apparent that there were other Genetic Factors. For instance, adequate neuromuscular control (particularly of the hands) seems necessary, and at least adequate eye function.

Given the characteristics which we have called the Genetic Factor, high level verbal and mathematical ability, and adequate sensory and neuromuscular control, did these give assurance of success in scientific research?

Here observation of graduate students was most useful. Assuming that acquisition of a Ph.D. in science is an acceptable index of commitment to scientific research—not necessarily high success in research—why did certain graduate stu-

The Gifted Student as Future Scientist 149

dents who apparently had high level ability fail in their work? Briefly, the factors which seemed to be responsible for failure were grouped under one head called the Predisposing Factor. (This excluded failure for financial reasons, where such reasons were readily demonstrable.)

The characteristics grouped under the Predisposing Factor consist of two major ones: the first includes a spectrum of traits which the writer places under the head of *Persistence*. This is defined as consisting of three attitudes.

1. A marked willingness to spend time, beyond the ordinary schedule, in a given task (this includes the willingness to set one's own time schedules, to labor beyond a prescribed time, such as from 9 to 5).
2. A willingness to withstand discomfort. This includes adjusting to shortened lunch hours, or no lunch hours, working without holidays, etc. It includes withstanding fatigue and strain and working even through minor illness, such as a cold or a headache.
3. A willingness to face failure. With this comes a realization that patient work may lead to successful termination of the task at hand.

The second characteristic placed under the Predisposing Factor has been arbitrarily titled *Questing*. *Questing* as this writer will use it is to be taken to mean a notable dissatisfaction with present explanations of the way the world works—in short, a dissatisfaction with present explanations of aspects of reality.

The writer's relative lack of extensive training in Psychology (rather, Psychiatry) is keenly felt here, but this may be just as well for the purposes of this work. Roe,[13] Kubie,[14] and other qualified workers are entering upon this most important phase. But the writer does not want to leave an impression that *Questing* has a necessary origin in neuroses or trauma in childhood. Possibly it has its origin in the mechanisms of sublimation. Suffice it to say that this aspect had best be left to others more qualified to speculate on the scientist's inner life.

From our observations, it may be useful to describe what Questing is not. The "so what" attitude is not characteristic of questing; the general acceptance of authority in a given field of scholarship without question and without ascertaining the reliability and validity of the authority is not characteristic of questing; the belief that all is well in this best of all possible worlds is not questing. Questing is not Panglossian. Questing arises in a dissatisfaction. Questing, indeed, results in curiosity, in asking "Why?" "How?" or in the three perennials as the writer sees them:

What Do We Know?
How Do We Know What We Know?
How Well Do We Know What We Know?

---

[13] Roe, Anne, *The Making of a Scientist,* Dodd, Mead & Co., New York, 1953.

[14] Kubie, Lawrence, in the *American Scientist:* "Some Unsolved Problems of the Scientific Career," pp. 596-613, October, 1953; and "Socio-Economic Problems of the Young Scientist," pp. 104-112, January, 1954.

150   P.F. Brandwein

That spectrum of attitudes involving dissatisfactions which energize the individual's activity and give him purpose is what the writer means by *Questing*. Persistence and Questing, then, make up what we shall call The Predisposing Factors.

It will be readily accepted that these factors are not all or nothing in nature. Verbal, mathematical, and neuromuscular abilities have their curves of distribution. And so have Persistence and Questing.

But from the observations of working scientists as well as from common sense observations, it seemed clear that Genetic and Predisposing Factors were not all that operated in the making of a scientist. Opportunities to get further training and the inspiration of the individual teacher were clearly factors to be considered in reaching a working hypothesis on the nature of high level ability in science.

Knapp and Goodrich[15] have studied the place the college teacher has in stimulating individuals with high level ability in science. Time and time again, without exception, the working scientists whom the writer observed stated their indebtedness to one or more teachers and cited the opportunities these teachers made available to them. On this basis, it seemed reasonable to assume that another factor needed to be added to the ones here stated. This one we shall call the *Activating Factor*. This factor is concerned with opportunities for advanced training and contact with an inspirational teacher.

As a result of these preliminary excursions into the problem of what makes a scientist, this working hypothesis on high ability in science was put forth tentatively:

High level ability in science is based on the interaction of several factors—Genetic, Predisposing, and Activating. All factors are generally necessary to the development of high level ability in science; no one of the factors is sufficient in itself.

The succeeding chapters describe attempts to determine whether this large-scale working hypothesis is useful in identifying and predicting high level ability in science.

The ambiguity and tentativeness of the hypothesis are admitted. It is ambiguous precisely because we are trying to investigate an element in social behavior—high ability in science—which has not been clearly defined. Here we are caught on the horns of the usual dilemma of the investigator of social events. We need to be able to define the trait so that we may observe it. On the other hand, if we define the trait arbitrarily, then we limit our observations. If we do not define it, we run the risk of not knowing what to observe.

Cureton[16] has made the point that we are on relatively safe ground to begin with if we admit that we are operating in an ambiguous area, that an investigator may at least state his hypotheses to his own satisfaction, provided they are not ambiguous to him, and proceed to test them in the light of clearly understood operations. As he learns more and more, his hypotheses and observations clarify themselves, one in counterpoint to the other. It may be assumed, however, that once the trait is opera-

---

[15]Knapp, R. H., and Goodrich, H. B., *Origins of American Scientists*, University of Chicago Press, 1952.

[16]Cureton, E. E., *Educational Measurement*, American Council on Education, 1951.

The Gifted Student as Future Scientist   151

tionally definable, and operationally observable, it may be possible to test it; that is, it may become operationally scorable. As investigations of high ability in science continue in scope, this objective will be approached with ever-increasing assurance.

Chapter 2: **Who They Are—*Self-Identification***

Any teacher who is part of the effort of helping children grow to the fullest potential wants to know *how* best to do it.

In the writer's experience, it is in the *how* of doing things that the teacher on the firing line finds himself in a quandary.

Early in this work, it became clear that to test the hypothesis presented in the Chapter 1 it was necessary to invent an educational environment which would give the widest opportunity to those with high level ability. This, of course, within the limitations, personal and otherwise, which exist in school situations. What is here to be described as an Operational Approach, is the environment, still imperfect, ever being improved, which we have attempted to develop so that youngsters of high level ability may flourish. It is necessary to emphasize that equally searching attempts have been and are being made to meet the requirements of youngsters of all kinds of ability. For instance, Forest Hills High School has a useful and successful program in Science (and other areas as well) for what has been called the "slow" learner.

An Operational Approach in Blueprint

Descriptions of this operational approach have been made by the writer[17,18] at various times. It is well, however, to give a description of the essential aspects of the Forest Hills program here because it is these aspects which appear to be necessary and sufficient ones in the attainment of a respectable degree of success in predicting high level ability through the use of the operational approach. Let us follow a freshman class of approximately 400 students through four years of work.

All of these students take a course in General Science. This course is designed to further understanding of the common phenomena in their environment, the problems in areas of life and living which science might help to solve, such as the use of natural resources, the use of energy for doing the world's work, the problem of getting along with other men, the conquest of disease. The first step in the identification of the "prospective" scientist occurs here.

---

[17]Brandwein, P. F., "The Selection and Training of Future Scientists," *Scientific Monthly*, 54 : 247-252, 1947.

[18]Brandwein, P. F., "Developed Aptitudes in Science and Mathematics," *The Science Teacher*, XX : 111-114, April, 1953.

The progress of students who show any signs of distinguishing themselves in science (those who show great interest, willingness to do extra and co-curricular work, whose achievement in class is high, etc.) is noted during the first few months of the General Science course. Attention is given not only to these students but also to those who may have a "hobby" interest in science as well as to those whose grades are low. These first distinctions are vague and we would have them so, for our purpose in the Operational Approach is to give each and every student the environment which will enable him to make the utmost of his gifts and opportunities.

At the end of the first term, those who have shown an interest as well as an ability to work in science are given an opportunity to work in our laboratories during their free periods before, during, and after school. Those who are interested (whether they have ability or not) generally ask for this opportunity. Others who have ability—as shown by their work in class—are invited to work. Some of the activities they may choose include the following:

1. Preparing teaching materials in chemistry, physics, or biology. (In laboratory squads under the direction of our laboratory assistants, specially equipped people of high ability.)
2. Assisting a science teacher in his field of special interest.
3. Maintaining a large school museum of a wide variety of living and preserved specimens.
4. Maintaining a vivarium of forms particularly useful in biological work. Here students learn to maintain insects and mammals as well as cultures of the common protozoa and algae.
5. Engaging in science work in a variety of activities such as the science journal, the Chemistry, Physics, Biology, or Engineering Clubs. Other activities embraced in such names as Bio-Arts Club, the Museum Curators, Tropical Fish, Science Research, Cancer Committee, the Photography and Laboratory Technicians' Club offer other opportunities. In addition, many teachers help small groups of students who are working on special projects of one sort or another. This club program is broad, having been made so purposely in order to attract and hold those interested in science. There is also a Science Honor Society which engages the energies of those who have decided to make science a life work.

During the second semester this program of guidance continues. The result is that many good students enter into this so-called extracurricular science work. It is our belief that this work is not extracurricular in its usual sense, but is an integral part of the school work in our science program.

Our students generally go on to a second year of science work. There are many factors responsible for this continuing study. Two not inconsiderable ones are the sympathetic viewpoint toward the study of science which the administrative officers and the guidance department of the school hold and the favored economic position in general of the student body. But we also feel that the wide program of science activity offered is in itself a factor.

In the second year of science, selection of students for special science training begins. In the *present* stage of development, selection is based on these criteria:

The Gifted Student as Future Scientist

(1) any student who plans to make science a career may apply, (2) those students who have shown ability in science are invited.

In practice it has been found that the students who do apply are among those in the first quartile of the entering class; generally speaking, the average I.Q. is 120, the average reading comprehension score is 13 in the ninth year, and the average mathematics comprehension score 12+ in the ninth year. Some students with 100 I.Q. and even below are to be found in the group. No attempt is made to induce students of high level ability (as shown by high I.Q., high reading score, and high mathematics score) to enter upon training in science if an interview shows their interests to lie elsewhere.

Out of 400 entering students, some 30 to 40 apply or are invited to do special work. They are placed in one class. Those students whose programing difficulties make it impossible for them to enter the class are nevertheless given opportunities on an individual basis for similar work in other classes.

After their first year in high school, these science honor students (the classes in Biology, Physics, and Chemistry are designated administratively as "honor" classes) enter upon three years of enriched science and three years of mathematics work each. The work in these next three years is not only enriched but proceeds at a more rapid pace. For instance, students will not only read a high school text but will use a college text as well. This is in addition to various special references and journals such as *The Scientific American, Science, Scientific Monthly, Science News Letter*, and special journals fitting the special project which these students select (see descriptions later).

Not all of the students in this group will continue with this program. For one reason or another—change in interest, lack of success—*some students drop out*. For one reason or another—change in interest, heightened success in science—*other students apply for admission*. Generally the class remains with a core of some 25 students who remain together during 3 years of the program, while some 10 or so drop out and are replaced by others.

Several purposes are served by having these students in one class. Capable of attaining a high appreciation of scientific method and its social implications, they are given not only a different course of greater difficulty, of more advanced material, but work which stimulates them to develop high efficiency in the laboratory and field. More important than this, perhaps, *is that considerable time is spent in personal guidance*, so that the opportunities and advantages of entering fields of science are opened to them.

It should not be assumed that these are the only students who enter upon further work in science. Our responsibilities in science education are twofold.

1. We are responsible for a program of *general education*. This attempts to give those who will not become experts in science the basic understandings, skills, and appreciations that will enable them to co-operate closely with those who will become, or are, experts. This program must also, we feel, have special and practical application to the problems of living, so that the student sees scientific methods at work in solving his personal and community problems.

2. We are responsible for a program of *special education*. This attempts to select and give special training to those who may become our scientists of the future.

The relationship of this program to the other components of our science program may be diagramed as shown below.

Thus, while practically 70 to 75 per cent of our student body of 3,400 is taking science work each term (many of the students elect four years of science), approximately 160 to 240 students are in this special science program. Are we justified in making this selection? We have permitted students with scholastic averages of 80 to 85 to enter this classification and have given them the same training. The student with a special hobby such as radio or the collection of insects is also selected, regardless of his grades. As many students as we can reach are given the opportunity to avail themselves of special training.

The students selected for this class have the opportunity to:

1. Engage in some "original" research work on the high school level. Each student is under the close guidance and observation of a teacher. (This work will be described briefly later.)

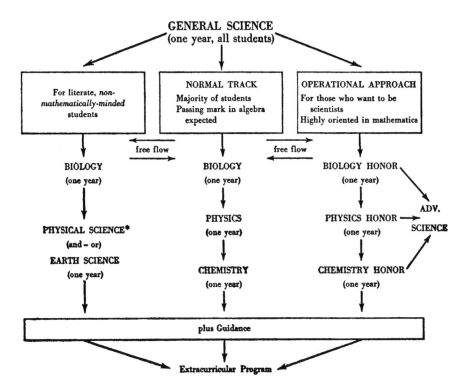

*Physical Science deals with selected aspects of both chemistry and physics around problems met in daily life. For Advanced Science, see pp. 155 and 156.

The Gifted Student as Future Scientist

2. Learn laboratory techniques (histological, bacteriological, analytical chemistry, work with glass, etc.).
3. Learn the expert use and operation of laboratory equipment of all types (analytical balance, microscope, electric oven, autoclave, etc.).
4. Gain special skills in shopwork, including handling of common materials, wood, metal, etc.
5. Engage in library research. Required reading includes college texts in biology, physics, and chemistry and other pertinent materials.
6. Take adequate training in mathematics. This includes a special class which may go as far as the calculus.
7. Prepare exhibits of their work for demonstration before other students, at science fairs, or local exhibitions.
8. Prepare reports of their work (in their senior year) in writing for the school science journal or for other journals.
9. Engage in seminar activity at regular meetings of the Forest Hills High School Science Society and the Mathematics Honor Society. Students who have shown competence in science are eligible for election to the society. Students who offer the best reports of their work will in turn be invited to submit their work at a Biology Congress sponsored by the New York Association of Biology Teachers or to exhibit their projects at the Science Fair sponsored by the Federation of Science Teacher Associations of New York.
10. Engage in the Annual Science Talent Search of the Westinghouse Educational Foundation.
11. Take a course in college physics (using the calculus). This course has more recently been organized by a group of the students who are part of this program.
12. As this goes to press, plans are being completed to organize a first-year college program in the different areas of Science, in Mathematics, Social Studies, English and Languages. The students eligible for this program are, however, those who can maintain high scholastic standing in all areas of high school work—not only in Science and Mathematics.

After a *half* year in this honor Science program, the students who wish it are given the opportunity to enter an Advanced Science class. This is in effect an additional period of science. However, it is not spent in classwork per se *but in the laboratory*. It is a period where a student may select his own project and solve it in the laboratory. Some students continue these projects at home; many work in the school laboratory before and after the regular school schedules. It is a gratifying sight to watch young people tackle a scientific problem and use their own abilities (without outside interference) to solve the problem and emerge with a solution.

What is the teacher's function in this class in Advanced Science? Generally speaking, when a student asks a question, his question is returned with another. The intention is to put the student "on the track," but only if he uses his brain and other resources, including the library. If a student seems to be way off the track, the teacher's questions may bring him nearer to a clearer analysis of the problem. The procedure is *to give the student no answers which he can discover for himself by painstaking work.* In addition, the teacher, by his observations in the laboratory (looking over shoulders) and by evaluating the monthly reports the students give him, is constantly aware of the extent of each student's progress in solving the problem selected.

In this Advanced Science class, the mature scientist could see a picture of the scientist-in-embryo. And this is especially true of the "research" activity in which these youngsters engage. In most cases, the student faces a problem he has never faced before. No solution is available in textbooks, and it may take two or more years of work to reach even a tentative conclusion. For instance do zygospores of *Rhizopus nigricans* germinate? Many textbooks assume such germination. Several of our students find no such evidence on the basis of their investigation. They find their conclusions supported by authorities in the field.

How long does digestion take in the food vacuoles of different protozoa? Why does *Chaos chaos* appear to have only a regional distribution? What factors does the gas produced by *Tribolium confusum* have on other insects? Other work includes a study of the embryology of Physa, diapause in Cecropia and other insects, development in vitro of certain plant embryos, studies on a modified method useful in the recovery of silver in photography, the structure of soils in the vicinity of Queens, N. Y., studies on aberrant electrostatic effects, meteors, background radiation in the Forest Hills area, inversion in Volvox, sporulation in lager yeast, the effect of ultraviolet on flour beetles fed with buckwheat, the influence of sun spots on agriculture, a modified circuit breaker and various other studies in biology, chemistry, physics, geology, and other fields. (See Appendix F for précis of some typical interim and final reports of "research" by some of these students.) There is little doubt in our minds, as we observe these young people at work, that they are using the methods employed by the professional scientist in his experiments. This is a matter of direct observation not only by the writer but by visitors from schools throughout the country. They have also had the opportunity to observe these youngsters at work.

This early training also places emphasis in the social responsibilities of scientists in this modern world. These boys and girls are still plastic. Perhaps if we train our students in high school and even earlier to see themselves as citizens first and specialists later, we shall not have the situation which Dean Harry J. Carman describes in these terms: "In public life we are ruled by scientific ignoramuses, and in the scientific laboratory we have, for the most part, political and social illiterates." The elimination of this state of affairs is a prime objective of our program of general and special education.

At the end of the second year of high school we have, then, 40 to 60 youngsters who are ready and willing to embark on an extended period of work. Every day for two or three years they have one period of science, one period of mathematics, and one period of laboratory work on a personal project, if they wish it (with, of course, English, Social Studies, Language, and the required Music, Art, and Health Education). In doing their personal project work these students must read source material, plan experiments, order materials, construct equipment—in short, over a period of two years, carry out in a small way the procedures which serve the scientist. Their teachers stand by—to guide.

The Gifted Student as Future Scientist

As has been said before, during the following two-year period (sophomore to senior), many students drop out of the work. Some find that athletics and social events are more important to them than science; others are not fitted—through lack of even the simplest manual skills—to carry on the work; still others lack originality. A few are lacking in honesty or a sense of responsibility and are advised by the sponsor to seek other work. This advice is given only after failure has attended many attempts to produce desirable changes in attitude by the methods at our disposal. And, of course, there are those who cannot learn to work with others. Again we have never released any of these students until attempts have been made to make desirable changes, and in many instances we have retained these youngsters till the bitter end—especially when their basic qualities warranted it.

In any event, by the beginning of the latter half of the senior year we have perhaps 25 to 30 boys and girls who have participated in most of the first 10 activities listed on pages 154–155. In addition, these students are most skillful in grasping scientific concepts, projecting them, and using them to solve scientific problems. We believe most of these have the ability to be scientists.

From these 30 senior special science students, come the five or six seniors who, we believe, may be the research scientists of the future (categories 1 to 3, page 144). Where opportunities warrant, these five or six are given a sort of apprentice training in their senior year. For instance, where possible, they may be placed in industrial, college, or research laboratories to assist scientists at work. For this we are grateful to the many individuals in college and industrial laboratories who have given generously of their time and energy. These students distinguish themselves by winning extensive honors in the various activities sponsored by different organizations. The other 20 to 25 generally go into the applied sciences.

Forest Hills High School is now thirteen years old. Since February, 1945, when the first class which had been with us four years was graduated, one way or another this program has yielded 89 students whom we consider on the basis of observation (in the Operational Approach described here) to have promise as research scientists. More than 300 others have committed themselves to engineering, mining, geology, medicine, dentistry, to work as laboratory technicians and in other areas; there are, in great probability, others in this group, but information is lacking. Twenty-eight more students are known to be committed to science teaching in secondary schools.

This, then, is a brief description of the Operational Approach. It consists, as can be seen, of the same elements which go into the early professional training of a scientist in graduate school—except that it is done on the school level, and that all students can enter it freely. Hence it may have value in predicting later success in science. The data to be presented indicates that the Operational Approach can be used to predict success in science.

Possibly a better analogy is this. The Operational Approach is more like a training camp for future baseball players. All those who *love* the game and think they can play it are admitted. Then they are given a chance *to learn how to play well* under the best "coaching" or guidance (however inadequate) which is available in the situation operating. No one is rejected who wants to play. There are those who learn to

158 P.F. Brandwein

play very well. Whether they can play well (in the big leagues) is then determined after they have been accepted into the big leagues (college and graduate school).

When the Commitment to Science?

Early in this work several serious questions arose. Was it fair, was it good guidance to commit youngsters to science so early in their careers? The question answered itself. First, as has been said before, it is the obligation of a community and the school to provide the best education possible. A dynamic program in science, with the fullest opportunities for laboratory work in science, is obligatory in a modern society where science has such increasing impact on life and living. Second, no boy or girl is forced in any way to enter upon this program; there is full freedom of flow out of it as well as into it. And youngsters take the opportunity to leave it (about 10 out of the original class, or 25 per cent, leave the program during the three-year period, and about 25 per cent enter it). Third, careful measures are taken to assure the most sympathetic guidance; for instance, competent youngsters with other clearly expressed interests and abilities are guided into other areas. Fourth, the student's personal guidance counselor (outside of the Science Department) has a general interest in the boy or girl—and generally guards against unwise commitment. Fifth, the school itself has a rich program which appeals to the widest interests; its English, Mathematics, Social Studies, Language, Business Subjects, Music, Art, and Health Departments have vigorous programs, curricular, co-curricular, and extracurricular. Sixth, the students' total program (this is a special one for these students), four years of English, four years of Social Studies, four years of Science, four years of Mathematics, and four years of a Foreign Language, four years of Health Education and required Music and Art, forces attention to the widest opportunities. Seventh, parents are kept apprised of the state of affairs. Eighth, the student is under this constant precaution, a shibboleth of guidance carefully and continuously stressed, that: "High School is exploratory. Here you may find out what you like. Here you don't need to commit yourself to any profession."

The checks and balances listed above enable us to face the question, "When the commitment to Science?" with equanimity and security.

There are several other significant factors in considering this most important question of when youngsters should be committed to a profession. It is not thought to be strange to commit a youngster to music, or art, or athletics (e.g., baseball) as soon as signs of inordinate skill manifest themselves. Proud parents, proud teachers, and proud guardians who happen to be associated with a promising young artist do not hesitate to plunge him into hours of practice—in and out of school. Their reasoning is sound; he has the skill and promise; the world needs artists; *ars longa, vita brevis;* etc., etc. We are convinced, with reason we believe, that the intellectual life, that which occurs in the cerebrum, has all the qualities, rewards, and promise for the individual and for civilization that the arts have. There is as great a thrill or sense of intense pleasure in an intellectual discovery as there is in painting a picture, composing a sonata, or writing a short story. The writer's observations support the

The Gifted Student as Future Scientist

notion that youngsters take keen joy in making an "original" discovery; they add a cerebral cubit to their growth as well.

Hence this notion: Science may well join Music, Art, Writing as a field of early commitment. There are those who need to discover with their cerebrum, or as Percy Bridgman puts it, need to "do their damnedest with their brains, no holds barred"; there are those who need to do science just as there are those who need to do art or music. And just as we commit those with high level skills in music and art to early training in those fields, so may we with justice commit those with high level skills in science, mathematics, languages, to early involvement in those fields—or in any field to which they may be contributors. All this with proper safeguards for mental and physical health. Yes, all this with proper safeguards for mental and physical health.

From our own viewpoint we could do no less than organize the program described above, for our original working hypothesis forced the design of the operational approach. Furthermore, the original working hypothesis led us to subsidiary notions, among the-most important of which are these:

1. Future contributors to science may be identified on the school level.
   a. They may be identified by observation during a training program (the Operational Approach described here) to which admission and exit are freely guaranteed.
   b. They may be identified by a testing program.
2. Science "talent," better called high level ability or developed ability in science, is not a specific factor, but emerges out of general intelligence.
   a. High level ability in science is a combination of intellectual abilities, inherited and developed.
   b. The expression of high level ability in science is directly related to the early opportunities available for involvement in science.
3. The number of scientists in supply for future scientific operations is significantly related to the number involved in school science.
   a. Youngsters who like school science, and are successful in it on the school level, tend to make science a life work.
   b. The supply of a great number of *future* scientists for our greatly increased needs does not *primarily* depend on efforts by the colleges but on efforts on the school level.

This science program, literally a "social invention" (our Operational Approach), is the product of much study and observation of the best practices in the country and the efforts of the personnel of the school, especially the members of the science department. We have borrowed freely from the schools we have observed, the articles (see below) we have read, and the brains we have picked. We have referred constantly to the four volumes of *Genetic Studies of Genius* (Stanford University Press), written and edited by Lewis M. Terman and others. This invention is subject to all the flaws of social invention—flaws of intention, omission, commission, complicated by the interaction of all the human personalities

Other Operational Approaches

involved. Being human, we could not have it another way, and we do not at present have another way.

The reader will realize, as a matter of course, that there are a good number of operational approaches, or training programs, similar to the one described here. Each one has evolved out of a particular situation and in a particular community. Some programs are that of the splendidly conceived Bronx High School of Science, described by its principal, Dr. Morris Meister, in various articles and especially in Paul Witty's excellent summary of work on high level ability, *The Gifted Child,*[19] that described by Klinge[20] and others in *Selected Science Teaching Ideas of 1952,* as well as those included in Bloom's "symposium"[21] by Witty and Bloom,[22] by Alpern,[23] Meister and Odell,[24] Sumption, Morris, and Terman[25] as well as many others. Such programs indicate various approaches to the problem of furnishing incentives and suitable environments for the student with ability in science. Some of these programs afford opportunities to observe and describe the individual who will make the scientist. In the next two decades enough observations will have been made to enable us to reach tentative conclusions about the nature of high level ability in science.

It has been the writer's privilege to talk with a good number of the teachers throughout the country who are responsible for these programs.[26] Without reservation, these teachers believe that no observable harm comes to youngsters who commit themselves to science early in high school. In fact, as will be developed later in Chapter 3, commitment to science may possibly fulfill a need established early in life.

Equally important and clearly apparent is a rock-bottom essential to the success of all these programs. There are many limiting factors—equipment, time, etc.—but without this one factor the program is not conceived, does not develop, and is not

---

[19] Witty, Paul (editor), *The Gifted Child,* D. C. Heath & Co., Boston, 1951. (See Dr. Meister's account.)

[20] Klinge, Paul, "Scientist Training," in *Selected Teaching Ideas of 1952,* National Science Teacher's Association, Washington, DC. (edited R. W. Burnett).

[21] Bloom, Samuel, "Science Provisions for the Rapid Learner—A Symposium," *The Science Teacher,* XX : 161-163, 182-184, September, 1953.

[22] Witty, Paul, and Bloom, Samuel V., "Education of Gifted," *School and Society,* 78 : 113-119, October 17, 1953. (See also continuing issue.)

[23] Alpern, Hyman, "How Can the School Meet the Needs of Gifted and Superior Students?" *Bulletin National Association of Secondary School Principals,* 36: 110–117, March, 1952.

[24] Meister, Morris, and Odell, H. S., "What Provisions for the Education of Gifted Students," *Bulletin National Association of Secondary School Principals,* 35: 30-40, April, 1951.

[25] Sumption, Merld R., Morris, Dorothy, and Terman, Lewis M., "Special Education for the Gifted Child," in *49th Year Book of the National Society for the Study of Education,* pp. 259-280, 1950.

[26] See footnotes on this page.

The Gifted Student as Future Scientist   161

brought to fruition. *That essential is the teacher.* In Chapter 5 the characteristics of the kind of teacher whose self-fulfillment lies in part in developing high level ability in science will be examined. This kind of teacher is the key to our national problem of supplying sufficient scientific personnel. Unless his needs are met, the needs of youngsters with high level ability in science will not be met.

Observations on an Operational Approach

Raw Materials

In June, 1945, the first class of students who had been given the opportunity of participating in the O.A. (Operational Approach) program described were graduated from the Forest Hills High School.

From 1945 to 1953, several results of the program have been clearly observable by almost anyone who would care to look for them.

Selection Through Self-Identification

Of the 431 students who make up part of the group on which this study is based, 354, or 82 per cent (approximately), carried through the program for the high school period. There were approximately 18 per cent dropouts for the many reasons which operate in a school. The main reasons were concerned with:

> voluntary change of interest
> result of guidance procedures by teachers concerned
> transfer to another school

Of these 354 students who were graduated as participants of O.A., 89 were thought to have potentialities for high level contribution to science. How were the 354, and then the 89, identified?

As soon as O.A. had been discussed with all administrative and guidance officers concerned, and as soon as agreement had been reached and approval given, our science teachers announced it in all classes. At that time, Forest Hills was a four-year high school (soon it will be a three-year school). From the beginning, our ninth-grade students who were interested vocationally in science applied for admission to the program. Since four years of science, mathematics, and language were required, it is clear that only those really committed—not just interested—applied.

Beginning in 1942, and ending in 1946, the number of students applying in each year was 35, 38, 37, 40, and 39. In eleven years, ending with 1952, some 431 students had applied, and 354 had been graduated as participants. When teachers were asked to recommend other students for the program, those recommended by teachers included more than 90 per cent of those who applied. To put it another way, teachers could add very few to the list of qualified students. And as will be

162 P.F. Brandwein

seen from a study of testing devices used (Chapter 3), these students were indeed qualified.

In other words, when a program with clear intentions (O.A. as described in this chapter) is carefully explained to the student body, when its curricular outlines are clearly understood, students with sustained interests and adequate qualifications come forth and *identify themselves*. So it appears to us. Further, as will be seen from Chapter 3, the tests used would have selected much the same youngsters.

The question remains, why so constant a number of applicants? Two main reasons may be adduced. First, we did not open the program to all students with high I.Q., high achievement in Science, English, and Mathematics, and high reading and mathematics score (on standard tests). We should have had three to four times the number who applied. The program was opened only to those who wanted to become scientists—no matter in what field—not to any student. Second, there were curricular requirements to be met—four years of science, four years of mathematics, four years of one language. Parents' consent was required in all cases. Apparently these practices (O.A.) had the net result of limiting the number of students who came forward to take advantage of the program and, curiously enough, selected those who might have been selected by tests. However, there were many, many more who were taking science courses as such who did not enter the program. These could take as much science as they wished, selected from courses in Biology, Physics, Chemistry, Earth Science, and Physical Science. Nevertheless, interest in science even when coupled with high I.Q. and high achievement does not necessarily mean that the student is committed to science as a career. Nor does it mean, as we learned, that students interested in science sustained themselves in science. This is worth looking into because the notion presently exists that we should seek our future scientists from the ninth-grade students (and lower) who are interested in science and whose achievement in it is high.

The Place of Interest

In 1944 a freshman group of 263 students took part in a study which involved questionnaires and interviews to check on the answers to the questionnaires. Two weeks after the answers had been tallied, 40 students were interviewed and their oral responses checked against their written ones; the indications were that the responses were a fair description of reality.

Early in this study of interests, it became clear that many students in the first year of General Science at Forest Hills had spent more than 8 hours per week in a science interest (chemical sets, radio, model airplanes, etc.) prior to the age of 13 or 14, yet did not sustain their interest. Of 263 students subjected to a questionnaire, 40 students (33 boys and 7 girls) indicated such a major interest in science. Their parents confirmed this interest, as did interviews with the students to check their statements. Yet of these 40, only 18 sustained their interest in science throughout high school in a major way. The other 22 turned to different fields. Subsequent studies confirmed this notion. In our present society, youngsters before 14 are inter-

The Gifted Student as Future Scientist | 163

ested in science much as they are interested in sports, music, reading, or collecting stamps. It is a sign of our modern culture. Then as they get into high school and college, their teachers, the opportunities for prosecution of scientific work, and their success in science determine, at least in part, who will go into science. Zim's study adds further evidence along these and other lines.[27]

It seemed clear from this and other observations that interest itself, as defined operationally here, was not a sufficient index of commitment to science. This was primarily true because the environment in which the interest in science was to be cultivated was an important factor in any student's commitment in science.

Selecting 89 Out of 354

For the time covered by these observations, 354 students were graduated as participants of this program. Slightly more than 95 per cent of these entered college. More than 90 per cent of these are now committed to a science career—engineering, medicine, dentistry, geology, psychology, and related fields; this includes research in all these various fields as well. Yet, of all these students, 89 were judged by us to have the characteristics needed for success in scientific research. The 89 were selected on the basis of results gathered by careful observation of 62 experimental and 62 controls in the group of participants in O.A. between the years 1945 and 1951. The 89 include the 62 experimentals.

Why weren't all of the 354 selected for careful study? First we wanted to follow these students through college and postgraduate work. Lack of secretarial help (the scourge of schoolmen) forced us to limit our group to 124 students, 62 experimentals and 62 controls. Approximately 9 experimental and 9 control students were chosen in each of the seven years of this study. These 124 students were then part of a pilot study. The observations we have made on this pilot group seem to us to apply with equal validity to the 354 students. The years to come will certainly refine our operations, but at present we are reasonably convinced that in the situation we have been describing, our observations of the 124 apply to the majority of students with high ability in science.

The two groups of 62 students were paired for I.Q. achievement in science and mathematics, reading and mathematics scores (on standard tests). However, the 62 experimentals scored 4 and above in a Man-to-Man Rating Scale and Inventory of Predisposing Factors (see Chapter 3 and Appendixes B and D). The 62 controls were selected on the basis of their score of 3 and below on the Man-to-Man Rating Scale and on the Inventory of Predisposing Factors. These scales or inventories, if indeed they may be called that, are admittedly subjective in nature but exceedingly helpful as a check on what might seem arbitrary judgment. Clearly, our observations on these 62 pairs indicate the need for re-examining any notion that students with high verbal and numerical ability, when given a rich program in science, will enter

---

[27]Zim, Herbert, *Early Science Interests and Activities of Adolescents,* Ethical Culture Schools, 1940.

science professionally. The Genetic and the Activating Factor are not sufficient. The Predisposing Factor (Persistence and Questing) may indeed play a most significant part in commitment to life work in science.

The 62 and Their Pairs

Of the 62 students whom we have been able to follow in one way or another (visits, letters, other communications), 39 are already graduated from college (1 has dropped out). Of the 39, 22 are at present in graduate work in science or in research fields varying from genetics through nuclear physics, 6 are in medical and dental schools, 7 in engineering, and 4 in other fields such as accounting, business, and law. Eight in the group of 22 are Ph.D.'s. Fourteen of the 22 have already published papers in science or have assisted in the research involved.

Of the 62 "controls, " 33 are graduated from college (4 have dropped out), 6 are in graduate work in science (research fields), 13 are in the applied sciences (7 are in medical and dentistry school, 6 in engineering), 14 are in other areas such as law, accounting, business, social studies.

It appears that approxiately 22 of the experimental group have committed themselves to research and 13 to applied science, whereas 6 of the control group have committed themselves to research and 13 to applied science. Or for total commitment to science (research or applied) for the experimental group, 35; for the control, 19. This gives us further indications of the significance of the Predisposing Factor (Persistence and Questing) in committing youngsters to high level work in science. It deserves further investigation.

Other Observations

Aside from the tentative results reported above there are other results (not listed in order of importance) which may be reported as benefits of O.A.

1. O.A. has enlisted the active support of the majority of teachers in the department with a noticeable beneficial effect on *esprit de corps* and morale.
2. A tradition of scholarship in science has been established.
3. There has been a clear commitment to the improvement of education for all our students because one becomes conscious of the problems of all students when a search is made for a specific group. In addition, the additional equipment and library material purchased for the special group is used by all students.
4. Community support and approval are apparent.
5. We have had an average of 25 visitors per year. These have come from all over the country and from abroad as well. We have learned from them as well and reacted to their questions. Teaching also improves when teachers expect visitors to their classrooms.
6. Each one of the 62 experimentals was convinced that one adult or another had kindled and maintained his interest in science; 20 of 62 students named parents, relatives or friends, 42 named teachers in elementary and high school. This is clear not only from talks with these students and parents but from questionnaires and biographies as well.

The Gifted Student as Future Scientist

From a preliminary study it became clear that certain teachers are much more successful in kindling science interest than are others. The students in this study were consistent in mentioning the same teachers time and time again.

Certain observations continued to give meaning to this observation. During one term there were enough students to organize two honor classes (as described on page 153). One of the teachers had to relinquish the class; his substitute taught the class for one term. Of the 36 students in the class, only 11 continued their work in science (as a major for three years, Physics and Chemistry and laboratory work). In comparing this class with a class of 35 taught by the other teacher, 27 students of the original honor class continued their science studies for the remaining years (Chemistry, Physics, and laboratory work). The main difference between the two teachers was that one was experienced in dealing with able students. The other was very young—although willing—and most inexperienced in making rich opportunities available to able young people.

In still another case, an experienced teacher, warmhearted, respected, and loved by students, found it very difficult to supply the kind of rich climate of science activity which these youngsters demand. In this class of 36 students, 16 elected to continue science for 4 years—as compared with the 85 to 90 per cent expected. Strange to relate, not one of these 16 qualified for the Westing-house Science Talent Search two years later.

As we shall see later, it is not enough to organize a full program of science. It is necessary to have qualified teachers. These qualifications will be detailed in Chapter 5. We repeat the need, therefore, to re-examine carefully the notion that youngsters with high level ability can be left to their own devices.

An Inference for the Present

It seems to us that at present it is not necessary for a teacher to depend on tests which seek to identify students with high level ability in science in the early grades. If qualified teachers were to furnish sufficient opportunities in science to all students, those with high level ability in science would come forth and identify themselves. One such climate of opportunity in science has been provided in the Operational Approach described in this chapter.

Teaching is a personal invention. Each school can develop its own operational approach. Each school can develop its own particular invention adapted to its own particular situation. There does not appear to be a shortage of youngsters who could very well become our future scientists. There appears to be a shortage in our schools of opportunities for the fullest development of students with high level ability in science. The teacher, therefore, remains the key to the solution of the problem presented by the present shortage of scientific personnel as well as the key to the problem of furnishing scientists in the future.

All our scientists pass through our schools, whether public or independent. Our scientists of the future are there. The people who teach them are there.

166 P.F. Brandwein

It is in the secondary school, therefore, that a solution to the problem of supplying the scientists we need should be sought. And good success may be expected.

Chapter 3: **Who They Are–*Identification by Testing***

There is a basic notion existing in educational measurement, namely that if something exists, it exists in some quantity. Hence it can be measured.

Early in this work, a criterion was sought for high ability in science, or "science talent," as it was known. "Science talent" as a descriptive term was discarded because it was soon a matter of observation that whatever it was, "science talent" was not a fixed identifiable quality which was determined in early youth, say in the elementary grades. Rather, it was a developed ability, given certain factors.[28] Paul Witty's definition of the "gifted" came nearer to the criterion of high level ability being sought since it was operational in nature; e.g., Witty places among the "gifted" a child whose "performance in a worth-while type of human endeavor is consistently remarkable."

This writer's observation of 31 working scientists (Chapter 1) gave him a clue to the kind of performance necessary in high level work in science, and the nature of the consistency which Witty emphasizes. These observations, plus his own experience in scientific research, plus the reading of the autobiographies and biographies of working scientists past and present, were basic to the setting down of the three factors which were possibly behind the possession of *science potential*, the ability to investigate and contribute to further knowledge in science.

When, after university work, the writer taught in the George Washington High School in 1936-1939, he had occasion to make his first observation of young people with high level ability and to apply these observations against the hypothesis that the Genetic, Predisposing, and Activating Factors were indeed the operative ones in fostering high level ability in science. Indeed five of eight students then selected as having S.P. (Science Potential) on the basis of Predisposing Factors are now all in scientific research (two women of the three are married, one man remained in the armed forces). At Forest Hills, when the writer was enabled to organize a full-fledged program, the Operational Approach (O.A.) soon forced the assumption that these factors were indeed significant. Would any testing devices show them to be significant?

The Genetic Factor

Analysis of the 31 case studies of working scientists indicate that all had in one measure or another these Genetic Factors: Verbal Ability and Mathematical Ability. Furthermore, all of them had high achievement in high school and college science. The word "high" needed definition of lower limits.

---

[28] The Genetic, Predisposing and Activating Factors previously discussed.

The Gifted Student as Future Scientist

Early in 1942-1944, it seemed that students with a high I.Q. (130, Henmon-Nelson) and high interest in science as determined by the number of hours spent in science after school would tend to furnish us with a group from which those with science potential would come. Two such groups were accordingly organized in the second term of the ninth grade. It was soon discovered that science interests as admitted by students were not stable indicators of a pervasive and sustained interest in science in high school nor were they to be indicative of commitment to science (Chapter 2). Youngsters in the junior high school (through the ninth grade) seem to be exploring their environment in all directions and apparently embrace one interest after another. Mallinson[29] indicates that science interests in high school are not stable indicators of a career in science; a similar conclusion comes out of a conference held in Washington.[30]

While Zim[31] indicates that the scientists he investigated had science interest early in youth, the notion, in converse, that early interest in science leads to commitment to a science career is not valid per se.

In fact, science interest may only mean that science is interesting—not that it is a vocational commitment.

Acting on this assumption, it seemed desirable to organize honor classes (in O.A., Chapter 2) *after* the ninth grade, and base the organization upon the specifically expressed interest in a *science career*, not in *science* per se. Thus any youngster, of whatever I.Q., of whatever level of verbal or mathematical ability, who was determined to make science his or her life work was admitted into the course. Some of these youngsters wanted to be nurses, others engineers, others research scientists. They varied in I.Q. from 98 to 165.

In 1944-1951, therefore, 40 youngsters in the middle of the ninth year who had applied for entry into the honor classes (in O.A.)[32] were checked for I.Q., reading score (based on a standard reading comprehension test),[33] and arithmetic score (based on a standard arithmetic test).[34] In addition, in 1950-1952 three of these groups were given three tests of the Differential Aptitude Inventory of the Psychological Corporation, New York City. These were tests titled Verbal Reasoning, Numerical Ability, and Abstract Reasoning. They were also given a battery of four of the tests of the Primary Mental Abilities, Form A (ages 11-17), published by Science Research Associates, Chicago. These tests were titled: Verbal-Meaning, Reasoning, Number and Word Fluency.

---

[29]Mallinson, George Greisen, and Van Dragt, Harold, "Stability of High School Students Interests in Science and in Mathematics," *School Review*, LII : 362-367, September, 1952.

[30]"Education for the Talented in Mathematics and Science," prepared by Brown, K. E., and Johnson, Philip, United States Department of Health, Education and Welfare, Office of Education, Bulletin 1952, No. 15.

[31]Zim, Herbert S., "The Scientist in the Making : Some Data and Implications from the Junior Scientists' Assembly," *Science Education*, XXXIII :344-351, December, 1949.

[32]As has been indicated in Chapter 2.

[33]Nelson-Denny Reading-Comprehension.

[34]New York Arithmetic Judgment Test, Grades 7-12.

168 P.F. Brandwein

It is a matter of common observation that not all high I.Q. youngsters with high verbal and mathematical reasoning will go into science. Some have no interest in science in the tenth grade; the interest of others is not sustained. It is apparent, therefore, that a determination of the Genetic Factor in relation to Science Potential is important only as it is related to actual commitment to scientific research or any scientific endeavor as a career.

Of the 124 students (62 experimentals, and 62 controls previously mentioned) who were tested for I.Q., reading comprehension, and arithmetic ability, more than 90 per cent were in the upper 10 per cent of the school's distribution of I.Q.'s, reading, and arithmetic scores. Forty per cent of these ninth-grade youngsters were in the range of about 140 I.Q., had reading scores of 15 (above the fifteenth year), and arithmetic scores of 12.0 plus (above the twelfth year).

Furthermore, in the series of tests of the Psychological Corporation, and those of Science Research Associates, all the students were above the national median, and those who are now committed to scientific research but for one were at the 90th percentile and above. Also *all* of the 35 (of the 62 experimentals) now in graduate work in science, and *all* of the 19 so committed (of the 62 in the control group) had I.Q.'s of 135 and above[35] and reading and mathematics scores of 15 years and 12 respectively (in the ninth year). All of these achieved grades of 93 per cent or over in science and mathematics.

Similarly, these students (35 + 19) achieved quotients of 130 and over (95th percentile and over) at whatever age they were tested (the highest quotient is 140) when given the full battery of S.R.A. tests.

Hence, the base level of the Genetic Factor was set operationally at 135 I.Q., reading score of fifteenth year plus (based on the ninth year), an arithmetic score of twelfth year plus, and a scholastic average at the ninth year of 90 per cent or above.

If Thurstone's general notion of the existence of primary mental abilities be accepted, then it is fair to predict that most of these youngsters would generally score above the 90th percentile in the tests of Primary Abilities mentioned (Science Research Association). Similarly, this is to be expected as well of their performance in the tests of the Psychological Corporation.

What seems to be necessary for more adequate testing of these youngsters is a test which will have a wider spread; that is, one which will demonstrate more selectivity in the higher ranges of ability.

The Predisposing Factor

Any sustained observation of youngsters in the process of educating themselves—or indeed, any observation of the human in any endeavor requiring sustained effort— soon leads one to the observation that intelligence per se is not enough for success.

---

[35]Witty and Bloom, "Education of Gifted," *School and Society*, 78 :113-119, cite 130 I.Q. as the level for gifted children.

The Gifted Student as Future Scientist

The nature of the goals which predispose individuals in science has been commented upon and studied most recently by Knapp and Goodrich,[36] by Roe,[37] and Kubie,[38] among others. They point increasingly to the socio-economic and behavioral factors (stemming from unconscious drives and motivations) which play a part in determining the scientist's purpose and goals.

When observations of working scientists led the writer to postulate the operation of a Predisposing Factor in the success of certain scientists, an Inventory of Predisposing Factors (Appendix D) was devised, based on observations of the way this factor appeared to operate. The purpose of devising this crude scale was to furnish our teachers with one operation which they might use in studying their students.

The 62 students in the experimental group were selected on the basis that their scores were 4 and above in this scale, while those in the control group were selected from those who rated 3 and below (Chapter 2). It will be remembered that the experimental and the control groups were paired as closely as possible for Genetic Factors and socio-economic factors. The average I.Q. for the experimental group was 138; that for the control 139. Both were above 15 (ninth year) for reading score, and 12 plus (ninth year) for arithmetic score.

It seems a useful assumption to the writer, and indeed it would appear only logical, that those with high Genetic and Predisposing Factors as defined here would tend to be successful in the modern competitive endeavor in science—e.g., judgment based on success in college (necessarily based on comparison with others), judgment based on admission to graduate school, granting of fellowships and assistantships (also based on comparison with others), success in research (which is itself based on drive, persistence, and questing as ordinarily defined).

Studying the Students

Man-to-Man Rating Scale

It was realized, as soon as the hypothesis of Genetic, Predisposing, and Activating Factors was set down, that in actuality these factors were based on an inventory of characteristics—consciously considered—of the successful research scientist, e.g., the one who was contributing to science. As the writer began working with youngsters in the George Washington and Forest Hills High Schools, he realized that here indeed were prototypes of the successful scientist. As these boys and girls went into college and validated or invalidated his original highly subjective predictions of their success, it was convenient to set up five types of youngsters in a rating scale and to appraise succeeding students in high school against this scale which attempted to "fix" a picture of successful "types" in science. What developed was the Man-to-Man Rating Scale included in Appendix B.

---

[36]Knapp, R. H., and Goodrich, H. B., *Origins of American Scientists*, University of Chicago Press, 1952.

[37]Roe, Anne, *The Making of a Scientist*, Dodd, Mead & Co., New York, 1953.

[38]Kubie, Lawrence, in the *American Scientist*, October, 1953, and January, 1954.

170 P.F. Brandwein

In 1948, this scale became more or less fixed. The top of the scale was a youngster rating high in Genetic and Predisposing Factors; the lowest was a youngster with ambitions to be a scientist, but low in Genetic and Predisposing Factors. The 62 experimentals were chosen on the basis of rating in the Man-to-Man Rating Scale (4 plus), the control by a rating of 3 minus. (An example of a typical case study differentiating the two groups is to be found in Appendix C. C-1 is an example of a student rated 4+, C-2 one of a student rated 3–.)

The Man-to-Man Rating Scale is merely an additional check on the subjectivity of judgments based on Predisposing Factors. It proved, however, to be useful operationally.

Productivity

It soon became obvious that students who had the base level in Genetic Factors and a rating of 4 and above in the Predisposing Factors had high productivity. It was relatively simple to set up an operational test of this productivity. In the Advanced Science course described in Chapter 2 monthly reports were required of work accomplished. Over the years more than 90 per cent of the youngsters in the experimental group turned in their reports on the required day, while less than 50 per cent of the youngsters in the control group turned in their reports on the required day. Furthermore, subjective rating of the reports indicated almost always a greater amount of work by the youngsters with a Predisposing Factor of 4 plus than was found to be done by youngsters of 3 minus. In addition, more than 90 per cent of the youngsters in the experimental group completed a project by the first term of their senior year; less than 25 per cent of those in the control group completed a project. Curiously enough 31 of the 35 who are now in scientific research were in the 90 per cent group of students who completed their projects, and 15 of the 19 in the control group who are now in research were in the 25 per cent group of students who completed their projects.

"Reliability" Criterion

Over a period of three experimental years (1948-1951), youngsters in the Advanced Science class were given the opportunity to take their own attendance, and also to note on the attendance blank where they were to be found.[39] At intervals, checks were made on the reliability of their reporting. In over 95 per cent of 100 distinct checks over three years, more than 90 per cent of the 62 youngsters in the experimental group were to be found in their designated places at work. Similarly, no more than 60 per cent in the control group were found in their designated places at work in 95 per cent of the checks. At other times they were dawdling in the laboratory or just talking idly.

---

[39] 1 indicated library; 2 indicated they were in the laboratory; 3 indicated they were in a special room given over to writing reports.

Test of Developed Ability

As our work went on, it became apparent that high level ability in science was not a limited, identifiable characteristic at an early age. It did not seem to be inherited. It seemed, as has been said previously, to be a developed characteristic dependent on Genetic, Predisposing, and Activating Factors. Perhaps a test of developed ability in science could be developed. When the writer had the opportunity to become a member of the Sub-Committee in Science of the College Entrance Examination Board, it soon became apparent that the members of the committee were thinking along the lines of testing what came to be called "developed ability in science."[40] "Developed Ability" as used here will be taken to mean an ability derived as a result of training. It assumes that this ability is a skill in the use of the materials particularly related to the area being examined (in this case, science) and that it differs from skills usually tested by achievement tests which involve mainly recall and recognition. If this be vague, it is because we are not yet at the point of clear definition.

Be that as it may, a test of this sort was the goal in mind, and in 1950 such a test was written. One item from such a test is de-tailed in Appendix A. From this item the nature and direction of the test is observable. Whether or not this test is one of developed ability in science, some definite relationships were discovered in testing a group of students in the Major Group at the end of their tenth year.

It will be remembered that these students had what purports to be a year in General Science and a half year in Biology, based on their science work in Junior High School. The first 40 items of the Test of Developed Ability in Science, Forms A and B (as it is now called by the College Entrance Board Committee on Science Testing), are general in nature, covering areas in science which would be recognized as Meteorology, Astronomy, Biology, Chemistry, Physics, heavily interlarded with the methods—manipulative, experimental, mathematical, and speculative—that the scientist uses. Five successive groups who took the first 40 general items in Form A of the test (in a testing period limited to one hour) in the middle of the tenth year and 30 of the specific items (making a total of 70) were compared as to their success in the Westinghouse National Science Talent Search as well as to their place in the Inventory of Predisposing Factors and in the Man-to-Man Rating Scale.

1. No youngster who scored below 30 in the first 40 items of Form A of the TDAS got Honorable Mention or was one of the 40 throughout the nation selected for a trip to Washington. Those who scored above 50 in the entire test successfully placed on the "Search" either as Finalists or were Honorable Mentions.
2. Only two youngsters who scored 3 or below in the Inventory of Predisposing Factors scored 30 or above.
3. Only one youngster who scored 3 or below in the Man-to-Man Rating Scale scored 30 or above.

---

[40]Brandwein, Paul F., "Science Teaching and the Board's Science Tests," *The College Board Review,* November, 1951.

172   P.F. Brandwein

When one considers that this test was given in the tenth grade, it is obvious that it was selecting youngsters who had high Genetic and Predisposing Factors. For it is precisely these qualities (among others not clearly discernible here) which will send youngsters along the road of seeking and knowing more, reading more, doing more—in short, experiencing more. There is also some indication that the Test of Developed Ability in Science given in the senior year (Form B was used) has also some predictive value. That is, 19 youngsters out of 20 who scored 50 and above (out of 70 items) in the TDAS were successful in the Westinghouse National Science Talent Search; all were part of the group of 89. Further use of the test may yield more useful data. But it seems that we have here a useful approach to the development of a test which will identify those with high ability in science.

The Westinghouse National Science Talent Search Examination

Every year since 1941 an examination has been held, with the objective of identifying and selecting youngsters with high level ability in science. This is the examination subsidized by the Westinghouse Corporation, administered by Science Service, with offices in Washington, and titled as above. For reasons previously described the term "Science Talent" in the title is not a useful one because the word "talent" has certain connotations in usage. Nevertheless, Edgerton and Britt[41] have shown high correlation between success in the examination and success in college science, as well as in contributions to science.

Of the 62 experimentals (rated 4 plus as to the Predisposing Factors) 15 were "winners" (a trip to Washington for final examination), and 32 achieved Honorable Mention. Of the 62 in the control, 10 were among the Honorable Mentions, none in the Winners' group. It will be remembered that the prediction of success of these 62 in science was made prior to the time they took the "Search" examination. Hence the "Search" examination record may be used as a check against the validity of the various devices described here as useful in the prediction of high level success in science.

This is not surprising to the writer, for the various parts of the Westinghouse examination place strong emphasis on the possession of the Genetic and Predisposing Factors described here, and certainly on their interaction with those of the Activating Factor. For instance, as part of the "Search" there is:

1. Appraisal of Record. Achievement of the youngsters in science is recorded. (In our experience the youngsters who have succeeded in the examination have around I.Q. 135, reading score of 15, and arithmetic score of 12 plus (in the ninth

---

[41] Edgerton, Harold, and Britt, S. H., "The Science Talent Search," *Occupations,* 22 :177-180, December, 1943 ; "Third Annual Science Talent Search," *Science,* 99 :319-320, April 21, 1944 ; *Educational and Psychological Measurement,* 7 :3, 1947.

The Gifted Student as Future Scientist

year). They also generally place within the first 15 in ranking in the graduating class. This is within the scope of the Genetic Factor.

2. Appraisal of Personal Characteristics. An appraisal of personal characteristics of youngsters, including honesty, inventiveness, curiosity, and persistence, e.g., a check list which includes such items as "has to be thrown out of the laboratory." (This is within the scope of our Predisposing Factor.)

3. An essay on "My Project"—a report of the student's project—stressing ability to report, reliability in reporting and originating. (This is within the scope of both Genetic and Predisposing Factors described here and, as will shortly be seen, is related to the Activating Factor. Those youngsters having Predisposing Factors of 4 plus turned in many more satisfactory projects than those with 3 minus.)

4. The Westinghouse Examination[42] is a 2 1/2-hour test of scientific reasoning and what the writer would call developed ability. Recently Watson[43] has indicated that part "B" of the examination produces the greatest spread in the final scores. It may have value then in discriminating among those in the higher ranges of ability. Part B is similar in structure to the items of the TDAS mentioned previously, but tests more of the ability of the student to understand the language of the specialist in science than does the TDAS—or so it seems to the writer.

5. Not part of evaluation within the "Search" examination per se, but an integral part of the success or lack of success of a student in the examination, is the Activating Factor, e.g., the opportunities offered to students in the school, including the science program and the kind of teaching. For without the Activating Factor students would not be stimulated to do a project, would not have the kind of superior instruction in science which the examination demands and, indeed, would not have the opportunity to take the examination at all.

This high success in the "Search," whether or not the Search itself will be shown finally to have the validity expected, indicates at least one possible conclusion: that the Operational Approach as we have described it here leads to high success in certain types of scholarship examinations in science. Furthermore, as has been said, it is curious to note that the record in the search compares very favorably with that of the "science-oriented" schools, the leading schools in the Search, the Bronx High

---

[42] The Westinghouse National Science Talent Search, started in 1941, includes a battery of tests—a written examination testing for science understanding, a personal data blank dealing with certain personal traits (as inventiveness, originality, etc.), an essay describing a project, and an examination of achievement record in high school. If an individual is adjudged a Finalist, he is one of 40 chosen to go to Washington for final interviews for the awarding of scholarships. Three hundred students in addition are given Honorable Mention.

[43] Watson, F. G., "Analysis of a Science Talent Search Examination," *The Science Teacher,* November, 1954, pp.274-276.

174   P.F. Brandwein

School of Science and Stuyvesant High School,[44] even though these had a full-fledged program going at the time Forest Hills High School had its first entering students (1941).

If, then, the "Search" is to be taken as any measure of high level ability in science, it may follow that a school such as Forest Hills High School can offer a general program of studies and yet, by offering such opportunities as have been described (Chapter 2), develop students with high level ability in science.

The Westinghouse examination requires processing; that is, some 12 hours per student are required in just filling out records, giving examinations, completing personal data blanks, correcting essays for English This aside from the year or two of guidance in accomplishing a project (getting equipment, reading material). Hence the Westinghouse Examination tests for the existence of the Activating Factor as well as the Genetic and Predisposing Factors.

It is clear that the data reported here, such as it is, confirms the usefulness of the Westinghouse National Science Examination in selecting youngsters with Science Potential in their senior year in high school.

Inferences

1. Assuming opportunities for development in science (the Activating Factor), students with a base I.Q. of 135 (Henmon-Nelson), a reading score (Nelson-Denny) of 15 in the ninth year, an arithmetic score of 12 plus in the ninth year (Arithmetic Judgment Test, N. Y. City), are at the base level of the Genetic Factor postulated here as being a factor basic to high level contribution in science. It seems also to be essentially true that youngsters who place at the 90th percentile and above in the Tests of Primary Mental Abilities, specifically those titled Verbal-Meaning, Reasoning, and Number (developed by Science Research Associates, Chicago), or the Differential Aptitude Tests, specifically those titled Verbal Reasoning, Numerical Ability, and Abstract Reasoning (developed by The Psychological Corporation, New York), are also at the base level of the Genetic Factor which is postulated here as necessary to high level contribution in science.

---

[44]To the best of our knowledge, in 1954, the records of the three leading schools in the "Search" were:

	Finalists	Honorable Mentions
Bronx High School of Science, New York	18	84
Stuyvesant High School, New York	17	54
Forest Hills High School, New York	17	57

The next three, Evanston Township High School (Illinois), Brooklyn Technical High School and Midwood High School (New York), had 8 finalists.

The Gifted Student as Future Scientist | 175

2. Assuming opportunities for development in science, students with a 4 plus on the Inventory of Predisposing Factors (Appendix D) and a 4 plus on the Man-to-Man Rating Scale are at the level of the Predisposing Factor which is postulated here as playing an important part in successful contributions to scientific research. Students with a 4 plus in this Rating Scale appear to be highly productive and highly reliable as well.

In addition, there is confirmation that the Westinghouse National Science Talent Search examination given in the senior year is an indicator of science potential.

There is also an indication that a test such as the Test of Developed Ability in Science (sample item, Appendix A) may be useful in predicting science potential when given early in the high school career.

Chapter 4: **Who They Are—*Certain Characteristic Behaviors***

As students work over a period of two years on their "research," as they come for guidance, or as they just "talk," ample opportunity is given to observe their behavior. There is also room for specific questions, casually introduced. These observations of behavior were noted in case studies, an example of which is given in the Appendixes.

An attempt was made to determine whether there were clearly discernible characteristic behaviors of youngsters with high level ability in science—as they worked in school.

Responsibility

How early can youngsters work by themselves in the laboratory? How soon can they be left on their own? It is worth recounting our experiences.

Since 1944 attempts have been made to discover the best way of guiding youngsters to fruitful use of time in the laboratory. Two small research or project laboratories, each capable of holding at least six students, were used. Students were scheduled throughout the day over a 6-hour-and-20-minute day in an attempt to limit the use of each laboratory to 6 students per hour.

In 1947-1949, youngsters were divided into two groups to determine whether students could indeed work on their own. One group was placed in a rear laboratory (away from surveillance); this group, the *far* laboratory group, was left to its own devices. An equated group (for Genetic and Predisposing Factors) was placed in a laboratory which could be easily kept under surveillance; this group, the *near* laboratory group, was visited once a period. In addition, members of the "far" group were permitted to go on their own once they had submitted and obtained approval of an outline of their project. Each student of the "near" group, however, was required to submit a monthly report, and advice and guidance were available to the students at all times. Each day the writer visited the "near" group to offer help.

176   P.F. Brandwein

The "far" group invariably accomplished little or nothing. The students were noisy, left the laboratory in a messy state, did their class assignment instead of laboratory work, and were constantly in difficulties with the school's laboratory assistants (who worked in a room in between the two laboratories). The "near" group had reasonable accomplishments, kept the laboratory clean, and had reasonably few disciplinary problems.

When the laboratory "far" and "near" groups were exchanged in the tenth year, they exchanged behavior; that is, the present "far" group (formerly the "near" group) became noisy, accomplished little, and vice versa.

This is reasonable confirmation of Lewin's work;[45] a laissez-faire policy does not appear to operate successfully. Guidance, permissive in nature, not autocratic, appears to be necessary.

In 1949, final confirmation (for the situation as described) was obtained. Twelve eleventh-year students (16-17 years of age) equated for Genetic and Predisposing Factors as well as sex were placed "under laissez-faire" (a group of 6), and under "guidance" (another 6). Although each one of the 12 had a project pretty much under way by September of that year, only the six in the guidance group (even though placed in the "far" laboratory) had completed their project reports by December. The six in the laissez-faire group had advanced somewhat but had not completed their reports.

A short but brief experience convinced the writer that an autocratic environment was as unproductive as one of laissez-faire. Throughout the years this work had been going on, few of the youngsters remained with a certain sponsor who tended to be decidedly autocratic. Time and again (and unfailingly) these youngsters would ask for another sponsor or would drop the work.

The Effect of Training

When the sponsor was available, that is, when he "dropped in" regularly so that he could suggest further references and a way out of difficulties (yet without giving the solution) ; when goals were arrived at in full discussion and in common agreement; when a permissive (not coercive, autocratic, or laissez-faire) attitude prevailed, there was a noticeable growth in the ability of the youngsters to work effectively. Thus youngsters placed in such an environment during the tenth year needed less guidance in the eleventh year, and in the twelfth were self-directed and self-reliant in the best and mature sense of the word. They were assured workers who used time well. Yet they could drop their work as well, and invite the teacher-sponsor to take part in just "talk" without the over-the-shoulder apprehensive look so characteristic of youngsters who are fearful of being caught at what they should not be doing.

Finally, instructors in certain colleges have been good enough to remark on the mature and effective behavior of these youngsters in the laboratory.

---

[45]For a brief description of Lewin's work, see pp. 453-456 in *Educational Psychology*, Lee J. Cronbach, Harcourt, Brace & Co., Inc., 1954.

Planning Ability

In 1947-1949, experimental observation of the laissez-faire group (far) and permissive-guidance group (near) indicated, as expected, that the latter planned its work better. An operational indication of this was the weekly "order" for material. The members of the laissez-faire group, in general, did not have complete orders; almost without fail they asked for additional material throughout the week. The permissive-guidance group soon learned to plan work and time so that a weekly order of materials sufficed. This is not strange, since these students were shown how to plan.

Be that as it may, the main observation to be made after 1949 was that in the eleventh and twelfth year, youngsters not only learned to plan but took pride in their ability to do so. In the twelfth year, youngsters both in the experimental and control group had learned to plan, not only in the sense of getting sufficient material for experiment but also in the general planning of a design and the carrying out of an experiment; some of these were quite complicated in nature. (See Appendix F for summaries of reports; note particularly F-1 through F-6.) Little difference could be seen in this ability of planning between the 62 in the experimental and the 62 in the control group (Chapter 2). Rather it was in the quality of work done that the difference between the experimentals and the controls was noticeable.

Leadership

It was interesting to observe the position of leadership which these youngsters accepted. In the 62 experimentals, practically every one of the boys and girls except for six had become chairmen of a subject class—Science, English, Foreign Language. Rarely did they seek or ask to serve in "political" office—e.g., elected officers in the General Organization, the Student Government of the school. It was the writer's observation that they were not interested in such posts, seeking instead posts which involved intellectual pursuits primarily.

On the other hand, the 62 youngsters in the control group had a total of 37 posts (some of them two) of a political nature—secretary, treasurer, president in the student government, manager of the school paper, track team, etc. Furthermore, almost without exception, youngsters from this group were elected as directors of the student laboratory (to manage laboratory procedures as head of a group of six in the "near" or "far" laboratory).

Attendance

There was little difference between the experimental 62 and the control 62 in attendance. Both had high attendance records as a group; they liked school. Furthermore, they were in as good general health as were their peers. A comparison of records of absence shows 2.2 days per year (illness, personal reasons) for 20 of the

178   P.F. Brandwein

experimental 62, and 2.1 days for a similar number of the control 62. In the former group, 11 had perfect records of attendance as compared with 9 in the latter group.

Trends in Personality Traits

In general, there was one personality characteristic which seemed almost obvious. The youngsters in the experimental 62 as compared with the "norm" of behavior at Forest Hills might be said to be more quiet, more reflective, more inward-looking; in short they exhibited, in general, a tendency to introversion, as compared with the norm. Twenty boys and ten girls, who not only had the Genetic and Predisposing Factors (previously described), who were also at the 90th percentile and above in the S.R.A. and Psychological Corporation tests (named in Chapter 3), who had scored above 30 in the Test of Developed Ability, and who in addition had placed as one of the fifty finalists (8) or had won Honorable Mention (22) in the Westinghouse Science Talent Search, were compared with their age-mates (controls) who were not committed to science vocationally although very successful in high school in other areas. These age-mates were of similar socioeconomic level but were not paired with the experimentals for Genetic and Predisposing Factors. The purpose of this was to obtain a trend or tendency[46] in behavior of these youngsters who had committed themselves to science. A comparison of behaviors observed is shown in the tabulation on pages 179–180.

From these main observations a picture, with hazy outlines, it is true, may be built up of these youngsters who may be our scientists-in-embryo. There is every indication that most of our 89 students fit this picture. It seems possible that the high rating in Predisposing Factors of the experimental 62 is related to their introversion (as indicated in the tabular description). MacCurdy[47] recently has made a detailed study of the characteristics of superior science students (Science Talent Search winners). His findings extend over a greater area of the background of these youngsters than does the description tabulated above. Nevertheless, there is essential agreement on the major points made. As this goes to press Terman,[48] whose studies on the general nature of giftedness are classic, writes (on the basis of a quite recent study of 800 gifted males now committed to the sciences, humanities, social sciences, and business among other fields):

> At any rate, in our gifted group the physical scientists and engineers are at the opposite pole from the businessmen and lawyers in abilities, in occupational interests, and in social behavior.

---

[46]Information obtained through interview. The word *tendency* describes an activity of more than 50 per cent of the group.

[47]MacCurdy, R. D., "Characteristics of Superior Science Students and Some Factors That Were Found in Their Background," Ed.D. Dissertation, Boston University.

[48]Terman, Lewis M., "Are Scientists Different?" *Scientific American*, p. 29, January, 1955.

The Gifted Student as Future Scientist

A curious observation bearing on item 19 deserves comment. Of the 89 students mentioned in Chapter 2, 47 were the only children in the family, and 24 were first children. How significant is this? How general is it?

A singular item was also noted for the girls. Except for one case, rarely was a girl who would ordinarily be judged to have great physical beauty found in the groups

Thirty within the 62 experimentals (with science potential)[49]	Thirty age-mates (without science vocational interests)
1. Outside of school the majority tended to individual sports—tennis, cycling, fencing, walking. Very few in team sports—basketball, baseball, football.	1. General involvement in team sports as well as individual sports. Most of the boys mentioned playing football, basketball; the girls tennis, dancing, etc.
2. A major part of the time spent in reading and other intellectual activities, homework, listening to music, school club activities. Minor, although significant, amount of time in social activities such as dancing.	2. A major part of the time spent in social activities—dancing, parties, theater, movies, group activities. A minor, although significant, amount of time in homework.
3. A major part of time in self-initiated, individual projects— astronomy, "ham radio," stamp collecting, classical music, learning foreign language, musical instruments.	3. Relatively minor part of time spent in self-initiated projects of sort described opposite. Tendency to group activity.
4. Tendency to classical music, chess, bridge, and serious reading of classics (Dickens, etc.), do crossword puzzles and acrostics.	4. Tendency to popular music, dancing, magazine reading, popular novels. Less tendency to crossword and other types of puzzles.
5. Tendency to read "serious" magazines: *Harper's, Time, Scientific American, Saturday Review.*	5. Tendency to read story type magazines, e.g., *Reader's Uigest, Saturday Evening Post.*
6. A tendency to go to movies less than once a week. Tendency to go to the theater.	6. A tendency to go to movies more than once a week. Prefer movies to theater.
7. Activities joined in school more of discussion type—Language Society, Problems of Civilization, Science, Chemistry Club, school paper, school magazine.	7. Clubs joined in school more of "doing" type—Glee Club, Orchestra, Intramurals, school paper, school magazine.
8. Approximately equal (to column 2) tendency to earn extra money in after-school work—but mainly in baby sitting (girls).	8. Approximately equal (to column 1) tendency to earn extra money in after-school work—in delivery work, baby sitting, delivering newspapers.

---

[49]For other descriptions of the characteristics of gifted children in science see—

1. Meister, Morris, in *The Gifted Child*, Paul Witty (editor), D. C. Heath and Co., Boston, 1951.

2. Subarsky, Zachariah, "What is Science Talent?" *The Scientific Monthly*, LXVI, No. 5, May, 1948.

3. Other definitions, descriptions of characteristics of the gifted will be found in the four volumes of *Genetic Studies of Genius*, edited and written by Lewis M. Terman and others, Stanford University Press, Stanford, Calif.

(Continued)

Thirty within the 62 experimental (with science potential)	Thirty age-mates (without science vocational interests)
9. Strong tendency to do social-service work—e.g., read to blind, collect for Red Cross, church activity, messengers, Civil Defense.	9. A minority tend to be in Boy Scouts or Girl Scouts, church activity.
10. Tend to be conservative in clothing, although within norm, e.g., most boys wear ties, rarely take on "fads."	10. A good number (not over 50 per cent of the group, however) take on "fads" of teenagers, e.g., plaid shirts, jeans, etc.
11. Vast majority buy books for personal library.	11. Only a minority buy books for personal library.
12. Tend not to smoke till senior year—or not at all.	12. Tend to smoke early.
13. All plan to go to college.	13. Most plan to go to college; some plan to go into business, minority into professional life.
14. Almost never get into difficulty with teachers or are disciplinary problems in school over school work. May disagree with teachers over interpretation of subject matter.	14. Almost never get into difficulty with teachers over school work or are disciplinary problems in school. May disagree with teachers over interpretation of subject matter.
15. Almost all the parents in this group possessed a post-high school education. More than half are graduates of colleges. A high number from graduate schools(Ph.D., M.D., Law, Engineering, Accounting). A high minority in professions.	15. Almost all parents in this group with a post-high school education, with a high majority in business. Slightly more than one-fourth are graduates of colleges, a number in medicine and engineering.
16. A tendency for parents to own a substantial library (500 books or more). A vast majority with 200 books or more.	16. A minority with parents having a substantial library; most with less than 200 books.
17. Vast majority of parents with ambitions for professional life for their children.	17. Vast majority of parents with ambitions for financial success for children.
18. Average of children per family, 1.2.	18. Average of children per family, 2.4.
19. Twenty-four of the 30 were the first child, 16 were the only child.	19. Eleven of the 30 were the first child.

who were in the experimental 62, or for that matter in the control 62. This is not to say that there were not attractive girls in the group. The "cover girl" or "model" was, however, generally not in the group. But we are dealing with such small numbers that no definite statement should be made here. As more investigations of the type Roe[50] and Kubie[51] are doing on the adult level are duplicated on the high school level, we shall know more about the individual characteristics of these scientists-to-be. Hence, the observations described here must be considered tentative.

---

[50]Roe, Anne, *The Making of a Scientist*, Dodd, Mead & Co., New York, 1953.

[51]Kubie, Lawrence, *American Scientist*, October, 1953, and January, 1954.

Ways of Work

One cannot work long with these youngsters without wondering how they get their unusual ideas. One wonders whether they have a different way of thinking than others. Of one impression the writer is certain: these youngsters get a joy (one called it a "thrill of the brain") when they get an "idea" or make an "original discovery." Bridgman in his *Reflections of a Physicist* develops the idea that the discovery of a new relationship, that is, activity within the brain, furnishes as great a thrill as composing a piece of music or completing a painting. These youngsters give the impression that they enjoy thinking. Their "I've got it" lights up their faces and seems to give them great satisfaction.

But is there any systematic approach that these youngsters have to a solution of their "intellectual" problems? How do they go about discovering "new" relationships, doing what would ordinarily be called "discovery"?

It seemed to the writer that there were several stages in their outward behavior as they solved a problem. At first, the writer used the following phrases, only to find that Wallas[52] had given a similar description and analyzed, in a general way, the nature of the "process." Poincaré, in his *Science and Method*, also deals with several stages to be described here (e.g., Preparation, Incubation, and Verification). Wallas has dealt so well with the present writer's independent observation that it is a pity that the writer came upon Wallas's formulations so late in this work.

It appears that when these youngsters first become interested in any problem area, there is just random *Exploration*. In this period of Exploration there is no focus, just a mental meandering. Then as part of this period of Exploration there seems to be a period of *Clarification*. This seems to be a period of focus when youngsters narrow down to a special activity, project, or problem.

Once this *Clarification* occurs, it seems as if the youngsters begin a period of *Preparation*. They read, they explore techniques, they discuss problems with their advisers, they draw on previous experience. They work *consciously*, it appears, to solve the problem.

But here is the nub of it. Rarely do they appear to get the "new idea," the "solution, " the "approach to the solution" while they are *consciously* at work on the problem. Almost uniformly they seem to admit getting *the idea* in a "flash, " a "flash of insight, " if you will. They get this flash while they are engaged in another activity, e.g., reading literature, listening to music, or walking—mainly walking—and when they are *not thinking*, or rather *not conscious of thinking of the problem* per se. This "flash" is what Wallas calls *Illumination*. Wallas and Poincaré both put forth the notion that a period of *Incubation* is necessary before this "flash" or *Illumination*. So too it seems to the writer.

During *Incubation*, the cerebrum seems to take over, seems to feed back the problem to itself as in a computer. Of all this the student seems not to be conscious. Then

---

[52]Wallas, Graham, *The Art of Thought*, Harcourt, Brace & Co., Inc., 1926.

the "flash"—the "That's it" — the "I've got it." Then the solution of the problem, the approach, the new road, is laid bare.

Then the youngster can barely wait to get to work, to embark on a period of *Verification*. Reading, experiment, planning, discussion are borne lightly, for the "way" seems to be clear. The youngster often says, "Now I know where I'm going."

These stages—Exploration, Clarification, Preparation, Incubation (unconscious), Illumination, and Verification—seem to be the vague yet perceptible stages through which these youngsters work. The Incubation period, with its subsequent Illumination, seems to be especially characteristic. Is it different for scientists than for others? Is it characteristic of all thinking? Is there anything to it at all? Does this seem profitable for further investigation?

Inferences

A brief recapitulation is necessary. It will be remembered that it was indicated, although not with the precision that the writer would like, that the majority of the youngsters in the experimental 62 had the high level ability thought necessary for contributions to science (science potential).

It will be remembered that the experimental 62 and control 62 were similar in the Genetic Factor but differed mainly in the extent of their rating, subjective though it was, in the Predisposing Factor. It is inferred here that the Predisposing Factor is mainly of environmental causation; that is, environmental influences (parental, school, church, and community) are partly responsible, at least, for the Predisposing Factor.

A picture of tendencies in behavior has been given of high school youngsters who have science potential. There is an indication that there are differences in the general behavior of these youngsters as compared with age-mates who have little science potential. Their behavior in the laboratory has been described mainly with regard to the problem of determining the amount of responsibility for self-guidance which they can reasonably be expected to assume. Also a description has been presented of the approach of these youngsters to a solution of a problem.

Whether the picture of behavior sketched here is valid would depend on observations gathered from further investigation.

The Predisposing Factor, which is probably based on psychological factors which this writer is not equipped to analyze, seems to be related to the strong tendency of inwardness (or introversion) which is strongly characteristic of these youngsters and their mode of activity. Curiosity and creativity as ordinarily used would seem to stem from one of the factors within the Predisposing Factor, namely, Questing.

It will have occurred to the reader who has charitably read the past four chapters that high level ability in science appears to be very similar to high level ability in other fields. And indeed it is the writer's strong belief that this is so with this modifi-

The Gifted Student as Future Scientist 183

cation. Three factors—Genetic, Predisposing, and Activating—have been proposed as profitable centers of identification for the individuals with science potential. To embrace art, music, athletics, literature, this hypothesis need be altered mainly in the description of the Genetic Factor.

It will be remembered that the Genetic Factor for science concerned itself with high I.Q., high verbal and mathematical ability. It would seem to the writer that it is mainly in the kind of Genetic Factor (not Predisposing or Activating) that the youngsters with other high level abilities differ. For instance, youngsters with ability in music need not *perhaps* have high verbal and mathematical ability; their special talent lies in their ability to cerebrate with sounds and symbols.

From a teacher's point of view, these youngsters in science are fine young people with whom it is a joy to work. From this it should not be inferred that the normal trials and tribulations of adolescence are not theirs. Be that as it may, among these youngsters are the scientists of the future. They are the people upon whom our nation depends so much. They are part of our human resources. And what happens to them in their school days is of exceeding importance to the future of this country. The past is prologue.

Chapter 5: **Who Teaches Them—*The Key***

Teaching is a personal invention. It is a result of many years of work coupled with the interplay of factors of heredity, personality, and the larger forces of the environment—education, home, church, and community. Yet there is no doubt that some teachers have invented a better teaching method than have others— better in the sense that they affect the growth of youngsters in a wholesome and desirable way. Furthermore, these youngsters are aware of this stimulation and acknowledge it.

Previously, it has been noted that high Genetic and Predisposing Factors alone were not sufficient to commit a youngster to science. Similarly, interest in science prior to the tenth grade was not stable—hence the postulation of an Activating Factor in high school. The teacher—and the opportunity he makes available—is the Activating Factor; the other—the financial situation of the student—operates more importantly in the college situation.

Early in the writer's exploratory period, in relation to the problem "What makes a scientist?" he could not help but note that many of the working scientists spoke of the great influence of a teacher, or of two or three teachers, in their lives. Not so much of teaching in general, mind you, but of the effect of one teacher, or two, sometimes three. Knapp and Goodrich[53] have published a work which deals with the characteristics of such teachers. This work, in a sense, substantiates the position that the teacher is the key, probably the single most important factor, in the training

---

[53] Knapp, R. H., and Goodrich, H. B., *Origins of American Scientists,* University of Chicago Press, 1952.

of a scientist, or what is as important, in stimulating young people in school to turn to science.

What are the characteristics of the teacher who inspires youngsters *on the high school level* to commit themselves to a career in science? In 1931-1939, the writer began making notes of the characteristics of the teachers who inspired him (the writer was then in college and graduate school completing bachelor's, master's, and doctorate degrees).

Since 1936, whenever he found a student who had committed himself to a career in science before entry to high school, that is, when the period of commitment was in junior high school or prior, he asked the student to write a brief description of the teacher or teachers who had helped him decide. Particular reference was made to a description of "how" the teacher went about stimulating these students.

Also since 1944, the writer has interviewed students who were exhibiting in the Science Fairs regularly held in New York City up to 1949. Five students were interviewed at each fair with the view of inquiring how they were stimulated to enter the fair. It was not necessary to ask about their teachers; they talked freely and enthusiastically of the help they had been given. The same was done at the Paul B. Mann Biology Congress held annually by the Association of Biology Teachers of New York. Here youngsters doing projects or "research" in biology report to a congress of high school students from all the schools in the city. Inquiry was not made of the names of teachers. Usually, however, they were freely and enthusiastically given.

The Westinghouse National Science Talent Search examination requires the student to answer certain questions relating generally to hobbies, activities, aspirations to college, etc.; among these questions is one asking the student to name a person who has influenced him most to enter upon science. At Forest Hills High School certain teachers were regularly named. The characteristics of these were studied through direct observation. Similarly, a study of the characteristics of 17 science teachers in New York City who had had good success in the Science Talent Search was also made; this was through direct observation, since all were the writer's colleagues and were active in teacher' organizations. Finally, through visits to national meetings in Pittsburgh, Cleveland, Chicago, Philadelphia, Washington, Boston, and Cambridge, and visits to a number of schools throughout the country (during the past ten years), the writer was able further to study the characteristics of teachers who were successful in stimulating youngsters to enter science.

Success in stimulating youngsters to enter science is defined operationally as follows. The teachers studied generally had youngsters in one of the following groups:

1. Youngsters who were taking the Science Talent Search (at Forest Hills, mention of teachers by more than 10 students over 5 years).
2. In junior high school, regular entry of one or more students with prize-winning projects in three or more science fairs.
3. A national reputation including two or more winners in the Westinghouse National Science Talent Search. (Eleven such teachers were known to the writer.)

The Gifted Student as Future Scientist 185

4. Reputation among colleagues and former students outside of school (e.g., in college); that is, frequent mention by others of success in stimulating students to enter science.

All in all, 82 teachers in junior high school and senior high school were studied. Among these, 22 high school teachers were studied carefully through direct observation—regular conversation, regular meetings over a period of a week, observation of all 22 in the classroom, at least one observation for 6 teachers, two observations for 9, eight observations for 3, twenty observations for 4.

The major observations which filtered out and which seemed constant features of teachers who were successful in stimulating high school students to commit themselves to science were these:

1. More than 90 per cent of the 82 (65 men, 17 women) had a Master's Degree in Science, in addition to the requisite work in education. They were exceptionally well versed in the subject matter of science.
2. More than 50 per cent of the 82 had at one time or another matriculated for a doctorate in science or in education. There were 11 Ph.D.s (6 in science and 5 in education).
3. More than 50 per cent of the 82 had taught in college (liberal arts or in schools of education) at one time or another. This varied from full-time teaching to one or more courses during the week.
4. More than 90 per cent of the 82 had published at least one paper in science or in education.
5. All but one had been an officer in a local or national organization of teachers.
6. All had, at one time or another, been members of a committee to formulate courses of study or a curriculum for the school district or township in which they taught.
7. All were in general good health and had a remarkably high attendance record; very rarely were they out of school for reasons of personal illness.
8. The 22 whom the writer knew personally had hobbies which ranged from expertness in chess to expertness in collecting antiques. All of these had some athletic activity—walking, tennis, gymnastics, handball, or baseball in which they took part regularly.
9. The 22 were invariably vigorous in their personal manner and, in the writer's judgment, were people of decisiveness.
10. Twenty of the 22 were judged by the writer and at least one colleague as having a sense of humor.
11. All of the 82 participated in extracurricular activities; not only did these include science work, but work in music, publications, athletics, etc. The 22 who were observed directly often gave up lunch and some evenings to work with students. As a regular occurrence they often came to school long before their first class to meet with students or prepare for special work with them.
12. The average age of the 82 was 40 years (plus or minus 2). The "plus or minus 2" is necessary, since in a number of cases the exact age could not be ascertained.

186 P.F. Brandwein

Of the 22, the average age at this writing is 39.2 years. This appears to be especially significant and will be discussed in the summary below.

13. The writer would judge 8 of these observed under teaching situations (in public and private schools) as outstanding and inspiring master teachers, 10 as superior teachers, 4 average. The 8 superior or outstanding teachers were dynamic personalities both in classroom and extracurricular activity. Of these 18, 17 were experts in discussion techniques, and their classrooms were centers of student activity; they rarely lectured. One lectured (in a private school). All were splendid demonstrators and experimenters; they had at one time or another given demonstrations before teachers' groups.

14. All 22, however, were dissatisfied in one way or another with their progress in teaching, with the state of knowledge of the learning process, with the professional status of teachers. All 22 were in some way (committees, officers of associations, editors of journals) associated with one effort or another to improve instruction.

15. All 22 intended to stay in teaching. All seemed to like children (19 expressed their "love" for them). All were vitally interested in science and in other intellectual pursuits.

16. Although the writer's training in psychology is meager, he could not fail to note that these teachers were in the relation of the "father" or "mother" image to the youngsters who were interviewed. These teachers were, in short, not only admired and respected as teachers of subject matter, but as teachers in the ways of life. They were guides, counselors, friends, guardians, father-confessors.

Aside from this, it was very clear that they held up to their students firm standards of competence in scholarship as well as in behavior. In brief, there seemed to be an element of accepted "coercion" (since it was accompanied by sympathetic, even warm, treatment, by the student—without fear or threat of punishment). Knapp and Goodrich have amplified this point to considerable length in their work, *Origins of American Scientists.*[54] A distinctly personal impression remains that these teachers were first of all fine human beings; on the whole they were sensitive to human problems, they were considerate of others, and they were plucky in the face of the frustrations teachers must face. They had considerable respect for the goals of children and for their dreams.

Inferences

From the summary of observations listed above, it would seem that the very traits which characterize students with high level ability in science identify the teacher with the qualities to furnish opportunities for youngsters with science potential.

---

[54]Knapp, R. H., and Goodrich, H. B., *Op. cit.*

The Gifted Student as Future Scientist   187

These are High Genetic Ability (vaguely, intelligence), the Predisposing Factor (Persistence and Questing as defined here). In the case of teachers, it is the Activating Factor which appears to be some what different.

The teachers described here are very well trained in subject matter. They like children, particularly at the adolescent stage. Teaching has enabled them to bring their two loves together, the student and science. The first Activating Factor for them appears to be the opportunity to teach.

It is well worth repeating that the children who have been interviewed almost always refer to these teachers in terms of the "father" or "mother" image. These teachers are people they trust; they are sympathetic to their problems. But they hold up firm and high standards of achievement and behavior as well.

The observation that these teachers are, on the average, 40 years of age, seems significant.[55] Two inferences, among others, are possible. One is that it takes almost that long for a science teacher to become competent. The inference to which the writer inclines is something like this: these teachers left college during the depression years. They were competent scientists who but for the economic failures of the depression years might have found their way into scientific work per se (research in a university or in industry). These opportunities were closed, however. They found their way into teaching in secondary schools, liked it, and became successful. And they stayed.

The Activating Factor for these "successful" teachers of students with science potential seems, then, to be a combination of three factors: High Training in Science, Opportunity to Teach, and Successful Relations with Children in a Teaching Situation.

Can such teachers be trained?

Can teaching be made attractive to them so that they will remain to develop the students with high level ability in science, students so desperately needed by a country approaching its epitome in world leadership?

These are two fundamental questions to which answers must be found.

The booklet "Critical Years Ahead in Science Teaching"[56] spells out a state of affairs now existing: teachers who are willing to work unremittingly in their profession but who must work after school to make ends meet; teachers without equipment; administrators, desperate in the face of increased enrollments, forced to hire personnel untrained in science to teach science; teachers who do not have the confidence of, nor help from, their communities.

Industry pays an individual trained in science more at the beginning of his career than do the schools; industry affords more opportunities for advancement than do the schools; hence industry attracts the individual who might prefer to teach but must

---

[55]Hugh Templeton, Supervisor of Science for New York State, made a similar observation and has communicated it to the writer.

[56]"Critical Years Ahead in Science Teaching," a report on nation-wide problems of Science Teaching in the secondary schools. Held at Harvard University, Cambridge, Mass., summer, 1953. Free copies may be obtained from Elbert Weaver, Phillips Academy, Andover, Mass.

also think of his family and his physical well being. There is this fact as well: unless the schools offer opportunities to youngsters with high level ability in science, the manpower resources upon which industry and science must draw will dry up at the source. That source is the secondary school.

Chapter 6: **Some Proposals—***Local and National*

These proposals are based on the notion that the schools, not the colleges, are in the long run the pool from which will be drawn the people of high ability in science, the arts, and the humanities. The proposals are submitted with diffidence.[57]

Any solutions to the problem need to be considered against the background of this situation:

1. Less than one-half of the most able youngsters (similar to those we have called youngsters with high level ability) ever reach college; many never finish high school.
2. About one-half of the able students who enter college are graduated. It is estimated that about 100,000 able students drop out of college each year.

The reasons for this dropout fall into several categories:

a. Most school programs are still aimed at the average student. The able student is not sufficiently motivated (i.e., the "Activating Factor" is missing).
b. Many able youngsters, who come from families of low socioeconomic status, do not consider going to college. Their pattern is to go to work as soon as possible.
c. Able youngsters are not identified early enough or are not given the opportunity to identify themselves in a school program geared to their abilities.
d. In a good number of cases, the financial situation of the family does not permit the able youngster to stay in high school, then to enter college and stay there.
e. The salary of teachers is so low that it does not attract able people—those who can motivate able youngsters, guide them, and produce the curriculum these youngsters need.

The categories a, b and c, d, and e are really facets of one picture—our failure to use our human resources wisely.

---

[57]F. L., Fitzpatrick, "Scientific Manpower: The Problem and Its Solution" (Teachers College, Columbia University, mimeographed) offers one solution.

On the Local Level

There are no two high schools alike in this country. However, there are certain similarities in the way the best schools meet their obligations, in the way they aid the development of all the different youngsters whose growth is partly in their care.

1. They have a rich curriculum.
2. They have extracurricular (or co-curricular) activities which extend the curriculum.
3. They have a number of specially skilled teachers who *want to* and like to work with youngsters *and know how to do it*.

The curricular and extracurricular activities are the opportunities, and the teachers make the opportunities come alive for all youngsters, including the able ones.

There are, of course, many, many ways of fashioning a science curriculum for the able student. There is the science high school, as exemplified by the Bronx High School of Science. This kind of solution to the problem is available mainly to large cities. The school has been described in good detail in Paul Witty's book, *The Gifted Child*;[58] there is no need to repeat that description here.

One major way of solving the problem of meeting the needs of the able student lies within the purview of the science department as organized in most high schools. One such program has been described here.

However, there are many small high schools of 500 students or less. For these schools, the science curriculums are generally based on four major courses. Enrichment and acceleration become truly individual, and as the writer has observed science work in small schools, work with individuals per se generally occurs outside of class in extracurricular or co-curricular activities.

The science curriculum in most schools means General Science, Biology, Chemistry, and Physics. The extracurricular activities are science clubs or similar activities held after or before regular classes. Some of the different types of clubs have been mentioned in Chapter 2. "The Science Sponsor's Handbook," published by Science Service, Washington, D. C., offers a great number of suggestions, as do the publications of Future Scientists of America, Washington, D. C.

Yet it is true that a number of smaller schools do not offer all the science courses; especially do they not offer Physics or Chemistry. In most of these cases, the size of the school does not permit sufficient yearly registration in these basic courses. In some cases the teacher of science has to teach all the courses in science and some courses in other areas in any given year—an impossible task as far as skillful preparation is concerned.

What can be done in the smaller school? These observations may be useful.

---

[58]Paul Witty (editor), *The Gifted Child*, D. C. Heath & Co., Boston, 1951.

Proposals Dealing with Basic Science Courses[59]

It has been assumed by many with whom the writer has talked that part of the problem, an exceedingly important phase, would be solved if only we could get more students taking Physics and Chemistry; if only we could introduce these courses where they are not now given; and if only we might increase the number of students in these courses where they are being given. This is an oversimplification of the problem—leading to an approach which may possibly be off the main track towards a solution.

The underlying assumption that more Physics and Chemistry courses would be a partial solution of the problem which concerns us reflects the fact that the students taking them must be among the better students. True. Generally speaking, Physics and Chemistry should be considered the *special courses* from which will come our students of high ability in science. But it is possible that the road to getting increased registrations in Physics and Chemistry involves more than just giving these courses. For these reasons:

1. In one state the registration in Physics is 4 per cent of the student body. There is little General Science given. It seems (and there is agreement among the teachers) that an interesting and vital course in General Science would stimulate youngsters to take more science, more Chemistry and Physics. Without General Science, there is little "seeding" of interest. This seems to be true of other situations where Physics and Chemistry registration is low.

   At the Forest Hills High School, a brief investigation showed that students coming from junior high schools where General Science was given in a functional way took more science than did others (about 30 per cent more). In addition, in these schools where substitutions for science were allowed for girls (home economics, cooking, etc.) their students hesitated to elect further science.

   In any event, this generalization is permissible on the basis of observation: where early courses in science are given (General Science and Biology), there tends to be more science elected in later years. Where General Science is required (and is taught well), the science registration in later courses is significantly higher than where it is not required. There is some evidence that where General Science is taught poorly—in a nonfunctional chalk-talk way, with emphasis on rote memory rather than upon investigation and experimentation—registration in later courses is reduced.

2. In mated schools observed, it was curious to note that in two schools of similar population (2,500), one had two sections in Trigonometry but none in Physics. The "teacher" of Physics there (who now taught other courses, Mathematics and History) stated that the reason for this was that Physics was a "hard" course

---

[59]The material on pp. 190–193 was originally prepared at the request of Charles C. Cole, Jr., of Columbia University for his study, "The Recruitment of Education of Youth with Scientific Ability," made for the College Entrance Board and the National Science Foundation (by permission, Charles C. Cole, Jr.).

The Gifted Student as Future Scientist

and there were very few students who could do the work. Yet the teacher of Trigonometry, a sympathetic, warm person, was giving what would be called a "hard" course in Trigonometry. (The principal of the school was also convinced that the school did not have students of high enough caliber to take courses in Physics.) This school also had no candidates for the Westinghouse National Science Talent Search.

In a school comparable in population and socio-economic background to the one above there were two sections in Trigonometry, two in Standard Physics, three in Modified (or Applied) Physics. There the teacher believed that Physics had application to life and it could be "hard" or "applied" and nonmathematical in nature.

In order to have a course in Physics one must have registration for it. And if there is no course in Physics one cannot meet the needs of

a. Those who will become experts.
b. Those who will become citizens in a world in which science has increasing impact, and must understand the expert and cooperate with him.

The paradox seems to be that to give Physics and Chemistry to the gifted, one must admit all students in order to have a planned course on the curriculum. Our thinking might be along lines somewhat like these:

a. Introductory courses in Science are important in "seeding" interest and as a method of selecting those who will contribute to science. These courses should be taught partly with this purpose in mind.
b. In general and only for the purposes of the solution of the problem which concerns us, students in the junior and senior years in high school might be divided into two groups: (1) The "literate" in Mathematics and English and (2) the "literate" in English but *not* in Mathematics. Courses in Physics should be given for both groups. If courses are given for both, the popularity of Physics will increase, registration will increase, and Physics will be taught, in many schools where it is not now taught or where it is taught to very few students. It may even be that the answer lies in organizing classes in Physical Science (a combination of Chemistry and Physics); in Applied Physics and Applied Chemistry; as well as in organizing standard, mathematically oriented courses in Physics and Chemistry. Since the problem in supplying scientific personnel is not only to supply research scientists but technicians, stenographers, typists, and helpers of all sorts, the latter might come from those students who might ordinarily take courses in applied science.

Physics and Chemistry remain the special courses to which we must give our attention. Why not Biology? A word on that. Biology registration holds its own and is even on the increase. Why? The answer lies in the course (which is related to life and living) and to the teachers who, in general, approach the subject as teachers of children as well as teachers of a course. Biology could be made as "hard" as Physics and Chemistry, but the thinking of teachers of Biology is, in general, that they must meet the needs of a variety of children—those who are literate in mathematics, as well as those who are not.

Special Groups

Most of the high schools in this country have enrollments of 500 or less. Many of them could have at least one class in Biology, one in Physics, one in Chemistry. Some of them, where the school registration is low, might conceivably give the courses in alternate years, General Science and Biology one year and Chemistry and Physics the alternate year.

Especially where there is only one teacher of Science, this is eminently to be desired, because it is impossible (from the viewpoint of the amount of energy solely) to give four laboratory courses in General Science, Biology, Chemistry, and Physics in a single year. It would seem better for this "single" teacher to offer two science courses only, with the rest of the program made up of courses which do not require a laboratory and for which the teacher has adequate preparation (e.g., Mathematics or English or History, but not a combination). It seems advisable that for the small school where there are *one* or *two* Science teachers the offerings in Science and Mathematics should be combined and the program for one teacher (for purposes of concentration of effort) might be as follows:

YEAR A. General Science, Biology—and Mathematics (lower math)
YEAR B. Chemistry and Physics—and Mathematics (higher math)

And the program for a school with two Science teachers might be:

TEACHER A. General Science, Biology—Mathematics (lower math)
TEACHER B. Chemistry and Physics—and Mathematics (higher math)
TEACHERS A and B. Alternate different science clubs and different groups

Even these solutions are not always possible in the very small schools. The writer, however, has observed these solutions in certain schools throughout the country:

1. In one small group (2 to 7 students) students meet twice a week after school for sessions lasting one hour and a half each. Thus a course in Chemistry or Physics is given. This the teacher does voluntarily and without pay. The Science teacher gives of his free time—generously and in good will.

One of these teachers compared himself with the coach of a team who gave some of his free time every day. However, many coaches are paid, while teachers who do as important work as coaching a team are not. Would it be possible (by subsidy by industry) to consider paying Science teachers for courses given after school? This could solve one aspect of the problem. It would not take an unnecessarily large sum for profitable results.

2. In one school only two students were involved. Again the teacher gave up time—before and after school—at irregular intervals. But the two students did do well in a standard Physics examination.

The Gifted Student as Future Scientist 193

In one school the work described in 1 above was done over a period of a year and a half. There is no apparent reason why the work must be done within a year.

In still another school, the principal scheduled classes for Physics even though there were only 3 or 4 students in the course. This was a small school where classes normally did not go beyond 18 to 20 students. By increasing the registers in other classes very slightly, the principal was able to maintain these special small classes in all areas, higher math, etc. In general, teachers apparently do not mind this arrangement if it enables them to take care of all students. The problem of giving adequate science training seems in many cases to be administrative, not instructional, per se.

On the Local and National Levels

There is, however, a grave need for a total nationally organized effort. (Otherwise, the individual community's contribution may well be lost.) Therefore, we must look into the possible development of national, state, and school organizations acting in collaboration. The proposal sketched here envisages three tracks.

Track I. Public Schools

1. A National Organization, the National Science Foundation, already exists, and it already has begun to develop basic policies.

The National Foundation, or an organization which serves a similar function, need not dictate policy. In the beginning, it might profitably serve as a clearinghouse. In the beginning, too, it might support research in the field and publicize existing procedures which seem successful. It is also possible for it to develop policies and practices which will result in making available "expert" advice to school systems throughout the country on a consulting basis. Such advice, however, is useless if it is theoretical and exhortative in nature; it must be practical. In short, the "experts" should be chosen from those who have worked intimately with the problem of identifying, selecting, and teaching youngsters with scientific potential. Such people now exist.

It would, for instance, be possible and practical to organize "task forces" (for each state) of two or three teachers who are equipped to deal with the gifted. These teachers would be free to visit the schools which invited their services. The purpose of the members of the task force would be to study the practices of the school and *demonstrate* practices useful in dealing with students of high ability in science. Funds secured by the National Foundation (from industry, perhaps) might be used to pay half the salary and expenses of these individuals; the schools which invited the task force would pay the other half. The writer feels confident that if such task forces were organized, gratifying results would be obtained within four to five years.

It may be that the National Foundation will choose to work through existing national organizations or set up a specific organization or organizations. Whether it does so or not, a national organization is still necessary to co-ordinate effort and give the project dignity and support. It may very well be that such support will come in a secondary way, through bequests to individuals and institutions from funds set up by industry; but there, too, co-ordination by a nonprofit group is essential and desirable.

2. An office or individual within each State Board of Education with primary responsibility for solving the problem of identifying, selecting, and improving the teaching of youngsters of high level ability in the areas for which the state has responsibility. This state office would be in close contact with the schools in the state with the specific purpose of stimulating the organization of teaching programs designed to deal with the problems of high level students in all areas.

3. A similar office or individual assigned to the "gifted" within each large city of over 200,000.

4. An individual assigned to the "gifted" within each school in the country— elementary, junior high school, high school. The individual or office in 2 and 3 above might conceivably deal with the problems of all exceptional children, gifted or retarded.

Track II. Colleges

1. The National Organization mentioned in Track I. Its special purpose might be to co-ordinate 2, 3, and 4 below.
2. An office or individual of some organization like the College Entrance Examination Board specially equipped to advise and guide colleges in the identification and selection of individuals with high level abilities.
3. An office or individual in each college with special training in identifying, selecting, and improving the methods of teaching students with high level ability. The purpose would be clear: to avert the tragedy of great waste which now occurs in our colleges as they fail to recognize the special abilities of the students *already* equipped (at least to some extent) by the high school. The articulation of the feeding high school and the freshman year in college would be the main purpose of this office.
4. An industrial board made up of representatives from college and personnel management in industry, with the special purpose of supporting research in identifying youngsters with high ability as well as of furnishing initial scholarships. Industry can only benefit from such a co-ordinated effort.

Track III. Teachers' Organizations

1. The National Organization mentioned in Track I. Its purpose is to co-ordinate 2, 3, and 4 below.
2. A national organization or organizations designed to deal directly with teachers and students to organize national conferences for the exchange of procedures and to furnish publications to teachers and students. These might very well be organized in different areas—art, music, science, health, farming. The Future Farmers of America, Future Scientists of America, Science Service, and similar

organizations are now in existence. What is needed is elimination of duplication of effort with resulting concentration and co-ordination of effort.
3. A committee within each regional state and city organization designed to work with the national organization above.
4. An industrial board to furnish funds and projects for research, for publications, and for consultants.

In brief, thus:

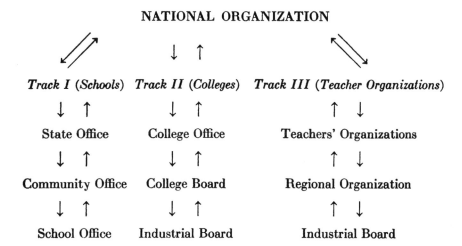

This plan is offered without presumption. Probably it is naïve. But there are able people in this country who have the organizational ability, the energy, and the will to set up workable organizations if the one above is wanting—as it must be.

We can no longer afford the luxury of unorganized effort. These are critical, even perilous, years. A reading of history has shown that there have never been enough people with high level ability. It is clear we are not doing our best for our boys and girls of such ability.

It is high time our intelligence and good will were directed toward the conservation of our human resources.

Appendixes

Appendix A

Sample Test Item

Test of Developed Ability in Science

In order to breed cows which produce large amounts of milk, one must determine the bull's transmitting ability—his ability to pass on the traits of good milk production. The chart below is a method developed by Heizer which he found useful in selecting bulls with high transmitting ability. These bulls were then mated with cows.

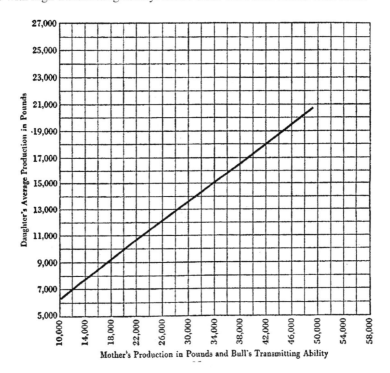

This chart is used for determining the transmitting ability of a bull in regard to milk production. The formula is based on measurements of the daughter's milk production.

To use the chart, locate the daughter's average production at the left, follow horizontally to the right until the diagonal line is reached, and then follow vertically down and read the figure on the base line. Subtract from this figure the mother's average production. The result is the bull's transmitting ability.

1-2. Bull A was mated with a cow and over a period of years produced six daughters. The average milk production of the daughters was 15,000 pounds. The average milk production of the mother was 14,000 pounds.

The Gifted Student as Future Scientist

Bull B was mated with a cow and over a period of years produced seven daughters. The average milk production of the daughters was 14,000 pounds. The average milk production of the mother was 16,000 pounds.

1. On the basis of the data, the transmitting ability of Bull B, in pounds is approximately

   (A) 14,000
   (B) 15,000
   (C) 16,000
   (D) 17,000
   (E) 18,000

2. On the basis of the data, a farmer wishing to buy a bull

   (A) should pay a higher price for A than for B
   (B) should pay a higher price for B than for A
   (C) should pay the same price for both bulls
   (D) should buy neither of these bulls
   (E) cannot tell which bull is worth more

3. The chart given is in reality

   (A) a hypothesis
   (B) a generalization based on experience
   (C) a result of theoretical study
   (D) a law of breeding affecting all animals
   (E) none of the above

4. "It can be predicted that a bull with high transmitting ability will have daughters with high milk production." This statement is

   (A) warranted by the data in the chart
   (B) contradicted by the data in the chart
   (C) too difficult to check
   (D) only partially correct because it does not contain data on the milk production of the mother
   (E) only partially correct because it does not contain data on the breed of the bull

5. The chart is extended in order to make a prediction of the bull's transmitting ability where the daughter's production is 23,000 pounds and the mother's production is 30,000 pounds. Which one of the following is true in regard to this prediction compared with a prediction where the daughter's production is 14,000 pounds?

   (A) Both predictions will be equally reliable since both are based on the same principle.
   (B) The prediction involving 14,000 pounds will be reliable since it is on the present graph, but the other prediction will be wrong.
   (C) The prediction involving 14,000 pounds will be reliable since it is read in a region where the data show that the principle holds, while the other prediction is less reliable because it goes beyond the region where the data are shown to hold.

(D) Both predictions are reliable since a straight line graph can be extended indefinitely.

(E) It is impossible to answer this without knowing more about the breed of cattle.

Appendix B

Man-to-Man Rating Scale (Refer to Case Studies 1 and 2, Appendix C)

Score

5   Like Student A (When in High School, similar to student described in Case Study 1, Appendix C) (Now Ph.D. in Physics)

4.5

4   Like Student B (When in High School, similar to student described in Case Study 1, Appendix C, except with an I.Q. near 130, Scholastic Average 90) (Now M.D.—Research Leanings)

3.5

3   Like Student C (When in High School, similar to student described in Case Study 2, Appendix C) (Now Engineer)

2.5

2   Like Student D (When in High School, similar to Case Study 2, Appendix C, except with an I.Q. below 130, Reading Score 14 (9.0), Scholastic Average 85%, sporadic work) (Now Pharmacist)

1.5

1   Like Student E (When in High School, similar to D above, except that he shifted goals repeatedly) (Quit College—now in business)

.5

00   Like Student F (When in High School, similar to D above—sporadic work although highly intelligent, liked "fun"—thought that "fun" was the goal in life) (Quit College—now in business—had several jobs in one year)

Appendix C

Case Study 1

Summary of a Typical Case Study: High School Student Rated 4.5 on the Scale of Predisposing Factors (disguised slightly)

Male, 11th grade, 15 years

Splendid attendance record (absent 4 days in high school career)

I.Q. 153, Reading Score 16 plus (9th year), Arithmetic Score 12 plus (9th year)

Scholastic Average 95.6

Intends to take Science, Mathematics, Language 4 years

Plans to go to Harvard, M.I.T., or Princeton

Extracurricular activities: Engineering Club, Research Club, Social Studies Honor Society, Science Math Honor Society, Fencing, School Newspaper

President of general recitation classes, and one club

Hobbies: Chess, Piano, Reading, Hiking (Boy Scouts), Woodworking

Works weekends in pharmacy—deliveries

Pleasant yet hurried, never thinks of himself, cannot say no, takes on activities on own initiative, always bemoaning lack of time, refuses to pity himself for any misfortune, high ambitions for public service, high sense of responsibility and integrity, a leader in scholarship, respected by his classmates, shy, great affection for father. Learning to accept himself and his environment

Ambition: Research Physicist

Case Study 2

Summary of a Typical Case Study: High School Student Rated 3 minus on the Scale of Predisposing Factors (disguised slightly)

Male, 11th grade, 16 years

Good attendance record (average 4 days' absence each year)

I.Q. 151, Reading Score 16 plus (9th year), Arithmetic Score 12 plus (9th year)

Scholastic Average 89.6

Intends to take Science, Mathematics, Language 4 years

Plans on Harvard, Michigan, Cornell

Extracurricular activities: Tropical Fish Club, Research Club, School Newspaper

Hobbies and after-school activities: Baseball, Basketball, Social Club, Orchestra, Tennis

Does not work after school. Often does not do homework. (Says he knows his "subjects" without doing it)

200 P.F. Brandwein

Pleasant, loquacious; likes to pass time of day, willing to serve but has to be asked, popular with students and teachers, a great organizer of parties, group activities (almost always takes on collection of dues, etc.), makes friends easily

Ambition: Dentistry, Medicine, or Science Writing

Appendix D

Inventory of Predisposing Factors (Based on Operations thought to Underlie Persistence and Questing)

SHEET 1

Score	0	*Attendance* 1	2	3	4	5
(Days' absence due to minor illness, colds, etc.)	More than 10 days per term	8	6	4	4	Less than 2 days per year
Score	0	*Interest* 1	2	3	4	5
No. of hours per week spent in science hobby work outside of school (out side of assigned work)	Less than 2	2	4	6	8	10 hours or more
Score	0	*Width of Interests* 1	2	3	4	5
No. of projects chosen from a mimeographed list of 25	none	1	2	3	4	5 or more
Score	0	*Tendency to Stick with Project* 1	2	3	4	5
No. of projects selected after commitment to one chosen because of interest	Drops out of project work	Does not complete last project chosen	5 projects chosen but completes last	4 projects chosen but completes last	3 projects chosen but completes last	Not more than 2 projects, and finishes last

*These operations seemed to be characteristics of those students with Persistence and Questing. An average of all the scores on sheets 1 and 2 was taken. This was checked against the average of scores on sheet 3.

The Gifted Student as Future Scientist

Inventory of Predisposing Factors (Continued)

SHEET 2

	Number of Free Periods Spent in Laboratory					
Score	0	1	2	3	4	5
Time spent in Laboratory in Advanced Science	Drops out	Attendance in signed period sporadic	Spends only assigned period	Spends free period	Spends free period and part of lunch period	Has to be thrown out of lab; spends every free moment, including part of lunch period, before and after school; also does work at home

	Scholastic Average					
Score	0	1	2	3	4	5
Average in all areas through 7th term includes $3\frac{1}{2}$ years science, $3\frac{1}{2}$ years math, $3\frac{1}{2}$ years English, $3\frac{1}{2}$ years Language	Below 86	86-88	88-90	90-92	Between 92-95	Over 95% (within first 4 of graduating class)

	Nature of Reading					
Score	0	1	2	3	4	5
Reported in interview	Reads little	Reads newspapers	Reading limited to school assignments	Reads novels, mainly modern	Reads history in addition to novels both modern and classic	Serious reading (includes philosophy) in addition to novels both modern and classic

In order to be rated 4+ or 3− in this inventory, ratings had to agree in general with those in the Man-to-Man Rating Scale (independently rated). Otherwise they were not considered.

Inventory of Predisposing Factors (Continued)

SHEET 3

Score	Type of Personal History					
	0	1	2	3	4	5
		Case study of student with I.Q. about 105, scholastic average 75-80 (termed average)		Like Case Study 2, Appendix C		Like Case Study 1, Appendix C

Addition or Drive

(Ratings based on case studies and studies of scientists)

Score	0	1	2	3	4	5
	Considered "lazy"			Normal ambition but gets over dis-appoint-ments readily		Intense desire to get ahead; does not get over disappointment easily

Tendency to Ask Questions

Score	0	1	2	3	4	5
	Rarely asks questions; accepts informa-tion			Asks questions only when stimulated by infor-mation given by others		Regularly asks questions; displays insatiable curiosity; questions tend to originality

Appendix E

Summary of Case Study 1 (Graduate Student)

Summary of a Typical Case Study: Graduate Student Rating High in Genetic and Predisposing Factors (Inventory of Predisposing Factors Built on This Prototype)

Male, Biochemistry major, 24 years

Working on thesis for Ph.D. Has passed his oral and written examinations and language qualifying examination. (I.Q. 161)

The Gifted Student as Future Scientist 203

Teaching Fellow, rarely absent from work or school; works some 15 hours a day

Brilliant conversationalist, mathematician, and researcher (has published three papers of high grade)

Hobbies: Music (violin), chess, reading, field work, landscaping

Tutors individual students

Intense, yet polite; reserved; makes friends slowly but is respected and liked; accepts assignments quietly and does them efficiently; fine organizer. Shy but warms up in discussions, writes well

Ambition: Biological Research
          Already had good offer from top university
          (Now one of top research men in field)

Summary of Case Study 2 (Graduate Student)

Summary of a Typical Case Study: Graduate Student Rating Low in Predisposing Factors, but High in Genetic Factors (Inventory of Predisposing Factors Built on This Prototype)

Male, Biology major, 26 years

Working on research for Ph.D., took written and oral twice, passed language exam (admits it was a narrow "squeak"). Extremely well-versed in literature, splendid mathematician

Teaching Fellow, lateness to class noticeable, sometimes absence due to illness or some other reason, works spasmodically

Has high social interest, likes parties; has published one abstract of work but not a full paper

Hobbies: Golf, mystery stories, cars, especially old motors

Doesn't like teaching particularly, and admits it
Affable, easy-going, makes acquaintances easily, does not keep appointments, is judged a "nice fellow" but hasn't found himself; likes to pass time of day; considered highly intelligent by all—but "lazy"

Ambition: Industrial Research or Professorship in large university
          (Now in business—selling cars)

Appendix F

Sample Reports in O.A. Approach

I. Interim Summary Reports on a "Research" Project*

(after four months' work)

1. Stability of a Phase Shift Oscillator

The first step in project "Stability" was the construction of a 600 cps tuning fork to be used as a frequency standard. This was checked on an EPUT meter and it was discovered to be accurate to 1 part in 6,000 and to have almost 0 drift. The next step was the construction of an electronically regulated power supply. That was to be capable of maintaining a constant regulated voltage over a considerable range of voltages. This was just completed and has not yet been checked. When this is checked and is in perfect operating condition I will begin the main part of my project, that is the construction of a phase shift oscillator and check on the stability of this against the beforementioned tuning fork oscillator. This is to be done by the observation of Lissajou patterns of the screen of a cathode ray oscilloscope. Then I will calculate the drift of the phase shift oscillator over a period of time. With this information I will then proceed to attempt to increase the stability of the oscillator and in doing this find out what things affect the variation of drift of a phase shift oscillator. I plan to test: One, the effect of a variation of the operating voltage by the use of the electronically regulated power supply. Two, the effect of load on the stability. Three, what happens if it is allowed to remain on for a long period of time, that is, giving the oscillator an extensive warm-up period. And four, the variation of components in the circuit. All of this will be checked by the observation of Lissajou patterns for an extensive period and, again, the calculation of the drift per unit time. This will then all be written up in the form of a report in which conclusions will be drawn as to the effect of certain factors on the stability of a phase shift oscillator.

> David Berkeley, 5th term
> (Beginning Junior)
> Forest Hills High School

2. Correlation Between Reaction Times and Natural Left- and Right-Handedness

Is there any correlation between precedence of reaction time and natural left- and right-handedness? Do right-handed people have faster absolute reaction times than

---

* Teacher sponsors for the projects reported in the appendix were Evelyn Morholt, Arthur Lazarus, Harvey Pollack, George Schwartz, and Paul Brandwein.

The Gifted Student as Future Scientist 205

left-handed people? These questions, and others along the same line, form the basis for my project—testing large numbers of students for correlations like these by means of electronic devices.

I feel that this project is especially appropriate for a school the size of Forest Hills High School, since I have a large number of students from study hall to work with.

Since September, I have been building the electronic circuits as well as the permanent frame for all the component parts. Before anything else could be done, a power supply was needed, and this itself took almost a month to build. I then started work on the heart of the project—the precedence circuit. I used an industrial circuit with some modifications, but after it was built, I found it did not work properly. A completely new circuit was designed and built, and is now working perfectly.

Before any actual work was started, other circuits were drawn up—a timing circuit to measure absolute reaction time in microseconds by use of condenser charge, and a stimulus panel, incorporating both visual and auditory stimuli. I saw that the project would have many component parts, so I designed a permanent frame on which all the chassis could be mounted. This frame has been built, and all mounting finished.

On one side of the frame is the stimulus board and the keys for the subject to rest his hands on. On the other side is a bank of switches for the experimenter, as well as several chassis. The chassis are mounted out in the open, making them easy to reach. This also solves the problem of ventilation.

During the coming vacation, I plan to finish the timing and stimuli circuits, which will complete the testing device. Then, from January until December, I will test as many students as I can, and compile and tabulate my data.

Stephen Slaton, 5th term
(Beginning Junior)
Forest Hills High School

II. Final Reports on Summary of Projects

3. An Ultraviolet Photosenitization in Para-Aminobenzoic Acid and Pantothenic Acid Fed to *Tribolium confusum*

I had read that when mice were fed buckwheat and were placed in a strong light they died, while mice lacking either the light, the buckwheat, or both, thrived. Lacking mice, I tried to duplicate the results on insects. I worked with the Confused flour beetle, *Tribolium confusum*.

The effect in mice can be duplicated on the flour beetle. I am reasonably certain that:

1. The ultraviolet rays of the light, acting with an agent (or agents) in the buckwheat, seem to cause the reaction known as a photosensitization.

206 P.F. Brandwein

2. When Pantothenic or Para-aminobenzoic Acids (in a concentration of 5 per cent and higher) are added to the diet of the flour beetle, the photosensitizing effect does not occur.

It may be that the photosensitizing reactions are caused by the conversion of either (or both) Pantothenic Acid or p-AB, both of which are needed by the cells to synthesize anti-metabolitic structural analogues. The cells seize upon these structural analogues but cannot utilize them; the cells thus suffer from a deficiency of these vitamins. Death may be the result.

Report: 1,500 words, diagrams

Michael Fried, Senior
Forest Hills High School

4. The Design of an Electrical Machine Performing a Multipartite Operation of a Mechanical and Mathematical Nature Previously Done Only By Man

The question of machines replacing man has been asked again and again during the last two centuries. As of now, machines have been developed that can completely replace man in certain jobs concerning mechanical operations and arithmetical computations. I was interested in knowing if a machine could be designed that could completely replace man in a more complex, multipartite job.

All jobs involve one or more of the following factors: mechanical operations, arithmetical computations, observations, reasoning, and the element of chance. I wanted to see if it was possible to design a machine which could perform a job which contained all five factors. Since the dealer in the game of Blackjack has a job which involves all five factors, I decided to try to design a machine which would take the dealer's place.

First I unified into a definite set of rules the various ways of playing Blackjack. Then, after arranging the controls of the machine for convenience of operation and prevention of confusion, I prepared an instruction sheet to tell how to operate the machine. Because of its multiphase operation, and its resulting complexity, the machine was considered as seventeen individual units, each being a block of cognate functions. All the functions of each unit were tabulated, and then wiring diagrams of each unit were developed. A method was developed for the machine to decide mathematically what the dealer would decide.

The designing of the machine was more than just a problem in designing circuits. It was also a problem in unification, tabulation, and classification of information, problem analysis, and probability.

My machine has the possibility of taking over the jobs of man. It could be assumed, however, that I am inferring that machines will replace man himself. If

The Gifted Student as Future Scientist 207

a machine can reason, it is following the reasoning that man told him to follow, and not reasoning by itself.

Report: 1,600 words and
diagrams (in large part)
and calculations

Carl Koenig, Senior
Forest Hills High School

5. The Optimum Concentration of Terramycin, in Conjunction with Nutrients, on the Reproductive Rate of *Paramecium caudatum*

James McGinnis first announced in February, 1950, that the growth of poultry was considerably influenced by residues of Terramycin production. Subsequently, many investigators experimented with Terramycin as a growth factor in animals, but its effect on other animal functions was ignored. This paper attempts to trace Terramycin as an influential factor in the reproductive rate of a protozoan, using as the experimental animal, *Paramecium caudatum.*

After first determining by experimentation that Terramycin was not lethal to protozoa, a general procedure for the remaining experiments was devised. A stock paramecium culture was thoroughly shaken and three sample. 05 ml. drops was were placed on individual slides. The number of these protozoa per drop was recorded. Then an average counting of the three drops was found and multiplied by twenty to determine the number of paramecia in each average milliter of the stock culture. The solutions to be tested were made up, 80 ml. being placed in each jar. To these solutions, 2 ml. of the stock culture containing a known number of paramecia were added. The average number of paramecia in each milliter and in each .05 ml. drop of the 82 ml. solution was then calculated. The exact time at which the cultures were started was recorded.

Seven days later, the cultures were examined. A procedure similar to the one outlined above was followed to determine the number of paramecia present in each milliter of the various solutions. Careful note was taken of the time lapse between the starting of the cultures and their final examination. On the basis of these facts, the number of reproductions that took place in each solution, and the length of time needed for one reproductions, were calculated.

Using these techniques, experiments were made with Terralac (a composite substance of Terramycin, skim milk, minerals, vitamins, etc.), Terramycin plus rice, and Terramycin plus skim milk. In each case, the optimum concentration was the 1:1000 solution of Terramycin plus the added nutrient. Each experiment was performed twenty times. Control groups of rice in water were carefully watched.

From these experiments it may be concluded that Terramycin has a decided influence in increasing the reproductive rate of *Paramecium caudatum.* The amount of increase is dependent upon the concentration of the Terramycin and the nutrient with which the Terramycin is combined. Thus a concentration of 1:1000 of Terralac in water had the most pronounced effect on the reproductive rate.

In light of these facts, it is probably true that Terramycin, rather than acting directly upon the protozoan to increase its reproductive rate, alters the selectivity of its membrane, thus changing the type and quantity of nutrient which it is able to assimilate. This alteration may cause the paramecium to grow larger and mature earlier, thus being able to reproduce more rapidly. The evidence which I have accumulated all seems to point to this conclusion. However, further investigation, including detailed studies of the structure of *Paramecium caudatum* before and after the addition of Terramycin to a culture, is necessary to prove or disprove the truth of this hypothesis.

Report: 1,200 words,
tables, graphs

Roberta Jane Fishman, Senior
Forest Hills High School

6. The Application of the Techniques of the IBM 701 Electronic Computer to the Harmonization of Music

A Study in Cybernetics

This project is the result of my combined interests in music and electronic computers. Work has recently been done on the application of computers to nonmathematical subjects such as the translation of languages (e.g., Russian into English). This study is another step in the same direction.

The specific problem dealt with in this project is the harmonization, in four parts, of a single melody line of music by an electronic computer. For purposes of this study only music with notes of equal time value to the measure was used. The rules and techniques of computing that are used are those of the IBM 701 installation.

In solving the given problem it is necessary to be able to change theoretical statements in music to mathematical equivalents and then to notations applicable to the 701 installation. These changes are achieved by letting given constants equal the different musical notes being dealt with and then using mathematical logic to assemble them in the proper equations. Translation into 701 notation follows quite simply from these equations as the computer is constructed to perform the operations indicated by them. An example of the procedure follows:

a. The theoretical statement: Eliminate all repeats in the bass notes.
b. Translated into its mathematical equivalent: Let r be the previous bass note. Let C_i (where i equals 1 thru 6) be the consonances being worked with. Then:

$$\text{If} \quad r - C_i = 0 \qquad\qquad \text{If} \quad r - C_i = 0$$

or

$$\text{set} \quad C_i = 0 \qquad\qquad \text{set} \quad C_i = C_1$$

The Gifted Student as Future Scientist 209

c. Then given its 701 notation:

$$R \text{ ADD} \qquad 4000$$

4000 is storage for r

$$\text{SUB} \qquad 3101\text{-}3106$$

3101-3106 is storage for C_i

$$\text{TR} \quad 0$$

If $r - C_i$ equals 0 then C_1 is eliminated by storing zero in C_i

$$\text{TR}$$

If $r - C_i$ does not equal 0 then continue computation.

In conclusion it may be stated that though this project is confined to a specific species of music it may be enlarged considerably. A change in the constants used to obtain the consonances would allow the 701 notations to be used for music in a minor key; at present only music in a major key can be used. A different rhythmic pattern and a larger number of parts of harmony could be obtained through a few more minor revisions.

Through the music written through the computation of the IBM 701 installation is theoretically correct, it cannot be expected to be creative. Only the human can create.

Report: 1,500 words,
diagrams, tables, computaions

David Reich, Senior
Forest Hills High School

7. Cannibalism: As Developed by Nutritional Variations in *Blepharisma undulans* in Microvivaria

It is known that within the species of protozoa, *Blepharisma undulans*, exists a certain variety of cannibal. The object of my study was to discover as much as possible concerning the nature of these cannibals, and to establish whether cannibalism in *Blepharisma undulans* is due to a genetic or an environmental factor.

From observation of an experimentation with cultures of Blepharisma (refers to the species *Blepharisma undulans*) I noticed that the cannibals are generally twice the length of the normal Blepharisma, feed on smaller ciliates as well as other Blepharisma, and have peculiar dark red food vacuoles. These last are a result of digesting Blepharisma, which have a reddish color.

To determine whether cannibalism in *Blepharisma undulans* is hereditary or environmental, I tried to develop cannibals by variations in their nutrition. If this could be done it would prove that cannibalism is environmental. I had noticed earlier that cannibals on a diet of rice only (feeding on decay bacteria) shrink to the size of normal Blepharisma. Upon further experimentation with nutritive variations, I developed cannibals by placing normal Blepharisma on a diet of smaller ciliates.

The Blepharisma first devoured very small ciliates, grew slightly larger, and their oral grooves became larger, enabling them to ingest still larger ciliates. This cycle repeated until the Blepharisma attained the average size of cannibals and found it easier to ingest the slower moving members of their own species. The rice was an indirect factor in the development of the cannibals because it enabled the Blepharisma to ingest the ciliates while they (the ciliates) were feeding on the decay bacteria surrounding the rice. Since Blepharisma are generally much slower moving than the ciliates upon which they feed they cannot ingest them unless they (the ciliates) have accumulated on a rice grain.

I therefore conclude that cannibalism in *Blepharisma undulans* is environmental.

Report: 1,200 words,
drawings

Peter A. Roemer, Senior
Forest Hills High School

8. The Effects of Various Micronutrients on the Embryos of *Planorbis* and *Physa* Gastropods

I got the idea for this project when I read of the experiments of Stockard, who had succeeded in producing a fish with one eye by adding small concentrations of magnesium chloride to their normal embryonic environment. I decided to go further along his lines after reasoning in the following manner: if chemicals were gradually added to eggs laid at the same time in different solutions, the observed effects might be an indication of when the cells of the affected organ were being formed. For example, if magnesium chloride was added in the fifth day of embryonic development, and a deformity was noted in the grown animal, it could be postulated that the cells forming the affected organ were formed in the fifth day of embryonic development. In all my experiments I used snails instead of fish, as their eggs are easier to find and work with.

I performed many experiments with various chemicals, and drew some conclusions. Most of these conclusions pertain to the time of the formation of the shell of the snail which is altered in different concentrations of magnesium chloride. Unfortunately, even extremely small amounts of chemicals added to the embryos usually proved fatal. Toward the latter part of my work I built my own apparatus for taking

The Gifted Student as Future Scientist

photomicrographs, in order to corroborate my observations, and I obtained some excellent results.

Report: 1,000 words, plus
photographs, tables

Martin Glass, Senior
Forest Hills High School

9. Problems of Ecology on the Myxomycete *Physarum polycephalum* with Special Attention to Preventing the Growth of Extraneous Organisms

Physarum polycephalum is one of a group of organisms, the myxomycetes, which taxonomists have placed either in the plant or the animal kingdom.

I decided to study the effects of changes in the environment of *Physarum*. Before launching actual laboratory work, I did some field exploration (in order to identify the myxomycetes in their natural surroundings); I read on the subject in the library, and I corresponded with several college professors working on related organisms. Then I began experimentation.

I worked with *Physarum* in the plasmodial stage, when the organism, as a large yellow mass of free, undifferentiated protoplasm, swarms over its surroundings on the hunt for food. During its complete life cycle, the slime mold undergoes a complete metamorphosis.

My first fields for investigation were the simplest of the ecological problems. I found the proper size of the oat substrate used, the optimum water concentration, the best amounts of light and heat. I investigated the relationship between change of color in the plasmodia and the *p*H.

Throughout these experiments, I was plagued by attacks on *Physarum* by extraneous molds. I tried to combat these molds by finding a new media on which *Physarum* would thrive but extraneous molds would not. After trying many foods, I found corn meal successful in place of oats as a medium in certain cases.

I tried to eliminate these molds by sterilization, by using a mold preventive, and by applying antibiotics to speed up growth of *Physarum*, but no one of these methods could be deemed completely successful.

At the completion of my work, I had acquired an understanding of the effect of various environmental conditions on *Physarum polycephalum*, and I had developed methods of dealing with unwanted molds.

But, much more important, I had learned something about the methods and the techniques of scientific research.

Report: 2,000 words,
tables, diagrams

Stephen A. Schuker, Senior
Forest Hills High School

III. Some Titles of Reports of Projects Finished in 1954

1. Brunschwig, Michael A Proposed Correlation Between "Rising" and Budding in *Wolffia columbians*
2. Fishman, Roberta Jane The Optimum Concentration of Terramycin, in Conjunction with Nutrients, on the Reproductive Rate of *Paramecium caudatum*
3. Fried, Michael An Ultraviolet Photosensitization in Para-aminobenzoic Acid and Pantothenic Acid Fed to *Tribolium confusum*
4. Glass, Martin The Effects of Various Micronutrients on the Embryos of *Planorbis* and *Physa* Gastropods
5. Goldwaser, Lillian An Experiment in Invertebrate Learning
6. Kaplan, Ellen The Effect of Pressure on the Growth of Elodea and the Germination of Radish Seeds
7. Koenig, Carl The Design of an Electrical Machine Performing a Multipartite Operation of a Mechanical and Mathematical Nature Previously Done Only by Man
8. Reich, David The Application of the Techniques of the IBM 701 Electronic Computer to the Harmonization of Music
9. Roemer, Peter A. Cannibalism: As Developed by Nutritional Variations in *Blepharisma undulans* in Microvivaria
10. Schuker, Stephen A. Problems of Ecology on the Myxomycete *Physarum polycephalum*, with Special Attention to Preventing the Growth of Extraneous Organisms
11. Stollnitz, Frederick An Ecological Study of the Marine Fauna of Nantucket Island, Massachusetts
12. Young, Stuart The Effects of Varying Dosages of Thiamin Hydrochloride on *Rhizopus nigricans*

Acknowledgments

Many of the ideas and much of the good will of the following have gone into this report.

Present Regular Members of the Science Department of Forest Hills High School: Elizabeth Gray, Alice Koelsch, Bessie Lumnitz, Lucy DeVivo, Nancy Feingold, Evelyn Morholt, Herman Gillary, Joseph Greene, Jerome Harmon, Arthur Lazarus, Charles Katz, George Schwartz, Bernard Udane, Morris Wigler, Ernest Wilson, Harvey Pollack.

Past Members: Sylvia Neivert, Mary Norton, Annabelle Osman, Ruth Katz, Lyndon Burton, Solomon Friedland, Brendan McSheehy, Fannie Liebson, Pearl Schwartz, Rae Silver, Dorothy Susskind, Doris Weinstein, Mark Yohalem, Margaret Kuhn.

To Dr. Leo Ryan, principal of Forest Hills High School, for sympathetic understanding, encouragement, and active leadership in planning the program.

To Martin Delman, Assistant Principal, Forest Hills High School, without whose constant help in organizing suitable classes the program described could not have materialized.

For initial encouragement in this early work at George Washington High School: Elizabeth Fitzpatrick and Dr. Frank Wheat.

For exchange of ideas related to "high" level ability: Professor Fletcher Watson of Harvard University, Professor S. R. Powers of Columbia University, Elbert Weaver, Phillips Andover Academy, Professor Hubert Evans, Columbia University.

For initial encouragement and sound advice in the early work concerning this program at Forest Hills High School: Dr. Michael H. Lucey, its first principal.

To Renée Fulton and Helen Gribben for early administrative advice and encouragement.

For permission to publish the test item in Appendix A: Dr. Henry Dyer, College Entrance Examination Board.

Science Talent: In an Ecology of Achievement

Paul F. Brandwein

The purposes of this analysis are fivefold. First, we intend to devise an environment in schooling in which the young seek and find opportunities to think and do science. The environment, structured in an ecology of achievement, may thus affect the course of change in the young as they fulfill their diverse powers in the pursuit of their special excellences. Second, we are obligated to devise environments that channel the interests of the young so that they learn the arts of investigation and, in a sense, become performing scientists akin to performing artists; that is, they demonstrate their talent in a work or performance.

Third, given the opportunities so devised, the young who participate discover that there is a need for individuals of different gifts, different talents, and different levels of achievement. Fourth, we shall describe an environment in which the young serve their apprenticeship to various kinds of well-ordered empiricism, a significant methodology of the performing scientist. Thus, we will probe essential questions: Does an individual's talent in science demonstrate a private gift? Or is an effective channeling and augmenting environment integral in evoking talent? Finally, is it possible to turn our experience and investigation into some form of generalizable knowledge? We think it is. But I anticipate.

A First Thesis: An Ecology of Achievement

The early environment of the child—home, family, community, and school—forms an extraordinary interrelationship of environments, ecologies that contribute to shaping the individual. David T. Suzuki, Anthony J. Griffiths, Jeffrey H. Miller, and Richard C. Lewontin (1986) furnish us with an intriguing model and a masterful analysis of the relationships of genes interacting with the environment. Consider, they say, two monozygotic ("identical") twins, the product of a single fertilized egg that divided and produced two sisters with identical genes, that is, with identical

---

Gifted Young in Science: Potential Through Performance, "Science Talent: In an Ecology of Achievement" (pp. 73–103). Copyright 1989 by the National Science Teacher Association. Reproduced with permission of National Science Teachers Association.

D.C. Fort, *One Legacy of Paul F. Brandwein*, Classics in Science Education 2, DOI 10.1007/978-90-481-2528-9_22, © Springer Science+Business Media B.V. 2010

complements of DNA. Say the two were born in England but separated at birth. Suppose one were raised in China, by Chinese-speaking foster parents. The other, in Hungary. The former will speak Chinese, the latter, Hungarian. Each will behave in accordance with the customs and values of her environment. But consider: The twins began life with identical genetic properties (equivalent and equal DNA, and identical genomes), but in the end the different cultural environments produce great differences not only between the sisters but also from their parents. Clearly, Suzuki and his colleagues maintain that "differences in this case are due to the environment and the genetic effects are of little importance" (p. 5).

It boggles the mind to consider the effects of the multitude of non-DNA differences in the environment that determine the actual course of change in the individual. Since the early work on genetics in 1900–1940, there has been no study in genetics generalizing findings of research on an organism's structure or function that does not base its hypotheses on the generality that an organism is the product of the interaction of its genes and its environment.

For example, the color of the fat of certain rabbits is changed from yellow to white, or vice versa, by the color of the mash they are fed; the arteries of the human are clogged or remain clear depending on the diet (an environmental effect), obesity aside. Inherited pink lungs turn gray-black from the smoke of tobacco; light-pigmented skin darkens in the sun; obviously, we learn and do not genetically inherit a knowledge of history. Further, certain inherited disorders can be treated as easily, or with as much difficulty, as those arising from environmental difficulties and accidents, before and after birth. By way of example, Wilson's disease is characterized by the steady degeneration of the nervous system and liver because the body cannot synthesize normal amounts of a certain copper-containing blood protein. Instead, copper atoms from food are deposited in the brain and other tissues. However, an available drug (an environmental factor) removes the copper atoms and prevents degeneration of nerve tissue.

Another example: The absence of a specific gene results in Phenylketonuria (PKU), a condition that pushes a child into mental retardation. The amino acid phenylalanine accumulates in the body, resulting in brain damage. A child may be spared the effect of its genetic defect when given a phenylalanine-free diet. In this case, as in Wilson's disease, the intervention of a changed environment reduces the deleterious effects of the genetic condition. Thus, in these two examples, we see an explicit interaction of gene and environment through medical intervention. The notion that some inherited traits are unchangeable or inevitable is no longer acceptable.

Clearly, the ability to do originative work or discovery in science, a sign of talent in science, must be strongly assumed to be a product of heredity and environment. As Liam Hudson (1966) suggests, the intellectual operations of the scientist depend on a huge accumulation of experience as well as on vast accumulation of knowledge. Neither can be ascribed to genes formed long before that knowledge was available. We seek then an environment (an ecology) which will give all our young the multiples of opportunity for development, allowing them to benefit fully from the consequences of the interaction of their genes and their environment.

Recall now that the term *ecology* describes a relationship, an interdependence, among organisms and their environment. Although an ecology seems a loose relationship, it is in effect a structure built on strong interdependencies among organisms in a *particular environment.* Together, organisms and environment form an ecosystem; if we alter the environment, the organisms may not survive. The interdependence or ecology we seek is that which leads to performance; we try to relate the manner in which the design of the ecology in schooling and education translates or metamorphoses potential in science into a performance: a *discovery* (the term I should substitute for "creativity").

As we shall see, the sole use of intelligence tests can result in a single reified judgment of ability. IQ, too often a dominant criterion in the selection of the "science-gifted" in isolation from the powerful environments that affect talent, is thus not central to our probe.

Note, however, that the transformation of an organism from one stage of life to another is a result of the unique interaction of its genes and its environment at each moment of life (Suzuki et al., p. 5). But this significant statement may be inadequate unless one considers two types of heredity—one *genetic*, the other *cultural*. We are aware that biological heredity (or inheritance) consists of the transmission and transmutation of DNA. Cultural heredity (or inheritance) consists of the transmission and transmutation of learned elements: knowledge, values, and skills. Put another way, the transformation of the human from one stage of life to another is a result of the unique interaction of its biological and cultural heredities.

In other words, a child is not the result solely of gene-driven factors. The very young child is already a complex of gene- and environment-driven factors interacting to form structure and function. Further, subsequent development stems from a newer base: the result of the up-to-this-point interaction of the two heredities, genetic and cultural.

Surely cultural heredity is within our control. Yes, but not historically—or certainly not in present cultural history. The life span of a child born in an underdeveloped country is an average 40 years; the inner uterine environment is as critical as the outer. In the United States, at present, the life span for those with optimum health is 75 years. What was the cultural history of a South Korean child in 1910, as compared and contrasted with one in 1988? Surely, the former did not enjoy a civilization centered in science and technology. Of a Japanese child, before the Japanese industrial and cultural revolution? Of a woman born to mature to her majority before, say, women were allowed to vote or readily accepted, generally, into the scientific professions? The word "environment" hides too much and is often too bland. For example, the brilliant young British physicist, H. G. Moseley, was killed at Gallipoli during a failed attack in World War I. Are not wars and their effects environment driven? You will know of other instances—dismaying ones, or those made glad—some by human and humane intent. You will find we have survived countless terrors.

For this reason, we shall be stressing throughout a diptych in human development consisting of two universes of factors or traits that are gene driven and environment driven. The simplest example of the former is height, which is mainly gene driven, but even here affected by diet and, perhaps, exercise. The most complex, perhaps, is intelligence, which is dependent upon many genes, or, as the term is used, is multifactorial, to an extent as yet unknown and surely affected by a host of environments also not fully known.

Environment-driven factors come out of life's experiences. Schooling is precisely that construct of the environment that is intended to nurture all. It is the environment that is intended to offer varieties of experience for multiple intelligences. In brief, schooling is the social construct designed to offer multiple channeling and augmenting environments for the varieties of young who make up a school—and by extension, the varieties of young who make up a democracy.

Richard H. de Lone (1979) in his *Small Futures*, a study for the Carnegie Foundation, furnishes impressive evidence of the effect of the socioeconomic environment on the futures of children in minority groups. The data he has gathered demonstrate powerfully how environmental disadvantages and injustices limit achievement. Kenneth Keniston's foreword to the study states this view clearly:

> For well over a century, we Americans have believed that a crucial way to make our society more just was by improving our children. We propose instead that the best way to ensure more ample futures for our children is to start with the difficult task of building a more just society. (p. xiv)

Simplistic consideration will yield the truth that, even if undertaken immediately, reconstructing society (and particularly the family) demands careful and incredibly intricate and elaborate plans over the definable short and long terms. Yet daily, each and every year, without halt, the young keep coming to our schools. The school environment, its effect on intellection, on personhood, on knowledges, attitudes, and skills—if you will, the ecology of achievement—is the heart of a school supported by a community that can stand some rebuilding. The impact of schooling is notable with regard to the opportunities for both the gifted (the seemingly advantaged) and those disadvantaged by the many social and economic ecologies that affect them. The term *environment* is thus best placed within the widest ecology of achievement, which includes development of the organism not only in its biological, psychological, social, and educational components but also in its full political and economic range.

Schooling per se, to be effective in our present technocratic society, with its techno-electronic technology—a society hastening into a postindustrial era—must accommodate an ecology of achievement. But a school is only a part, certainly not all, of this ecology. That is, schooling is reasonably a part of the society and the culture that mother it, just as the separate communities of the forest and the sea are parts and not the whole of the ecosystem. As we shall note again and again, a community furnishes a part of the ecology of achievement in which a school succeeds or fails. The ecology of achievement that nourishes a successful school is characteristic of a successful community. And conversely, it is difficult to find a successful school within a community that does not support the kind of schooling required in an open society, especially one entering the global economy of the postindustrial period (Brandwein, 1981).

The community of scientists within a given field also establishes an ecology of achievement. Scientific knowledge is cumulative knowledge; scientists cannot ignore precedent work: Indeed, they build upon it. The library and the computer's data base precede and endure during the scientist's work in the laboratory. Simplistically stated, "brains on" before and during "hands on." And, if you will, as the young mature and rush to learn, to do on their own, a certain dose of "hands off" is desirable.

It seems also that ecologies of achievement affect the kinds of problems accessible to the problem-solving activities of science. James J. Gallagher (1964) remarks in the 1970 preface to *Teaching the Gifted Child* that much of the material contained within was "nonexistent" 5 years ago. That is to say, studies on giftedness seem to be grouped in periods—in the early 1950 s to 1960 s and again in the early 1970s. Possibly studies of the nature of the cyclical change in interest in giftedness generally, and in science specifically, parallel closely the crises of society—particularly those in economic, social, and political ecologies. So Abraham J. Tannenbaum (1979), both a teacher of the gifted and a student of giftedness, remarks, "The cyclical nature of interest in the gifted is probably unique in American education. No other special group of children has been alternately embraced and repelled with so much vigor by educators and laymen alike" (p. 5). And again, Tannenbaum, quoting Spaulding (undated), writes, "A review of the state of research for the years 1969 to 1974 reveals a fairly bleak picture; only 39 reports on the gifted had been published in that period" (1983, p.33). On this oscillation by the culture or "change in signals" by society, see also Harry S. Broudy (1972) and Brandwein (1981).

Thus, there appear to be concerted efforts in a given field at a given circumscribed period, ecologies of achievement apparently responding to periods with similar problems. On this, Robert K. Merton (1961) remarks:

> I should like now to develop the hypothesis that, far from being odd or curious or remarkable, the pattern of independent multiple discoveries in science is in principle the dominant pattern, rather than a subsidiary one. . . . Put even more sharply, the hypothesis states that all scientific discoveries are *in principle* multiples, including those that on the surface appear to be singletons. (p. 306)

Merton's account is, thus, a reference in our terms to an interrelationship in effort of different individuals and groups of individuals, concentrating on achieving solutions to the problems of society (an ecology of achievement).

Derek de Solla Price (1961/1975) in *Science Since Babylon*, his thesis on the nature of "scientific civilization," refers to discoveries by foremost scientists: "Probably it follows that to double the population of workers in the few highest categories, there must be added eight times their number of lesser individuals" (p. 120). That is, lesser individuals who prepare the ground, or assist in the investigation, or add to the field. Among them are individuals who, given a certain a certain intellection and personality (neither yet fully understood) as well as the necessary opportunity and luck (also not yet fully understood), may become giants in their own right.

Yakon M. Rabkin's (1987) analogy of the contributions of those scientists in the "lesser" categories compared to those in the "higher" categories is found in his discourse in *ISIS* on "Technological Innovation in Science: Adoption of Infrared Spectroscopy by Chemists." There, he states,

> Among the most important new methods was infrared spectroscopy, which acquired remarkable popularity during the 1950s and 1960s. The number of infrared instruments, a handful before the war, rose to 700 in 1947, to 3,000 in 1958, and to 20,000 in 1969. The technique's use in scientific research, as recorded in a 1965 report issued by the National Academy of Sciences in Washington, D.C., skyrocketed correspondingly. (p. 31)

Obviously the corresponding rise in numbers of scientists working in infrared spectroscopy included those in both the "lesser" and "higher" categories, presumably the greatest number in the "lesser" categories. Question: Could those in the "higher" categories have done their work without the efforts of those in the "lesser" categories? May we not say all were involved in similar ecologies of achievement?

I refer here only to a very few studies that offer firm cognizance of the notion that science discovery (synonym: "creativity") does not come out of a single "Eureka!" or "Aha!" but from a network, a seamless fabric of the effort and work of many individuals—of all levels of ability and temperaments—over time. It is almost with regret that I must redact Newton's fabled statement that even a dwarf can see farther when seated on the shoulders of giants.[1]

In essence, the figure of dwarf seated on giant is one aspect of the methods of intelligence of the scientist, namely: Science as a field, and a scientist working in an area of special knowledge, depends on cumulative knowledge. Thus, I must revise the Lucan-Burton-Newton model (Virgil probably had an earlier hand in the aphorism) of dwarf on giant. That is to say, a dwarf may stand on a gigantic pyramid of bricks composed of the clay and straw of prior massive effort, and so even a dwarf may see farther. But if a dwarf may see farther, so may a giant. This is not to reject the truth that, in an ecology of achievement, one or two individuals may have knowledge, skills, and attitudes that make them clear leaders in a field. There surely are seminal thinkers, or giants, in the general attributes of intellection (say, Aristotle, Galileo, Newton, Kant, Einstein, the Curies, Mead) or in a special talent (say, Mozart, Kant, Beethoven, Stravinsky, Rembrandt, Monet, Van Gogh, Shakespeare, Dickens, Cassatt—select your own) who grasp a field entire and set it into a new context. However, they too have learned from others before them. Further, children and adults are not gifted, or talented, in *all* things, are they? They are talented in *some* things—even many, but certainly not all. The towering genius does not tower in all fields but leaves some towers to others. That is to say, the processes of work in any field encumber us with the methods of thought and the ways of work, as well as the knowledge and skills gained in time known and unknown, but *probably* prior to the time to discovery and *possibly* within the same period.

---

[1] Even here there is an ecology of achievement. Surely, among others, it occurred to Lucan (39–65 A.D.) to note: "Pigmei gigantum humeris impositi plusquam ipsi gigantes vident." (Pygmies on the shoulders of a gaint see farther than the giant.) Robert Burton (1577–1640) referred to Lucan's Didacus Stella—and it would not be surprising that Newton read Burton. I am indebted to Merton for bringing this to light in his *On the Shoulders of Giants* (1965/1985).

It seems, then, we cannot escape the ecology of achievement no matter how we examine it, any more than the talented scientist can escape the past. We may thus safely take a first look at the devices that channel the interests of the young so that they may undertake significant role exploration in the acts of discovery in science as coming out of a *dyad: genome interacting with environment.* Yet, we as teachers do not tamper with the genome, we tamper with its environment, a precious yet dangerous opportunity. Precious, because we are designing an environment for the young, who deserve an optimum environment in which to fulfill themselves. And dangerous, because in the sense of the Hippocratic oath we must "first, do no harm."

A Second Thesis: The School Environment, as Dyad

Are there any environments that would approach the condition that would permit us to use the term "identical environment" even as we use the term "identical genomes" for identical twins? We should expect to find such a home environment for twins or siblings reared in the same family in one household. Yet the consensus of recent findings is that, in a sense, each child within the family (a macroenvironment) is in an environment of her or his own (a kind of microenvironment). For example, siblings reared in the same home environment are generally different in significant behaviors. That is, the siblings are reared within different ecologies of achievement—their genotypes and initial environments interact to furnish them with a singular, not similar, ecology of achievement. Indeed, the parents are likely to react differently to children of different temperaments, behavior, and ability. Parents furnish offspring with different microenvironments, thus creating different ecologies. This should be expected if the siblings have different genotypes, and further, if, as is thought, as much as 30–40 percent of the traits of temperament are inherited, are gene driven.

Is there then a hypothesis that may guide our efforts to determine whether there is a trait we may call "giftedness" or "talent" in science? Forty years ago (1947) I tried to define one, based on my early work in scientific research and thus on my observations of scientists at work in their laboratories. As a participant in scientist's researches for 6 years, I learned their "methods of intelligence" (a term Percy Bridgman so often used). I used what I had learned in my own research. When I turned to teaching, I applied what I had learned from my observations of the laboratory environment into environments nurturing those whose wishes and intentions were to become scientists and who came forward (i.e., they selected themselves) to undertake the work open to them. The program was described in its initial plan in *The Gifted Student as Future Scientist* (Brandwein, 1955/1981) and more recently and briefly in "A Portrait of Gifted Young with Science Talent" (Brandwein, 1986).

The titles are significant. First I had noted that all the scientists (some 26) I had worked with were gifted students; that is, they had mastered their fields, could use the tools of "uncovery" in library and laboratory in order to come to their aim: a *discovery*, a new and meaningful work. What seemed to characterize the scientists who made new knowledge through that definitive, creative act, discovery?

First, being *gifted students*, they obviously had a certain level of intelligence. Recall, intelligence is a gene-driven factor; it is multifactorial (Guilford, 1968; Gardner, 1983) and thus is not to be reified in a single measure, IQ. But multifactorial intelligence should not be confused with *achievement*. While the former is strongly gene driven and, of course, highly factored by environment, the latter depends strongly on an environment channeled in schooling and education.

Not only were the mature scientists gifted students in their particular field in science (and apparently also in general intelligence as indicated in their verbal and mathematical skills), but they also were persistent in pursuing the solution of problems. Generally, their observed behavior may be expressed thus: Theirs was a quest, a search for verifiable knowledge. Embraced in the quest was their notable and observable dissatisfaction with present explanations of the way the world works— particularly in their chosen field of study. They often spoke warmly of certain experiences in school and university—and of their mentors in research. In short, their channeled interest was augmented by personal and activating attention of a memorable sort.

On the basis of these observations made in a laboratory setting, and the basic interpretations that sprang from them, early on I developed a "model" to guide our work in preparing an environment that would

- *Channel* the energies of those who wished to carry on the study of science (a channeling environment)
- *Augment* this channeling environment with opportunities to do such research (an augmenting environment is specially designed to furnish situations in which "original" problems in science would be found and an attempt made to engage in the kind of discovery characteristic of the scientist: Students would engage in problem solving, not the usual problem doing, of the scheduled lab)
- *Transform* potential into performance (once the young had performed with success and personal satisfaction, it seemed as if they had transformed themselves)

Our test of "creativity" in science was thus to be a test of ability to do an experiment or an investigation leading to a "new" bit of knowledge, a *work*.

Our first concern in the design of the "model" was to focus on the nurturing face of the environment. We would create a novel environment, and thus we could attempt to specify the competencies of our students in terms of performance. Ours was a kind of psychology or "unpsychology" (Michael A. Wallach's term) based on discerning and discovering conditions that stimulated field-specific, real-world performance in a particular area such as science. We would substitute the students' *work* for their scores on tests purporting to measure general "giftedness," "talent," or "creativity."

As my colleagues and I observed youngsters at work over the years, it became apparent to us that *questing* included a reaching out, perhaps an avid search for experiences that were stimulating, perhaps "experience in search of meaning," as

Einstein would say. These students seemed to want an environment fitting certain personal tendencies toward autonomy; they were bent on hollowing out a kind of capsule of freedom. Further, our interviews with a number of parents of these young who used novel approaches in problem solving led us to conclude that these youngsters had the privileges of early "independence training." (Anne Roe's phrase in Roe and Marvin Siegelman, 1964, p. 5). And, in their active search for experience, these young exercised a certain autonomy. Indeed, Lois-Ellen G. Datta and Morris B. Parloff (1967) suggest the possibility that the "main influence on early scientific creativity was autonomy versus parental control" (quoted in Tannenbaum, 1983, p. 295).

Our observations of mature scientists at work and of young aspirants to scientific careers at Forest Hills High School (New York City) led to the development of the working hypothesis summarized as follows: *"High-level ability in science is based on the interaction of several factors: Genetic, Predisposing, and Activating. All factors are generally necessary to the development of high-level ability in science; no one of the factors is sufficient in itself "* (Brandwein, 1955/1981, p. 12). This hypothesis guided us in our efforts to develop a channeled and augmented environment for those who *selected themselves* for the program we called OPUS—*Occupational Program Undergirding Science* (Brandwein, 1955/1981; 1986).

Clearly, what appears to be a *triad* rests in an interaction of gene and environment, a *dyad*. Probably the elements of the genetic factor display themselves in the early intellectual and physical development of children, as they interact with the people and things in their early environment. *Possibly* the predisposing factors, involving as they do personality, that is, temperament, are at least in part gene driven but are expressed in interaction with the environment. Thus Sandra W. Scarr and Kathleen McCartney (1983) state,

> We all select from the surrounding environment some aspects to which to respond, learn about, or ignore. Our selections are correlated with motivational, personality, and intellectual aspects of our genotypes. The active genotype-environment effect, we argue, is the most powerful connection between people and their environments and the most direct expression of the genotype in experience. (p. 427)

The predisposing factor I postulated comprised two identifiable traits: *persistence* and *questing*. Early on, Catherine M. Cox (1926) observed that high but not the highest intelligence, combined with the greatest degree of persistence, would achieve greater eminence than the highest degree of intelligence with somewhat less persistence. The trait questing (embracing a free-floating curiosity) demonstrated itself in the need to know and in a dissatisfaction with certain explanations of phenomena. Roe (1953), in her study of mature scientists, and Donald W. MacKinnon (1962, July) agree. MacKinnon, for example, states, "Our data suggest, rather, that if a person has the minimum of intelligence required for mastery of a field of knowledge, whether he performs creatively or banally in that field will be crucially determined by nonintellective factors" (p. 493). That is to say, factors of personality, even luck. Hudson, in considering the performance of creative work, states plainly that the work depends not so much on an individual's "intellectual apparatus but the use he sees fit to make of it" (p. 30). Hudson asserts as well that, given a certain level of ability, the personal, not the intellectual, factors are crucial.

From our observations of working scientists as well as from common sense, it seemed clear that genetic and predisposing factors were not all that operated in the making of a scientist. Opportunities for further training and the inspiration of the individual teacher and/or mentor were clearly factors to be considered in reaching a working hypothesis on the nature of high-level ability in science. Robert H. Knapp and Hubert B. Goodrich (1952) have studied the place the college teacher has in stimulating individuals with high-level ability in science. I recall that, without exception, the scientists who gave me my early training stated their indebtedness to one or more teachers and cited the opportunities these teachers made available to them. What we then called the *activating factors* turned out to involve the channeling and augmenting environments mentioned earlier. Indeed, Roe found the scientists she studied recalled those teachers who stimulated them to find things out for themselves.

It is important to emphasize that the hypothesis does *not* postulate talent, giftedness, or creativity in science per se—as rooted in the gene. We postulate that *high ability* (rooted generally in genetic factors) interacting with *predisposing* and *activating factors* (rooted generally in the effects of environment) are necessary to the development of scientists.

Why not use tests of creativity, for example, those of E. Paul Torrance (1966) and Jacob W. Getzels and Philip W. Jackson (1962)? Note that Tannenbaum (1983), in a major review of tests of creativity, states, "It remains to be demonstrated conclusively, however, that divergent thinking and creativity are synonymous and that so-called 'creativity tests' have strong predictive value" (p. 298). Richard S. Mansfield and Thomas W. Busse (1981) have also remarked that the diversity of definitions of creativity has produced a jumble of findings with only dubious applicability to real-life creative performance. On this, also see MacKinnon who concurs, "Our conception of creativity forced us further to reject as indicators or criteria of creativeness the performance of individuals on so-called tests of creativity" (p. 485).

We found that the young who selected themselves for apprenticeship in research and were thus faced with the tasks of the research scientist exhibited what we know to be the behaviors of the scientist. In selecting themselves for the demanding work, they opened for themselves a period of instruction in problem solving as a central part of an *augmenting environment*: research over a period of 6–18 months during and beyond the course. This experience, differentiated from that of students who preferred the accelerating enrichment of the *channeling environment*, may well be what is sometimes described as "differential education for the gifted." For example, those young who sought out the augmented environment were able to

- note discrepant events
- discover a problem situation within the event they wanted to investigate
- uncover the prior literature related to the work on the problem
- propose a hypothesis

- design an investigation involving observation and experiment on the basis of the hypothesis
- record their data (including error)
- design control experiments in an attempt to defeat their hypothesis
- offer a tentative solution
- propose new experiments in an attempt to defeat their solution
- state their solution in a systematic assertion
- present their work in seminars with other apprentice-scientists
- present their assertions and predictions in a paper
- present the paper to their peers in a science congress
- offer their work for critique to their mentors—scientists in the field
- enter their work in the Science Talent Search for further appraisal by scientists (if they wished; on this, see pp. 203–204)

True, the congeries of activities called forth by the individual research these young could and did undertake were entirely unlike the creativity tests Getzels and Jackson and Torrance proposed. However, for us and the two psychologists we consulted, the activities represented a decent test of the ability to *discover*. For us, the research constituted a test of creativity directly related to the lifework these young were contemplating. As such, the research was perhaps similar to the audition of the aspiring musician, the portfolio of the aspiring painter, the tryout of the aspiring athlete, the story or essay of the aspiring writer. Or, as Hudson put it, confirming our view, at least in part,

> When we ask a scientist to complete a verbal analogy for us, or a numerical series, we are asking him to perform a skill insultingly trivial compared with those he uses in his research: when he grasps a theory; reviews the facts for which it is supposed to account; decides whether or not it does so; derives predictions from it; devises experiments to test those predictions; and speculates about alternative theories of his own and other people's. In all these maneuvers he exercises skills of a complexity greater than we can readily comprehend. (p. 109)

We must content ourselves then, at this point, in rooting science talent in a dyad: genes interacting with environment. And the *evidence* of the presence of science talent was to be a *work*, which would necessarily come out of the interaction of heredity and environment undergirding the qualities of thought and action interpreted in our hypothesis stated earlier in this paper.

At this point, we may postulate strongly that high-level ability *in science specifically* is not to be conceived of as lodged in DNA. The gifted student's high-level ability in intellection (or critical thinking) and in numerical and verbal skills may be *initially* gene driven. But high-level ability *turned to science*, to the cumulative knowledge and the skills in inquiry that characterize science, are *environment driven*. The triads offered by Brandwein (1955), by Joseph S. Renzulli (1977), and by Robert J. Sternberg (1985) all seek a nexus within genetic and environmental expression as a sign of giftedness. The field-specific hypothesis expressed here for science, embracing high ability as emerging from the interaction of the *triad* of genetic, predisposing, and activating factors, nonetheless is subsumed by the reality of the *dyad*, the highly evidential interaction of an individual's DNA (genome) with the environment. Our thesis proposes that—as part of the ecology of achievement acting on the antecedent development of the young—carefully designed field-specific curriculum and instruction in science that encourage originative laboratory work (an environment) would catalyze the activity of the young as "performing scientists," albeit in the early stages of development. As early as 1957, A. Harry Passow had developed the essentials of a science curriculum undergirding a channeling and augmenting environment. (On curriculum, see also Passow, 1983; F. James Rutherford, 1985.)

A Third Thesis: The Student as "Performing Scientist"

How does one "select" those who have a potential for becoming performing scientists? Selection implies acceptance of some and rejection of others. Between 1947 and 1952, when we were engaged in researches to investigate whether there is such a trait as giftedness in science (also called "science talent"), measurement of "science talent" was in its infancy. Still, at Forest Hills High School, examinations were not a prelude to entrance. The school doors were open to all the students in the district. We were both required and glad to make opportunities available to all who wished to do the work; we could not, and indeed, we would not, make an examination requisite to entry to any program in science.

If we use the phrase "potential *to* performance," we seem to imply that first we find a potential and then turn it into performance. The actuality is that in giving young aspirants their opportunity to develop a personal art of investigation through performance, we seek out potential *through* (not *to*) performance. The strategy is to furnish students, their interest perhaps now channeled and augmented, with an environment, a problem-solving situation, in which they further augment interest and fix it in observable behaviors.

In 1947–1948, we gave students the opportunity to enter a program of individual work in science in the second half of the year in their study of biological science. They could select themselves for the opportunity, one that included individualized work and instruction (including mentoring) to solve a problem through modes of research resembling closely those of research scientists. Our model for the test was thus a simulated "real-world" process of testing an ability to solve a problem using the scientists' methods of intelligence. To a high degree, the "apprentices" would reveal how they faced problems that required an invented, that is, a novel approach. And yet the apprentice experience would embrace an unknown, approaching a decent tincture of the complexity found in adult research. Solving the problem would require sustained effort over 6 months to a year, perhaps even longer. A certain originality in insight and evaluation and, what is more, in overcoming countless failures, small and large, was required. In effect, the apprentice would face a paradigm of persistence in scientific critical thinking and ultimately, in origination, a *discovery*.

Science Talent: In an Ecology of Achievement

Is *performance* evocative of, even a *test* of, *potential* in science? It may well be a crystallizing experience masquerading within a test of talent in science, and it is possible that such an experience may exert powerful, long-term effects on the individual. Further, to quote Joseph Walters and Howard Gardner (1986), crystallizing experiences "are a useful construct for explaining how certain talented individuals may first discover their area of giftedness and then proceed to achieve excellence within the field" (p. 309).

Let me state one aspect of our program plainly: Once students had been accepted as members of the school, they were not to be excluded from participation in any activity devised by the school; we let each one of them have a try at any program with full attention to physical and psychological safety. Some 40–60 students were to apply (that is, select themselves for the science work) each year, and all students were accepted, no matter their IQ. The environment was so channeled and augmented to permit their performance to be the test of their ability.

True, the students in the research program tended toward 4 years of science, 3 years of formal mathematics plus one special mathematics course of their choice, usually the calculus. But it was clear to us that, in this act of self-selection, the vast majority who applied for individual work had come through a period of self-appraisal based on their achievement. Indeed, the oft-repeated research that, given the opportunity to do so, students can judge themselves by examining their own achievement was validated. In effect, to us, self-selection implied an active seeking of a channeling environment leading to an augmenting environment.

In any event, the students who selected themselves for the work displayed several sets of traits: high interest, generally high ability, persistence, and questing (the predisposing factors). But would they all be able to generate, pursue, and complete an investigation (a work performed) that would test their ability in discovery, in creativity? Similarly, but not in the field-specific context of science, Renzulli, Sally M. Reis, and Linda H. Smith (1981) utilize Renzulli's triad to describe giftedness as

> an interaction among three basic clusters of human traits—these clusters being above-average general abilities, high levels of task commitment, and high levels of creativity. Gifted and talented children are those possessing or capable of developing this composite set of traits and applying them to any potentially valuable area of human performance. (p. 27)

Further, Renzulli and his colleagues, in stressing task commitment, press the point that "whereas motivation is usually defined in terms of a general energizing process that triggers responses in organisms, task commitment represents energy brought to bear on a particular problem (task), a specific performance area" (p. 24). Possibly, by task commitment, Renzulli and his co-workers confirm the elements of the predisposing factor (questing and persistence) stipulated earlier, in our hypothesis. Nancy E. Jackson and Earl C. Butterfield (1986) are content with the following definition: "Gifted performances are instances of excellent performances on any task that has practical value or theoretical interest. A gifted child is one who demonstrates excellent performance on any task of practical value or theoretical interest" (p. 155). Robert S. Siegler and Kenneth Kotovsky (1986) propose that "... careful observation of the products children produce may prove to be the most practical way to improve on intelligence tests as assessment devices" (p. 432).

We are persuaded, then, by practitioners in research in the field of giftedness that performance in a field that has practical value or theoretical interest is a useful test of promise, or potential, for a given talent. However, one of these indicators is certainly the ability to perform an experiment or to do a theoretical investigation or analysis in an augmented environment in science.

One Search for Science Talent: Effects of Ecologies of Achievement

Why is it that some secondary schools, especially those apparently endowed with all the possibilities for developing a fruitful program for a body of gifted students who might seek out a lifework in science, do not offer an augmented environment in science—that is, one with individualized instruction that fosters and crystallizes experience in the arts of investigation characteristic of scientists? We found that the use of the Science Talent Search as an *instrument* to study various aspects of the ecology of achievement for the putatively talented in science was effective in dissecting out certain aspects of the augmented environment that seemed to be missing in these schools. The "Search" is instrumental in two ways: first, as a possible test of science talent and second, significantly for our purposes, as an indicator of the nature of certain ecologies of achievement that affect the demonstration of science talent.

Early in my observations I thought I had clues to the answer to this perplexing question. Apart from special schools whose practice was and is to select science-prone students for admission by a series of tests, why do certain schools with heterogeneous populations seem to succeed in developing a rich environment in science and mathematics in which the *potential* for science talent is expressed, while others with a similar student population are relatively devoid of such an environment?

Siegler and Kotovsky (1986), in considering the question "What will be the most fruitful approaches for research on giftedness in the next 5 to 10 years?" suggest,

> One useful approach would be to focus on people in the process of becoming productive—creative contributors to a field, for example, high school students who win Westinghouse Science Competition prizes; ... [or students] who publish articles in nationally circulated magazines, or who have their drawings shown in major exhibits. Members of these groups are of special interest for two reasons. They already have made creative contributions—they have not just learned to perform well on tests—but they are still in the process of becoming eminent. (p. 434)

And so too, Julian C. Stanley, Director of the Study of Mathematically Precocious Youth at Johns Hopkins University, states in the *Phi Delta Kappan* (1987, June),

> I firmly believe that a residential state high school of science and mathematics should follow the lead of those prestigious programs [referring to those in the Bronx High School of

Science and Stuyvesant High School, both in New York City] by preparing most of its students to compete in the Westinghouse Science Talent Search when they are seniors. To do less is to underdevelop the investigative scientific spirit of highly talented students.[2] (p. 771)

Further, Robert D. MacCurdy's study (1954) of 600 men and women who had been awarded honorable mentions or were finalists in the Science Talent Search further suggests its validity as a measure of manifest originality in science. E. G. Sherburne, director of Science Service, which administers the Science Talent Search, stated (1987),

The Science Talent Search is unusual among scholarship competitions in that it puts primary emphasis on the quality of a paper reporting an independent research project in some area of science, engineering, or mathematics and only secondary emphasis on academic achievement.
In short, the evaluation is on the basis of the student's ability to "do" science in a way that is analogous, though at a less sophisticated level, to what a professional scientist does [italics ours]. To use a sports analogy, one does not test a student's ability to play tennis by giving a paper-and-pencil test. One puts the student on the tennis court to play so the performance can be observed.

For the purposes of this book, then, it is useful to add to the literature the thrust of the Science Talent Search in respect to the place of the school in the ecology of achievement that results in the emergence of young who are manifestly originative in science and mathematics. As I see it, the ecology of achievement conducive to performance in science, and thus to demonstration of science talent, assumes a certain constellation. This ecology calls for an appropriate curriculum to include an individual investigation; a mode of instruction necessary to pursue independent investigation; and the mentors essential to guide and advise those aspiring scientists, who are willing to enter the Search, in the problems attending individual effort. (On this, see Brandwein & Morholt, 1986.)

In the period 1952–1962, I had the opportunity to study the science programs of 103 schools. I was able to select 22 of these schools for further study and to visit 17 of these heterogeneous schools two to four times during this period.[3] These 17 had a somewhat similar socioeconomic group of students and similar high levels of acceptance to colleges and universities; all, confirmed by my observations, had developed effective science and mathematics departments. All had developed channeling and augmenting environments; the latter gave opportunity for performance in an investigation requiring originative work on the high school level. In the period

---

[2]Note that Stanley recommends "preparation to compete"—not an insistence on competing. Preparation would mean giving students an opportunity to perform, that is, to do an investigation, the major requirement of the Science Talent Search. Note, too, that, as my investigations disclose, a number of students did not choose to compete.

[3]Certain of my observations during this period were part of my functions as chair of the Gifted Student Committee of the Biological Sciences Curriculum Study (BSCS) as well as those of a member of the steering committees of both BSCS and the Physical Science Study Committee (PSSC).

Table 1 Westinghouse Science Talent Search 1944 Through 1954[a]

School	Finalists	Honorable Mentions
1	17	79
2	17	57
3	17	53
4	8	34
5	8	19
6	8	8
7	7	48
8	6	38
9	4	3
10	3	74
11	3	0
12	1	17
13	2	3
14	2	1
15	2	0
16	1	1
17	0	5
18	0	0
19	0	0
20	0	0
21	0	12
22	0	2

[a]Grateful thanks to Dorothy Schriver and Carol Luszcz of Science Service, Washington, D.C., for their aid in supplying me with the data from past Westinghouse Science Talent Searches.

Key: Schools 1, 3, 7—select schools for science (based on entrance examinations)

Schools 9, 11—independent schools

School 13—a school of performing arts

School 2—Forest Hills High School, a school with a heteroge- neous population, furnished the students included in this 10-year study

The remainder are schools with heterogeneous populations.

1944 through 1954, the records of the 22 in the Science Talent Search were as follows (please refer to Table 1, above).

Recall that participation in the Westinghouse Science Talent Search is voluntary among hundreds of schools in the nation. The 22 schools in my study do not compose a statistical sample but are studied (aside from the select schools) because of their similarity to the school in which my teaching was done and the one that gave me opportunity to carry on work with the talented in science.

Schools 1, 3, and 7 had populations selected by tests specially designed to attract those with "science potential"; schools 9 and 11 were independent schools with selected populations and high socioeconomic levels; the remaining schools had heterogeneous populations, with somewhat similar socioeconomic levels. School 13, which selects its students for ability in performing arts, is especially interesting.

My observations and notes, based on checklists and honed by work for the Board of Examiners of New York City in selecting individuals seeking admission to the posts of science teacher or chair of a science department, show that all of the 22 schools offered exceptional course work. Their success in placing students in "sought after" colleges and universities was considerable.

Within the 103 schools, I also had opportunity to study another sample of 17 schools that served as a control group; these 17 schools did not enter the Search. Explicit statements by 14 heads of departments and/or deans of the 17 schools summarized their belief that the *training of scientists could well be left to the universities*. The explicit policy of their schools was the task of preparation for future study in universities where further extended preparation would be available for chosen careers, especially science, but also most areas of scholarship involving research activity. This view seemed acceptable and generally was found to be the prevailing one. Indeed, Lloyd G. Humphreys (1985) reports that "differences among chemistry, physics, geology, and engineering measures of attainment are obviously produced during postsecondary education" (p. 344).

In the schools that did not make the effort to enter the Search, there was a demonstrated absence of the individualized instruction and mentors necessary to help students do "an experiment," that is, to express their *potential through performance*. For example, a special sample of 18 of the 103 schools consisted of schools with populations under 600–1,000 students. There were 15 in rural areas that were unable to organize a program of four consecutive years of science and did not have the facilities for individual research (in an augmenting environment). However, 3 of the 18 attempted a mentorship program for those 6 students who aspired for careers in science.

Another Aspect of the Ecology

After I had left Forest Hills High School to take the post of Director of Education for the Conservation Foundation (now situated in Washington, D.C.), my colleagues in the school continued their work in the program with somewhat similar success. Then, three of them were promoted to the chairs of science departments in three schools in different areas of New York City, areas where the culture was not disposed toward intensive work in science and mathematics and where the teachers were not specially trained in experimental protocol and techniques. The records of these schools in the Science Talent Search were not noteworthy—three honorable mentions in 5 years. In spite of the efforts of these highly effective teachers and supervisors, the nature of the population and the environment per se (parts of our ecology of achievement) were not then conducive to the development of science-talented individuals, as measured either by their own standards or those of the Science Talent Search. That is to say, a channeling environment and augmenting environment in science in a given school may not function if the culture (a part of the ecology of achievement) is not disposed to support or interact with it.

Still another factor, seemingly minor but worthy of attention, was the availability of laboratory assistants educated sufficiently to supervise individual laboratory work. Without such staff, it was necessary for a teacher/mentor to be available to ensure safe use of equipment and substances while students were doing individual experimentation. This safety factor, rarely considered, acts as a block to individualized experimentation in science throughout the country. But it seems clear that necessary to place students in the finalist or honorable-mention categories in the Science Talent Search are a student body with a sufficient number possessing the genetic, predisposing, and activating factors and a supportive community cognizant of the significance of science in our culture. One aspect of participation in the Search depends on a certain performance in solving a scientific or mathematical problem and, therefore, on an environment that supports the activities facilitating the performance. It can also be demonstrated that whether or not an ecology of achievement undergirding science talent exists and persists in a school depends at least on these or similar factors:

- the early education of children in the home environment and the behavior necessarily reinforced in the community
- the early schooling and education of the student population within the school and community
- the policies of the school and community as they affect a decision to enter the Search[4]
- the preference of the school to offer special opportunities in an augmenting environment for the gifted, and, moreover, the ability of the school or community to provide for the *one* or the *very few* who aspire to a career in science
- the decision of students who, for various reasons, apply to enter or decline to enter the competition
- the recognition by the faculty, supported by the community, that it is the performance of the student in a *work* that is significant; its caliber and completion is itself a test, whether or not the student enters the Search
- the judgment that science is a collaborative enterprise, in which a given work reflects the antecedent and present efforts of many with different functions, levels of ability, and skills

All are essential to an ecology of achievement, which must be considered as a whole. Thus, the development of all—of high and modest ability—who can contribute on whatever level plays a part in sustaining the ecology.

On the other hand, in those schools that entered the Search, the belief prevailed that the young aspiring scientist or mathematician might well benefit from the judgment of others, just as writers, musicians, painters, and other artists, and,

---

[4]This is not, however, intended to convey the impression that schools need prove themselves through participation in the Science Talent Search.

Science Talent: In an Ecology of Achievement 233

of course, athletes, submit their work to the review of expert and peer. It was interesting to determine whether schools (Table 1) maintained a steady course in the Search over the years, say a quarter of a century later. Table 2 contains the 1942–1988 distributions of finalists and honorable mentions from the same schools studied earlier (1944–1954).

Table 2 Top High Schools in Westinghouse Science Talent Search 1942–1988[a]

School	Location	Winners
Bronx HS Science (1)[b]	New York, NY	106
Stuyvesant HS (3)	New York, NY	63
Forest Hills HS (2)	Forest Hills, NY	42
Erasmus Hall HS (8)	Brooklyn, NY	31
Evanston Township HS (6)	Evanston, IL	26
Benjamin Cardozo HS	Bayside, NY	25
Midwood HS (4)	Brooklyn, NY	20
Jamaica HS (22)	Jamaica, NY	19
Martin Van Buren HS	Queens Village, NY	15
Brooklyn Technical HS (7)	Brooklyn, NY	11
Central HS	Philadelphia, PA	11
Abraham Lincoln HS (5)	Brooklyn, NY	11
Hunter College HS	New York, NY	9
Lyons Township HS	La Grange, IL	9
New Rochelle HS (10)	New Rochelle, NY	9
Coral Gables Senior HS	Coral Gables, FL	9
North Phoenix HS	Phoenix, AZ	9
Phillips Exeter Academy (9)	Exeter, NH	8
Melbourne HS	Melbourne, FL	7
Newton HS (16)	Newtonville, MA	7
Ramaz HS	New York, NY	7
Niles Township HS West	Skokie, IL	7
Columbus HS	Marshfield, WI	7
Stephen Austin HS	Austin, TX	7
Woodrow Wilson HS	Washington, DC	6
Wakefield HS	Arlington, VA	6
Princeton HS	Princeton, NJ	6
Nova HS	Fort Lauderdale, FL	6
James Madison Memorial HS	Madison, WI	6
Alhambra HS	Alhambra, CA	6
McLean HS	McLean, VA	6
Eugene HS	Eugene, OR	6

[a] Grateful acknowledgment goes to Dorothy Schriver and Carol Luszcz of Science Service for furnishing the data in Table 2.

[b] The numbers in parentheses refer to the schools' coding in Table 1.

Note please, while only winners are listed, the order of listing may be interesting. What is significant in the above is the relative stability over 40 years of the ecologies of achievement of certain participating schools. (See below "Select and Heterogeneous Schools.")

Select and Heterogeneous Schools

It is clear from the data in Tables 1 and 2 that two of the select schools of science maintained their positions; they had maintained a steady ecology of achievement. Essentially, select schools accomplish this not only by the selection of their populations of students, but also by the support of parents (a gathered community), the continued support of the boards of education or boards of advisors, and the selection of teachers, as well as the maintenance of their channeling and augmenting environments. Others with heterogeneous populations retained their presence over the years. Some schools that were not present in the earlier study have now begun to make significant showings. One set of assumptions may explain some of these trends:

- Select schools have developed their own following, a dispersed but like-minded community, a kind of homogeneity. Therefore, they have maintained and sustained their niche in an ecology of achievement.
- Over a period of time, populations in heterogeneous communities may change economically, politically, socially, and, thus, in the objectives of their schooling. It follows that, in a given school where there are strong efforts to maintain an augmenting environment, there may be temporary changes in administration, faculty, student body, or community support. Such changes could affect not only interest in or commitment to science and technology and the school's own ecology of achievement but also the ecology of the community.

Nonetheless, over the years, within a given geographic area the relative total number of finalists and honorable mentions coming from heterogeneous schools (however they shift in identity) compares favorably with that of the select schools. The ratio of achievement of heterogeneous to select schools in a defined geographical area seems to be fairly steady.[5] And it is clear that the vast majority of students attend heterogeneous schools. As do the select schools, the heterogeneous schools (in communities that have developed notable ecologies) may also serve as models affecting the establishment of channeling and augmenting environments for communities in the surrounding areas. Thus, the Science Talent Search, reflecting as it does the achievement of certain students, is not only a demonstration of individual

---

[5]Discussions with several administrators of schools indicated that they were loath to enter a national competition in which the number of prizewinners was drastically limited. Is it possible that Science Talent Searches held annually in all the 50 states might serve to attract students at a variety of levels of ambition to seek a lifework in science? The National Science Talent Search might then be extended to draw on the combined populations involved in 50 distinguishable Searches.

However, many entrants have still another opportunity for recognition through State Science Talent Searches (33–43 in number depending on the year of the Search). The number of state talent searches has increased by approximately 25 percent over the 40 years of the Search. The Science Service duplicates the written entries and forwards them to directors of the State Searches. The states then conduct their own competitions, many of which offer numerous awards, including scholarships.

talent but a demonstration, as well, of the presence of an ecology of achievement in home, school, and community.

Contrasting Ecologies

Recall that, as we have defined it, an ecology of achievement affecting a child embraces the sustaining environments of the home, peers, and community in the *educational* functions that support the particular construct of schooling in that area. Thus, the performance of a child in schooling per se and, of course, in a particular area of performance—art, science, or athletics—is not solely an outcome of the particular curricular and instructional practice in a given school but is also a test of the family and community support of both the education and schooling of the young. In sum, this support not only is exemplified in the attitudes of parents and in their competence in rearing their children but also is an earnest of the traditions, attitudes, and practices of the community reflected in the school.

It is also necessary to remind ourselves that, even when there is evidence of a child's tendency to excel, superior—or mediocre or failing—performance is still not solely or even mainly an index of the child's potential. Performance on any given occasion may well be a reflection of the effect of an entire complex of prior environments within schooling as well as the educational opportunities (or lack thereof) and influences (good and bad) afforded by home, peers, and community. A case in point: hapless addiction to drugs of able young. We may not disregard, as well, the steady accumulation of disadvantages those already disadvantaged may not be able to set aside, that is, to overcome without constructive affection.

Recall Tannenbaum's reflection on the cyclical nature of interest in the gifted. In his paper (1979), he documents an easily verifiable observation: that gifted children as a special group have been "alternately embraced and repelled with so much vigor by educators and laymen alike" (p. 5). It is also observable that certain states or communities with a notable dropout rate may, in the same period, also have schools with notable records in admission to elite universities—and, in fact, notable records in the Science Talent Search. Nonetheless, the schools that I have observed promoting the kinds of channeling and augmenting environments that make provision for the gifted also provide well-planned programs for the disadvantaged. The effectiveness of these programs is reflected by the considerable improvement in the tendency of the disadvantaged young in these schools to continue there (in contrast to the high dropout rates elsewhere). It may well be that the philosophy and practice of most of these schools is to attempt to provide for all their young. In my experience, Tannenbaum's remark about the swings in affection and disaffection for the gifted also applies to the disadvantaged.

On the other hand, throughout the country, we appear to alternate in our attention to the schooling of our gifted and, then, to that of our disadvantaged. These cycles or the recurrent formula of our "crises" in schooling have been well documented. We find too often a periodic rise in concern about the effectiveness of our schools stimulated by the momentary prominence of some event such as rises and falls in test scores, or Sputnik, or the economic success of another nation, or possibly our failures at the Olympics. If serious enough, this concern is followed by a period of vigorous effort mounted, almost *entirely*, through "reform" of schooling—to ameliorate certain symptoms of the decline. In time, this is followed by another "failure" of the schools.

It is fairly easy to demonstrate that what is considered to be a general failure of the public schools is in effect a failure within the ecologies of achievement. To repeat, it is rare to find an effective school within a community that does not support its schools and teachers. However this may be, there remains undeniable evidence that an individual's traits are the result of the interaction of heredity and environment at any point in development.

To recapitulate, we understand the term *environment*, in respect to its various effects on the traits of the young, to mean precisely the effects of the ecologies of achievement that are at the heart of this discourse. At this point, it is observable that the environments conducive to developing our young's learning capacities and personal growth have not achieved the stability that reduces oscillations in the effectiveness of the communities—whatever their precursors in political, economic, or social events—which support schooling and education. In time, these oscillations may become intolerable. It remains to be seen whether an open society can maintain the requisite stability of a schooling system that can respond equitably to its supreme responsibility of attending to the future of *all* its young in their various capacities. It is obvious that in so doing we attend to our own future, as is required of a society that acquits itself so nobly in a document beginning with that epiphanous phrase, "We, the people. . . ."

The Need for the Talented in Science

However this may be, the world is in dynamic change. Daniel Bell (1973), among other observers, made the case early on that essential individuals in industry in the decades to come will be those centered in science and mathematics. The postindustrial society is here; it has global consequences; we now accept that the scientist and science-trained individual are necessary to the well-being of present and future societies. However, it is not the scientist of earlier years whom we seek; we need individuals educated in societal as well as scientific aspects. We need not only biologists, chemists, physicists, geologists per se; they are required, in a certain measure, to be ethicists as well. For scientists, through their discoveriers, have indeed made the planet a global village.

Perhaps it will now be necessary to consider whether it is not a general function of schooling to design the channeling and augmenting environments that evoke the potential of those whose driving interest is to think and do science and technology. If so, it will become clear that these environments *cannot* all be sequestered in a limited number of select science schools. The young who are to seek their lifework in science must come out of the schools that exist, and these, for the vast majority, are part of some 15,000 school districts across the country serving mainly heterogeneous populations. Should not all the young have available a mature ecology of achievement in which they can fulfill their powers in pursuit of their personal swath of excellence?

It has been shown in earlier periods of work on the gifted; and again in the most recent spurt of activity in research on the gifted young; and further in this paper in the accounts of a number of scholars; and it has been and is being shown through the validating data of the Science Talent Search that it is feasible to develop programs that bring forth the young who may become our future scientists at various levels of originative work. We do not, in my view, find the science talented through paper-and-pencil tests of creativity or necessarily in those whose abilities are reified in the IQ. The confluence of traits and competencies of the young secured in antecedent environments and reflected in a performance, that is, in a work, more nearly reflects the scientist's processes. The process is central to our thesis that the completion of such a work is a sign of talent in science.

A strong hypothesis may thus be advanced: The demonstrated opportunity of the *school* to develop the curriculum and instruction devised as the channeling and augmenting environment, enabling individualized work in science, and the demonstrated ability of the *student* to plan and complete an experiment or theoretical analysis, with the meticulous application required, may well be a valid test of science talent. The hypothesis is in support of Wallach's (1985) suggestion that creativity in a specific field, say science, might be a "by-product of field-specific instruction" (p. 115).

Schools are filters of feasibilities. In the environments they construct lie the seeds of the destinies in which the young and their various but remarkable capacities find perdurable ways of advancing the culture.

Reflections and Conclusions

Recall that Gregory Bateson remarked, when asked to define "scientific truth," that he contented himself with this definition: "What remains true longer does indeed remain true longer than that which does not remain true as long" (1979).

As one who aspired to spend his life in science and once was welcome and worked in various laboratories, I am keenly aware that in the field of study called "giftedness and/or talent," we are not yet in the area we may call science. In this area, we are still in a science of practice and not of the laboratory. We are beginning to know something of the kinds of environments that may bring forth the metamorphosis of potential into performance. We *may* know more of one part of the dyad, the environments that encourage potential, than we know of the other part, the genes that are basic to performance.

There are thus certain clusters of paired reflections and conclusions I am obliged to put before you.

First: We are teachers. We are obliged therefore not to use the young as implements of our particular sociopsychological warfare in measurement or method. Because we cannot change the genetic complexes in the young that come to us, we are obliged to take the only course available to us: to change the environment to fulfill the powers of the young so that they may pursue the best course of development within their capacity. For the facts are these: An individual is the result of the interaction of inherited genes and of environment; this interaction occurs at every moment in the course of development. Yet, although the concepts of the ecology of achievement and the dyad (in numerous diverse phrases) are, and have been, strongly evident, the culture generally and societies specifically continue to attribute to each individual aptitudes, attitudes, and achievements without reference to augmenting or suppressing environments. Indeed, the attempt is to reify an individual's history in a single measure—the IQ. *Not* in works, *not* in achievements, but in a measure of potential that does not describe the individual's advantaged or disadvantaged prior history.

Indeed, our study and numerous others support this hypothesis: *The ability to apprehend the arts of investigation and to complete an empirical study is, in conjunction with other indices, a better predictor of future entry into successful scientific work then are paper-and-pencil tests of creativity.*

In other words, *not only the work but also the doing* summarizes the originative traits found in the self. The painting, the musical score, a building's architecture, the scientific or entrepreneurial act of creation—these furnish pictures of the self in the act of creation. We are thus required to wait until geneticists have analyzed the DNA of a host of gifted and talented grandparents, parents, and young (for genetics is a study of families) to determine whether attributes of talent are inherited, and if so, in which gifts or talents.

I *conclude*, then, that we are obliged—no, required—to develop the most fruitful course of development in the schools: a channeling and augmenting environment fulfilling the most generous of auguries. These environments are to nurture in fruitful curriculum, instruction, and mentorship the aspirations of all young who wish to demonstrate a talent in science *through performance* in the arts of scientific investigation.

Second: We work in an admirable environment, the architecture of a long cultural history: the school. But the school is not an isolate; it is part of an *ecology of achievement* affected by and participating in all the activities accommodating the particular community, society, nation, and culture of which it is a part. Just as a gaunt deer is not generally to be found in a rich, capacious forest, so a poorly supported school is not found in an effective community, state, or nation. Put another way, the human is the result of interaction of a biological and social (cultural) environment of which one important environment is the school.

I am obliged to *conclude* that the environment of the young who will eventually act to conserve, transmit, expand, and correct the culture consists of *all* the advantages the entire school-community-nation can and should offer and afford to offer. We are acutely aware that the talented, whether scientists or not, whether giants or auxiliaries maintaining giants, or the young not yet tall enough do not—and cannot—complete their work in isolation.

We know enough about curriculum and instruction and administration; we are competent and compassionate enough to develop models of apprenticeship in science to fit all manner of competency. If we will it. Further, there is evidence that both select schools and those with heterogeneous populations design and utilize similar models of channeling and augmenting environments.

I *conclude*, then, that *both* select and heterogeneous schools can develop appealing and adequate models of curriculum and instruction that channel and augment the interests and abilities of all who wish to enter the various levels of excellence required in science.

Third: We know enough about the traits of scholars, whether scientists or artists, to afford them sufficient opportunity. Our society cannot speedily be made perfect, although we—in the United States—are fortunate enough in minds and resources to travel that long and tortuous road. If we would. But the pieces of the constructs of most excellent schools are everywhere; while society is not yet perfectible, the school, an instrument of society, is a small enough community to approach the perfectible within an appreciable, even predictable, time.

I am obliged to *conclude* that within a society intending perfectibility, the schools may be exemplars in the design of a perfectible social construct—an ecology of achievement—for developing youth. For in the schools lie our future resources of mind and personhood. It is entirely conceivable, then, if the perfectible school could become a prime objective in the humane use of human beings, then talent would indeed be a by-product of study in a field freely chosen by each individual.

In a sense, our schools are strong signs of our character. They are bulwarks against the enemies that plague us: ignorance and indifference, mindlessness and meaninglessness. They are signs of our persistent, character-rooted passions: It is better to know than not to know. They are signs of our intent—to construct an open society in which all of us, without coercion, can behave as if we could find out what is true.

References

Bateson, Gregory. (1979). *Mind and nature: A necessary unity.* New York: Bantam Books.

Bell, Daniel. (1973). *The coming post-industrial society: A venture in social forecasting.* New York: Basic Books.

Brandwein, Paul F. (1947). The selection and training of future scientists. *Scientific Monthly, 54,* 247–252.

Brandwein, Paul F. (1955). *The gifted student as future scientist: The high school student and his commitment to science.* New York: Harcourt Brace. (1981 reprint, with a new preface [Los Angeles: National/State Leadership Training Institute on the Gifted and Talented]).

Brandwein, Paul F. (1981). *Memorandum: On renewing schooling and education.* New York: Harcourt Brace Jovanovich.

Brandwein, Paul F. (1986, May). A portrait of gifted young with science talent. *Roeper Review, 8*(4), 235–243.

Brandwein, Paul F., and Morholt, Evelyn (1986). *Redefining the gifted: A new paradigm for teachers and mentors.* Los Angeles: National/State Leadership Training Institute on the Gifted and Talented.

Broudy, Harry S. (1972). *The real world of the public schools.* New York: Harcourt Brace Jovanovich.

Cox, Catherine M. (1926). The early mental traits of three hundred geniuses. In Lewis M. Terman (Ed.), *Genetic studies of genius* (Vol. 2, pp. 11–842). Stanford, CA: Stanford University Press.

Datta, Lois-Ellen G., and Parloff, Morris P. (1967). On the relevance of autonomy: Parent-child relationships and early scientific creativity. (Described in Abraham J. Tannenbaum [1983], *Gifted children: Psychological and educational perspectives*. New York: Macmillan)

De Lone, Richard H. (1979). *Small futures: Children, inequality, and the limits of liberal reform.* Foreword by Kenneth Keniston (pp. ix–xiv). New York: Harcourt Brace Jovanovich. (Carnegie Council on Children)

De Solla Price, Derek. (1961/1975). *Science since Babylon* (Enlarged ed.). New Haven: Yale University Press.

Gallagher, James J. (1970). *Teaching the gifted child* (2nd ed.). Boston: Allyn and Bacon. (Original work published 1964, with a new preface for this printing).

Gardner, Howard. (1983). *Frames of mind: The theory of multiple intelligences*. New York: Basic Books.

Getzels, Jacob W., and Jackson, Philip W. (1962). *Creativity and intelligence: Explorations with gifted children*. New York: John Wiley and Sons.

Guilford, J. Paul. (1968). *Intelligence, creativity and their educational implications*. San Diego: R. R. Knapp.

Hudson, Liam. (1966). *Contrary imaginations: A psychological study of the young student.* NewYork: Schocken Books.

Humphreys, Lloyd G. (1985). A conceptualization of intellectual giftedness. In Frances Degen Horowitz and M. O'Brien (Eds.), *The gifted and talented: Developmental perspectives* (pp. 331–360). Washington, DC: American Psychological Association.

Jackson, Nancy E., and Butterfield, Earl C. (1986). A conception of giftedness designed to promote research. In Robert J. Sternberg and Janet E. Davidson (Eds.), *Conceptions of giftedness* (pp. 151–181). New York: Cambridge University Press.

Knapp, Robert H., and Goodrich, Hubert B. (1952). *Origins of American scientists*. Chicago: University of Chicago Press.

MacCurdy, Robert D. (1954). Characteristics of superior students and some factors that were found in their background. Unpublished doctoral dissertation, Boston University, Boston.

MacKinnon, Donald W. (1962, July). The nature and nurture of creative talent. *American Psychologist, 17*(7), 484–495.

Mansfield, Richard S., and Busse, Thomas V. (1981). *The psychology of creativity and discovery: Scientists and their work*. Chicago: Nelson-Hall.

Merton, Robert K. (1961). The role of genius in scientific advance. *New Scientist, 259*, 306–308.

Merton, Robert K. (1985). *On the shoulders of giants*. New York: Harcourt Brace Jovanovich. (Original work published 1965).

Passow, A. Harry. (1957). Developing a science program for rapid learners. *Science Education, 41*(2), 104–112.

Passow, A. Harry. (1983). The four curricula of the gifted and talented: Toward a total learning environment. In Bruce M. Shore, Françoys Gagné, Serge Larivée, Ronald H. Tali, and Richard E. Tremblay (Eds.), *Face to face with giftedness* (pp. 379–394). Monroe, NY: Trillium Press. (World Council for Gifted and Talented Children).

Rabkin, Yakon M. (1987, March). Technological innovation in science: The adoption of infrared spectroscopy by chemists. *ISIS, 78*(291), 31–54.

Renzulli, Joseph S. (1977). *The enrichment triad model: A guide for developing defensible programs for the gifted and talented*. Mansfield Center, CT: Creative Learning Press.

Renzulli, Joseph S., Reis, Sally M., & Smith, Linda H. (1981). *The revolving door identification model*. Mansfield Center, CT: Creative Learning Press.

Roe, Anne. (1953). *The making of a scientist*. New York: Dodd, Mead.

Roe, Anne and Siegelman, Marvin. (1964). *The origin of interests*. Washington, DC: American Personnel and Guidance Association.

Rutherford, F. James. (1985). *Education for a changing future.*Washington, DC: American Association for the Advancement of Science. (Originally called *Project 2061: Understanding science and technology for living in a changing world*).

Science Talent: In an Ecology of Achievement 241

Scarr, Sandra W., and McCartney, Kathleen. (1983, April). How people make their own environments: A theory of genotype-environment effects. *Child Development, 54*, 424–435.

Science Service, Inc., 1719 N Street, NW, Washington, DC 20036. (Agency that administers the Westinghouse Science Talent Search)

Sherburne, E. G. (1987, January). Washington, DC: Science Service, Inc. (Announcement relating to the Westinghouse Science Talent Search)

Siegler, Robert S., and Kotovsky, Kenneth. (1986). Two levels of giftedness: Shall ever the twain meet? In Robert J. Sternberg and Janet E. Davidson (Eds.), *Conceptions of giftedness* (pp. 417–435). New York: Cambridge University Press.

Stanley, Julian C. (1987, June). State residential high schools for mentally talented youth. *Phi Delta Kappan, 68*(10), 770–773.

Sternberg, Robert J. (1985). *Beyond IQ: A triarchic theory of human intelligence.* New York: Cambridge University Press.

Suzuki, David T., Griffiths, Anthony J., Miller, Jeffrey H., and Lewontin, Richard C. (1986). *An introduction to genetic analysis* (3rd ed.). New York: W. H. Freeman.

Tannenbaum, Abraham J. (1979). Pre-Sputnik to post-Watergate concern about the gifted. In A. Harray Passow (Ed.), *The gifted and the talented: Their education and development, Part I* (pp. 5–27). (Seventy-eighth Yearbook of the National Society for the Study of Education.) Chicago: University of Chicago Press.

Tannenbaum, Abraham J. (1983). *Gifted children: Psychological and educational perspectives.* New York: Macmillan.

Torrance, E. Paul. (1966). *Torrance tests of creative thinking.* Princeton: Personnel Press. (Norms technical manual).

Wallach, Michael A. (1985). Creativity testing and giftedness. In Frances Degen Horowitz and Marion O'Brien (Eds.), *The gifted and talented: Developmental perspectives* (pp. 99–123).Washington, DC: American Psychological Association.

Walters, Joseph, and Gardner, Howard. (1986). The crystallizing experience: Discovering an intellectual gift. In Robert J. Sternberg and Janet E. Davidson (Eds.), *Conceptions of giftedness* (pp. 306–331). New York: Cambridge University Press.

Science Talent in the Young Expressed Within Ecologies of Achievement: Executive Summary

Paul F. Brandwein

Introduction

This study was undertaken to attempt to define some of the educational ecosystems that encourage the young to discover, first, proneness toward science and, then, if they choose, to begin actively to express that interest through science talent. It defines "education" in the manner of Bailyn (1960) and Cremin (1980, 1990) as being made much more than schooling, which is but an exceedingly important part of the ecology. The ecology of education comprises three intereffective ecosystems—that of the family-school-community, the culture, and the postsecondary systems. When these three ecosystems interact harmoniously, they form an ecology of achievement that offers all the young opportunity for their special endowments—both intellective and nonintellective—to flourish.

While the ecology of achievement is essential to all the young seeking to fulfill their individual powers, it has special implications for those students who may eventually decide on careers in science or technology. In the preindustrial agricultural world, men and women struggled primarily with nature; in industrial society, they worked with machines; in "postindustrial" (Bell's term, 1973) society, minds contend with informed minds. In this world, literate, numerate, and scientifically productive citizens are profoundly necessary. New industries, based in what Schultz (1981) calls "human-made capital," concerned with the knowledge and processes coming out of biology, chemistry, physics, space, and environmental science, call for inspiring teaching and learning in an educational ecology of achievement.

An observable model of an ecology of education—*not an organized nationwide educational system*—appears to exist in the United States. Although individuals, groups, and organizations strive to create such a skein of achievement-centered

---

Research for this report was supported under the Javits Act Program (Grant No. R206R00001) as administered by the Office of Educational Research and Improvement, U.S. Department of Education. Grantees undertaking such projects are encouraged to express freely their professional judgment. This report, therefore, does not necessarily represent positions or policies of the Government, and no official endorsement should be inferred.

This document has been reproduced with the permission of The National Research Council on the Gifted and Talented.

D.C. Fort, *One Legacy of Paul F. Brandwein*, Classics in Science Education 2, DOI 10.1007/978-90-481-2528-9_23, © Springer Science+Business Media B.V. 2010

environments, a number of environments limiting achievement and unfavorable to self-discovery of science proneness hamper many of the young. It seems necessary, then, to forge programs where instructed learning becomes, first and foremost, a system of discovery of abilities through achievement, through the self-identification of capabilities by all the young in their increasing variety. Envisaged and conceivable are such programs that will validate themselves as a means of natural assessment of growth in science talent. When endowment projects itself in enriched opportunity through doing science, through performance, the young will find their own capabilities, learning how to discover for themselves and revealing portraits of intellective and nonintellective abilities. Science potential may then be discovered or confirmed not only through performance in programs in instructed learning, not only from the varieties of evidence gleaned through assessments of science proneness and talent, but also—and most importantly—through the originative work that is their criterion sample.

Limiting Environments

Before proceeding to a consideration of the qualities of environments that enable achievement in science and technology, it is necessary to look at the factors that account for the crisis that has hampered America's students' success in those fields. In 1983, the National Science Teachers Association's yearbook summarized a crisis in instructed learning in science. The syndrome of 10 it defined described a state of affairs in science education that largely continues. The yearbook's conclusion: "A wide variety of writing and reports, current projects, and research converges in a characterization of current science as plagued by 10 common recurring problems." [They follow:]

1. The textbook is the curriculum.
2. Goals are narrowly defined.
3. The lecture is the major form of instruction, with laboratories for verification.
4. Success is evaluated in traditional ways.
5. Science appears removed form the world outside the classroom.
6. A shortage of science and mathematics teachers has led to the widespread use of un- and underqualified teachers.
7. The outdated curriculum neglects the needs and interests of most students.
8. Current science instruction ignores new information about how people learn science.
9. Supplies, equipment, and other resource materials are severely limited or obsolete in most science classrooms and laboratories.
10. Science content in the elementary schools is nearly nonexistent. (Yeager, Aldridge, & Penick, 1983).

In teaching and learning, what is not *open* to children early on may be closed to them later. And science talent is both a general category and an amalgam of personal traits and abilities focused in specific fields. While giftedness is general, talent comprises the specific aptitudes required for the subsets of a field. Individuals with various talents and exceptional competence can begin to make significant career choices even during precollege and freshman years.

Perhaps, the family-school-community, college-university, and cultural ecosystems would contribute to the brilliance of the world if, in their interconnectedness, they would lend their collaborative resources to *all* young who aspire and are capable of achieving. Then, students who acquire the trained intelligence—in whatever capacity—desiring to enter the sciences prized in the United States would fulfill their powers in the pursuit of excellence. And, as they shaped their own opportunities, they would begin to define their self-concepts as well. They would know, from the beginning, that the massive achievements characterizing scientific research generally result from the works of scientists in all categories: From artisan to novice to eminent scientist.

The mutualism of the three human ecosystems acting intereffectively within an ecology of education is, however, not a matter of course. Because they exist within a total framework, their interaction is generally not mandated but lies within the sphere of choice, except when a specific function is dictated by law. No matter; their acts in support or neglect affect the totality of American education within an ecology of achievement.

Nothing in this study calls for a curricular and instructional experience composed of a stable set of experiences to fit all abilities and predispositions, thus attempting to ensure a steady progression through the grades. Quite the opposite, this study presses the invention of programs that encourage *differences* in expression and performance, and the inclination to seek special excellences and worthwhileness through a family-school-community program. In this sense, the limiting ecologies discussed here can stand in the path of the expression or attainment of desired abilities.

When barriers, such as limitations in instruction as summarized in the syndrome of 10, inadequately prepared teachers, and inadequate funding, combine with other factors to prevent the creation of an ecology of achievement, the results can be serious. Their consequences in the wide educational environment—especially the socioeconomic conditions affecting home, family, school, and community—can contribute to a reduced supply, first of young with interest in science and then of scientists and artisans.

Women and minorities, though making some headway currently, are particularly affected. The fall-off continues through misuse of what the Government-University-Industry Research Roundtable called the "weed and seed" approach in many of the nation's college–university ecosystems (1987). The National Science Foundation, along with other institutions concerned with the fullest representation of contributors in science, finds the origin of the present underrepresentation in early schooling, particularly in inadequate preparation in science and mathematics.

Granting that some young take the challenge of limitation and overcome it, research emphasizes that supporting environments, particularly those from early childhood through the grade school years, are generally necessary to prepare the young for the course they take in securing competence and performance.

This study aims to define an environment in schooling and education designed to encourage self-identification and self-selection of science-prone and science-talented young. This ideal was and is a necessary intervention (or invention), since the ecologies of both school and culture intereffect the development of abilities and predispositions, thus attempting to ensure a steady progression through the grades.

An ecology of achievement allows the intermeshing relationship of heredity and environment to encourage the full, direct expression of talent, whether in science or in another area of value in human and humane prospect. First, however, these data give rise to certain important assumptions. They follow:

1. Almost all American and foreign immigrant young who will become scientists in the 21st century are presently in our schools.
2. It is apparent for the present and possibly for the near future that a sufficient number of American young are unavailable to fill the need for the scientists of the future. Foreign scientists are now being trained here, but there is no guarantee that they will not return to their countries of origin.
3. The frequent premise that the thrust of practice in curriculum and instruction for the science talented should aim at the apex—the research scientist—requires reexamination. A visit to almost any research laboratory dispels the notion. All competent laboratories prize the contribution of skilled artisans and/or technicians. Practices in guidance and during early schooling, as well as programs, should be developed for those whose inclination is to artisanship. At present, the well-formed American system of community colleges makes available opportunities for credentials in a variety of skills.
4. Stressing achievement and self-concept at the beginning of a career in science is as necessary as stressing the history of achievement of the eminent. The latter holds up a vision of greatness as stimulus, the former, the high probability of a worthwhile lifework (however hidden from public view) and a significant contribution.
5. This construct's emphasis on limiting factors brings to mind only half the case, only part of a human ecology: The environments that make up this ecology are not severable; seeming opposites interpenetrate and, eventually, a natural ecology heals itself. In the communicable human ecology, the significant factors of materials, energy, and information engage purpose and action to introduce enabling environments to offset and replace limitations on a productive ecology.

Enabling, favored environments in intervention and invention may be able to neutralize, offset, and replace the limiting environments characteristic of flawed educational ecologies.

Enabling Achievement: An Instructional Approach for Self-Identification of Science Proneness

Here are presented some models of teaching and learning enabling expression of learning toward science in the primary school years: First, through examination of theoretical constructs, from which are drawn clues to instructional practices that help to identify and define early science proneness; and, second, through study of practices of science instruction that encourage children to identify themselves as science prone and demonstrate their awareness prior to high school. The clear purpose: To ameliorate, if not to annul, the syndrome of 10.

By casting a wide net for excellence and equity, the family-school-community ecosystem can enable the search for and by the young for competence in general or specific performance in science. A significant improvement in science teaching may empower a larger pool of talent than selection based on IQ alone.

A model of instruction in science is offered through which children may identify themselves as science prone before the talent pool develops. Certain science lessons both demonstrate characteristic behaviours of elementary school young in various contexts and lead to self-identification of potential. This approach concurs with Havighurst's (1972) aim to design programs to meld with the potential of children early and so to increase the numbers of them who develop it.

A curricular-instructional base promoting reciprocal interaction of child and environment is essential. (It is, however, important to remember that *curriculum* is a plan for teaching in classroom and laboratory; *instruction* is what happens in these environments, the field, or in independent study [at home or library] that stimulates learning through interaction between teacher and student.)

These objectives call for "instructed learning" (Bruner's phrase, 1966). If differentiated programs are developed during the course of schooling, gifted young should have the opportunity to identify themselves early as science prone. Their path should be through personal activity in instructed learning and independent study available in a gifted environment, planned in curriculum and instruction that nurture science proneness.

At this point, formal testing is unnecessary for either self-identification by the young or as a means to prejudge their capacity. A number of field observations of the young in instructive learning situations in interdependent-independent environments are presented.

The hypothesis is that evocative instruction, consistently stimulating idea-enactive, inquiry-oriented behavior in the classroom, laboratory, or in individual work, may be used as a mode for the young early to identify in themselves a tendency to science proneness. The aim to have the young, if they so choose, do science is and was the essence of the idea-enactive, inquiry-oriented approach. And this self-definition may be followed by self-selection for further participation in differentiated curricular practice in science and in its supportive verbal and mathematical knowledge and skills. Because evidence of self-identification and self-selection of science proneness takes careful observation, the teacher becomes also a researcher and an interpreter.

A science curriculum built around conceptual schemes is flexible and responsible for children's needs and interests. Such a program, far from being rigid, permits a consistent organizing principle, one that encourages incidental learning from the media or in special environments. Such a curriculum reflects both the ways of scientists and those of growing children as they progress into and retreat from the vastness of their universe. It permits the teacher to interpret the child's questions in a manner relevant to the kind of inquiry that results in individual activity.

Equal opportunity opened up through instructed learning may result in a seeming paradox: Namely, equality of opportunity may lead to situations where differences in expression of abilities appear. Such differentiated self-expressions through early study and work may become the first instruments through which peers, teachers, parents, or others contemplate differences among students in scope and in interests. These observations may lead to a common consent that a certain child may or may not be science prone.

If idea-enactive, inquiry-oriented teaching as a strategy of instructive learning becomes general practice, *then,* in the revolving-door instructional model (Renzulli, Reis, & Smith, 1981), it may become a mode of early self-identification of the young. Their individual responses to multiple stimuli may advance a program of self-identification and self-selection through performance for the beginnings of a science talent pool. Later, early instruction may be modified into more sophisticated experimental procedure and well-ordered empiricism in the classroom and laboratory.

Idea-enactive, inquiry-oriented instruction becomes a first procedure in observing the young in early achievement in science. The complex of such behavior plus ability and achievement testing can then become part of a cumulative record, which can be compared and contrasted with *field-specific* demonstration of ability in science and mathematics. Formal testing per se is *not* to be the gate to entry into differentiated programs in science and mathematics. *If* final judgment on selection for differentiated instruction in science and math is withheld until late middle school, *after* the young have had the chance to identify themselves for it, *and* their choice is followed by consistent science-specific *works,* then we have a better picture of in-context potential signaled through performance.

National, federal, state, and local, nonprofit and proprietary, industrial and post-secondary groups are joining American's schools to advance the skills of teachers and the quality of instruction for K–12 science and mathematics. The pool of well-schooled and educated young may then increase and so too the science prone.

Enabling Achievement—A Curricular Approach for Self-Identification in Conjunction with Instruction

Siegler and Kotovsky (1986) posit that "the fit between the individual and the field is important for both intellectual and motivational reasons. A superior fit allows the individual to learn quickly and deeply the material in the fields" (p. 419). An essential element in this "fit," which in turn is necessary to the creation of a significant science talent pool, is the function of curriculum and its congruent instructed learning as valid identifiers of science talent. This study concludes that the science talent pool is incomplete until those at promise are assessed through several exemplars. The science prone give evidence of two qualities: They early show exceeding competence in acquiring knowledge in a specific field, and they early perform excellently demonstrating their powers of originative inquiry in a work. Because gifted young can begin to demonstrate heightened capacities in earliest schooling, they should be given opportunity to fulfill them in pursuit of excellence. In the particular terms of this study, they need a chance to demonstrate their science proneness.

At present, however, high school is mainly where further expression of science proneness and/or talent is empowered. Three major exemplars in the design of augmenting high school environments designed for those with promise in science are identifiable: a pervasive exemplar, another fast-paced in content, a third based in originative inquiry and enriched in acquisition of knowledge. In spite of the efforts of current reformers, most high school science programs still follow the traditional, pervasive mode in curriculum and approaches to instruction not only in the United States but also in most of the Western world. Practiced in different intensities in various high schools, this pervasive mode is based in a lecture/prepared laboratory mode with foretold conclusions generally accompanied by limited discussion. The lecture-textbook mode remains basic to instruction.

This exemplar held, in most of the observations I made in 600 schools from the 1930s to the 1980s. It is still the road to the credential to enter college and university as well as to graduation from the university. In turn, this credential opens doors to further participation of the novice scientist in the originative inquiry that adds to science and technology. As the United States tries to make its students first in the world in mathematics and science, a number of curricular constructs have emerged and are emerging. What is new now, however, is instruction not curriculum. *If*, however, changes in design are introduced for the succeeding years of study—in the complexities of mathematical treatment, in computer-related inquiry, or by the science prone's compacting of subconcepts or using college textbooks in rigorous high school programs—*then*, the curriculum would *actually* be augmented in content.

In a modified philosophical approach (and, therefore, possibly a changed epistemic or axiological emphasis) in curriculum and instruction, these *stable conceptual schemes* (Kuhn's "paradigms," 1970) remain in context within a newer view predicated by the culture. The emphasis on science, technology, and society would offer a different face to the curriculum, however. In an overall updated approach to science, the nuances of a changed philosophy and, thus, a new view of the function of science in culture and society would call for an innovative instructional stance.

Besides the science, technology, and society curriculums, the American Association for the Advancement of Science (1993) is at work on *Science for All Americans*, revising approaches to science K–12. The National Science Teachers Association *Scope, Sequence, and Coordination* (1992) offers a complementary approach for the middle school–high school years. Both these curriculums are influenced by the

National Council of Teachers of Mathematics' (1991) groundbreaking *Standards*. And the National Research Council is currently writing standards for precollege science. A number of federal initiatives are also underway.

All these approaches to reform furnish at least three clear positions to those who frame explicit curricular and instructional designs:

- First, curriculum and instruction should advance the scientific literacy of the young. The imperatives of this issue are stated in clear, unmistakable aims and ends.
- Second, teachers and learners should be involved in activities that join science and technology to relevant social issues. In this, the newer technologies of science education—calculators, computers, interactive videodisks—are vital.
- Third, the needs of various populations of students—namely females and underrepresented minorities—often lacking scientific literacy are brought into focus.

All the frameworks stress the idea-enactive, inquiry-oriented mode of teaching and learning—postulated here as central to instruction for *all* young—that enables the science prone to identify themselves for advanced study. Particular refinements of course content and approaches for the science prone fit readily in fast-paced and originative augmenting frameworks. Fast-paced subject matter in elementary school can lead to originative inquiry in the high school years.

My research has shown that precollege science instructional materials, whether formed in textbooks, computer programs, or initial inquiry procedures, have been cloned in conceptual structure from the curriculum structures (made into textbooks) created by the various committees at work during the curriculum reform period (1958–1962). Approaches created by scientists and teachers of the Sputnik era still appear in the textbooks of present publishers. The additions concern new discoveries and cycles of crises; the rigorous treatment has diminished, however.

This pattern holds K–12, except where videodisk technology and, at times, computers and hand-held calculators have been introduced. Future changes in design for the science prone may occur in great part by augmentation through the new possibilities of integrated mathematics and science made possible through computer-assisted instruction and inquiry. Such enrichment could also take place through the compacting of subconcepts or through college textbooks used by the science prone in rigorous high school programs.

In sum: Differentiated programs are necessary for evaluation and identification of science-prone and science-talented young because special curricular and instructional devices are favorable to cultivating and evoking desired abilities. Whatever the mode of selection of qualified students, their *performance* in an enabling environment differentiated to fit various abilities and skills is the most valuable identifier of future ability in science, whether expressed by the scientist or the artisan to be.

In the case of the science talented, the teacher and students reinvent the curriculum as they proceed. The dyad of curriculum and instruction as enabling environments for talented young then needs to be as innovative as are the young who will benefit from it. For they may change its future form and function. An environment in which the young discover for themselves, whether through the guided discovery of teachers or the initiative of science-prone learners, is part of idea-enactive, inquiry-oriented teaching and learning, an approach that counteracts the syndrome of 10 inhibiting enabling curricular and instructional practices. Further, the idea-enactive, inquiry-oriented teaching model engenders activities that can and do serve as identifiers of science proneness in the young.

Three inferences follow:

- First, the structure of curriculum and the mode of instruction in classroom and laboratory serve to identify science proneness, an understanding that suggests a significant way to increase the science talent pool.
- Second, the widest net ought to be flung to open opportunity for all young in an idea-enactive, inquiry-oriented learning curriculum and instruction. This generous cast offers access to equal opportunity for self-identification, along with but not exclusively through ability and achievement testing, as composite factors for entry into the science talent pool.
- Third, exemplars distinguishing three schools of thought indicate science proneness and/or science talent: (a) fast-paced instruction (earlier than usual exposure to courses) with abilities measured in achievement testing; (b) originative inquiry as an in-context measure resulting in a work considered to be a criterion sample of prospective science talent; (c) the pervasive exemplar of curriculum and instruction in U.S. high schools, with augmenting modes in acceleration and enrichment, scholarships, and rewards.

This last (college-preparatory) model now furnishes most of the cohort composing the science talent pool and remains the matrix for present innovations in schooling. The fall-off of young with interest in science before graduation from high school and after the freshman year of college, however, is a definite cause for concern.

A newer model suggests itself. It modifies the pervasive exemplar, making provision for a differentiated curriculum and mode of instruction suited to the needs of the science prone and leading to the expression of science talent. Select science schools are increasing in number as are select programs for the science prone in heterogeneous schools.

New frameworks in curriculum, as well as new technologies, are available, but all will require modification and augmentation to fit the abilities of the science prone on their way to demonstrating talents. New technologies in science education promise certain advances in independent study and inquiry-oriented teaching and learning.

The preparation of present programs, defined by the National Education Goals (*Building a Nation of Learners*, 1991, 1992, 1993, and *The National Education Goals Report*, 1994) and designed to augment abilities in science and mathematics as well as to secure an increase in the science talent pool by the turn of the century, is only beginning. Noted throughout this study are national and local initiatives calling for an increase in resources to support the capital expenditures needed for the teaching of science—as well as the need for a full complement of teachers skilled in science and mathematics.

Enabling Early Self-Identification of Science Talent

The penultimate section of this study proposes

- to suggest a mode by which students identify and select *themselves* to participate in differentiated programs of demanding study culminating in long-term originative inquiry
- to report observations of the young in the activities of inquiry to identify certain correlative behaviors
- to argue that, by submitting their work to examination and external evaluation by qualified scientists, students experience the peer review and tests of validity to which works in science are traditionally subjected

This section will also define a working exemplar encompassing these purposes. Originating in the late 1930s, this exemplar has gained support through usage and has accumulated a weight of evidence through constant evaluation. Study of this exemplar's analysis, synthesis, observations, and findings supports recent theories and findings.

When the young enter into the climate of science, they should benefit from at least two resources as gifts of schooling: First, they deserve access to the *substance* of science, a rich, even massive, conceptual structure of cumulative knowledge. Second, they deserve opportunities to participate in problem finding and concept seeking and forming—that is, to experience the *style* of science—its particular modes of inquiry and explanation. With these twin thrusts in mind, in the 1930s, 1940s, and 1950s, I organized curriculum and instruction encouraging the acquisition of advanced, rigorous, structurally organized knowledge, along with its companion, originative inquiry. Students solved unknowns through commitment to long-term individual probes.

My convictions about the essential value of originative inquiry programs to high school instruction in science grew from my own early experience in scientific research. The disparity between school science education and the working world of the scientists who taught me when I was young and brought me to the adventure of inquiry was apparent. I tried to set up a secondary science program close to the reality of working scientists and found that certain young—not all—were eager to give it a try.

At George Washington High School in upper Manhattan and at Forest Hills in Queens (both heterogeneous New York public high schools accommodating all students in their residential area), I made trial-and-error attempts to develop a differentiated curriculum and mode of instruction to give full opportunity to the capacities of a variety of students attending a general high school. We outlined the program at George Washington (1937–1940), later took it on a dry run there (1942–1944), and then used it experimentally at Forest Hills (1944–1954). Our program saw its fullest development at Forest Hills, and I was able to offer a first hypothesis (1947), a theory I developed more fully in the ensuing years as a result of continuing study (1951, 1955/1981, and 1988).

Science Talent in the Young Expressed Within Ecologies of Achievement

In such programs as that conceived at George Washington High and maturing at Forest Hills High, the young undertook research-productive, originative inquiry resulting in new knowledge, testable and falsifiable by the template of processes and procedures of mature scientists. Their achievements, written with the signature of the scientist-to-be, reflect the philosophy, the observable behavior, and the methodology of science.

In a paradigm evoking science talent, the three intereffective elements—students, teachers, and the other individuals and entities making up ecologies of achievement—support curricular and instructional methodologies that allow self-selection and identification through the methods of originative inquiry. These elements cannot be considered apart: They are an inseparable, entwined, connected, and intereffective whole. The paradigm then describes the "methods of intelligence" (Bridgman's phrase, 1945) within the "human ecological structure" (Tannenbaum's phrase, 1983). The behaviors of the scientist-to-be emerge in certain processes and procedures demanded by the constructivist experimental mode of originative inquiry and suffused by the processes of critical thinking.

Almost *never*, in my personal work with some 26 scientists prior to teaching, with 14 more during the Sputnik crisis, and with the 354 young doing originative exploration between 1944 and 1954, did I note their paths following the procession of steps of the so-called "scientific method." On the other hand, often with Bruner's "effective surprise" (1966), I saw brilliant mental breakthroughs—evidence of methods of intelligence beyond the capacity of published tests of creativity. Bridgman made this point decades ago when he wrote that the scientist, in attacking a specific problem, suffers no inhibitions or precedent on authority but "is free to adopt any course that his ingenuity is capable of suggesting to him ... In short, science is what scientists do, and there are as many scientific methods, as there are individual scientists" (1949, p. 12). In teaching and learning, students may see the limitations of the "empirical approach" (Conant, 1952).

But scientists seem to value knowing what's wrong as much as what's right: both spur them on.

The young at Forest Hills who presented their experiments in scheduled seminars faced penetrating questions not only from the apprentice scientists and their teacher-mentors but also at times from visitors from nearby colleges and universities. These seminars evoked critical examination of problems, hypotheses, processes, and led to next steps. And finally, if the young experimenters wished to present their papers to the Westinghouse Science Talent Search, panels of practicing scientists probed their defenses of processes, of explanations, of—in fact—the caliber of their thinking. Their papers were at times published by the Talent Search, which often followed-up with reports on the careers the students eventually chose after winning the competition.

VanTassel-Baska (1984) pointed out that "the Talent Search focuses much more sharply than most identification protocols on self-selection or the volunteerism principle. The commitment to the Talent Search and to follow-up procedures must be made by students and parents in order for the identification to occur" (p. 175). Former principal of Bronx Science, Kopelman, explaining why—of all the awards his

254　　　　　　　　　　　　　　　　　　　　　　　　　　　　　　　　　P.F. Brandwein

students won—he announced only the Westinghouse said, "A young person has to involve himself for a prolonged period in a piece of work and then do a *research paper* on it. Then the work is judged by *research people*. That's very special" (Phares, 1990, p. 53).

The special science schools, with their students selected for entry by examination, and heterogeneous schools, with differentiated programs within a curriculum open to their residential populations, did about equally well until the late 1980s in producing Search winners and runners up. And, as Search Records amply show, many finalists went on to significant careers in science, mathematics, and technology. Their awards include, for example, five Nobels, two Fields Medals, and nine MacArthur awards. Seventy percent of the winners earned a Ph.D. or M.D.

In short, select science schools and heterogeneous schools constitute different ecologies of achievement, both capable of encouraging significant originative work in science. The paradigm of originative inquiry is a way of identifying promise in students who might tend in the future to choose a career in science. As such, it deserves a firm place in differentiated curriculums in science.

My direct observation of the behaviors of the young undertaking originative inquiry in the environments of teaching and learning led me to discard the Cartesian concept of one-to-one correspondence of cause and effect and to develop a triad as a working hypothesis:

> High-level ability in science is based on the interaction of several factors—genetic, predisposing, and activating. All factors are generally necessary to the development of high-level ability in science; no one of the factors is sufficient in itself. (1955/1981, p. 12)

Originative inquiry calls on *general* and *special* abilities. One of the nonintellective factors is persistence, which Roe (1953) noted in selected working scientists and I (1955/1981), in the young. Tannenbaum (1983) pointed particularly to dedication and will. Environmental factors are also important, including, of course, the chance to attend a school whose opportunities included originative inquiry.

If the evidence here supports the studies of Renzulli (1978) and Sternberg (1985), both asking for reality-based intelligence tests, as well as Tannenbaum's (1983) psychosocial theory, then producing a work through originative inquiry may well measure science talent. Perhaps this finding has broader applications. Perhaps the procedures of originative research by adolescents could also indicate talent in other domain-specific fields open to originative inquiry.

Within an Ecology of Achievement— A Conception of Science Talent

Scientific judgments, concepts, and findings of fact must be testable, and thereby verified, falsified, or amended through commonly accepted processes within a community's structure. Thus, scientists and scholars seek to transmit, correct, conserve, and expand the substance of a field to achieve a continuity of cumulative knowledge.

The community is usually tightly knit, given over to a particular subset of a domain (say, astronomy, biophysics, zoology, ecology, organic chemistry, ophthalmology, computer science, psychology, genetics, and the like).

Talent in science is not general. Even in the young, it may be centered in biology, physics, or chemistry, and later it is almost always shown in works undertaken within matrices—often extremely specialized ones—in given fields. Then, as required, the findings are communicated to a body of scientists through specific modes: journals, associations, and meetings. These procedures are self-energizing: The substance in all scientific works coming out of originative inquiry is subject to a well-understood style.

One of the most striking features of science talent identified in the acts of discovery is the scientist's unrelenting persistence over time. Succeeding generations create their works in part through building on prior findings. Scientists stand *on* the shoulders of others even as they stand shoulder *to* shoulder within the life-sample of a generation of discovery.

In the spirit of Bridgman's "methods of intelligence" (1949), then, this operational definition follows: Science talent in high school students is demonstrated in originative works rooted in the self-testing and self-correcting code of scientific inquiry.

The definition stems from the essential methodology of the scientist: Originative inquiry leads in its successful end state to a work that encompasses the methodologies that inspirit it—and quarrels with none. This is the premise that has affected practices within 48 states and a large body of teachers and their colleague-scientists and 50-odd years of judging by the many panels of scientists who have evaluated submissions to the Westinghouse Science Talent Search.

Talent in science is unlike that in music, art, or mathematics—where specialized aptitudes can be readily recognized in the young (Csikszentmihalyi & Robinson, 1986). Science proneness begins, I believe, in the base of a general giftedness and develops its component skills in verbal, mathematical, and, in time, the nonentrenched tasks of problem seeking, finding, and solving in specialized science fields. Eventually, given favorable ecologies, science proneness can shift to an expression in a work showing science talent.

This definition of talent in science calls for identification through in-context evaluation in long-term inquiry without reference to IQ or standardized tests of achievement. It provides for testing of science talent through a criterion sample of work of the young as predictive of their future accomplishments (Feldman, 1974, 1986; McClelland, 1973; Renzulli, 1977; Tannenbaum, 1983; Wallach, 1976).

The following sequence shows a portent of science talent in young demonstrating focused high-level ability in both acquisition of knowledge and a capacity for inquiry:

> ***First,*** during the early school years, some children exhibit raw, unfocused giftedness: Their amorphous potential seems in search of a purposive expression of talent.

Second, like others' signs of a preference for music or art, some students exhibit a definitive focus toward science. Thus the science prone may shift from showing raw ability to demonstrating domain-specific interests, not necessarily excluding their attraction to other fields.

Third, given a choice later in high school (without pretest), such young may select themselves for participation in a course of study that calls for rigorous acquisition of knowledge and offers opportunity for research-productive originative inquiry.

Fourth, such young may complete an originative work and submit it to a definitive test: The scrutiny of a panel of scientists.

Such students have a solid conception of *themselves*, are secure in their *self constructs*, and employ *transformative power* (Gruber's terms, 1986). They make a choice among the potentialities claiming their recognition within self. Further experience may highlight other choices—for there are talents still to be discovered in individuals seeking excellences as yet unknown or untested. This conception embodies giftedness not as a freefloating, generality-seeking definition but as an end state in a domain-specific talent. *It is easier to measure talent expressed in a work, talent that presupposes a certain giftedness, than to try to infer from general giftedness raw traits that will project a specific talent.*

A *powerful program of teaching and learning can be—or should be—a transforming experience and engage as catalyst the young in the shifting from gifts into talent*. This conception lies within the postulates of Feldman's stage-shifts in the development of talent (1982), Gruber's formulation of "transformative power" (1986) as comprising giftedness into creativity, Renzulli's enrichment model (1977), and Borland's (1989) and Tannenbaum's (1989) conception of curriculum as identifier of talent.

We are not limited by inherited behaviors. Learned behaviors can engender connection and interpenetration of seeming opposites; the brain can hold alternatives (Bateson, 1979; Toulmin, 1977). Human behavior cannot be posited either as pure hereditarianism or as pure environmentalism; the two mingle inextricably (Gould, 1981, among many others).

In sum: A triad of inseparable factors can result in the expression of science talent:

1. students with promising intellective and nonintellective factors (MacKinnon, 1962; Tannenbaum, 1983)
2. teacher/mentors with high-level abilities and personalities necessary to develop the optimum instructional and curricular environment
3. the three ecosystems that support necessary curriculum, instruction, and physical facilities

Human talent leaps out of its definition and redefines itself in more formidable expression. In time, the community of scholars engaged in research will probably decipher the human genome, particularly in its specificity in identifying the

DNA components of intelligence. In time, the newer insights of the neurosciences will uncover how the meshing of physical, chemical, and physiological functions of neurons, synapses, and neurohumors function in intellection and how they create a thought, an idea, a letter, a musical notation, or a concept. In time, scientists will unearth how the three-pound brain with its 10^{12} or 10^{14} neurons and, possibly, 10^{24} synapses creates the encompassing mind. In time, researchers will develop a social invention that assures equitable access to fulfillment of human worthwhileness to unimpeded limits in pursuit of individual powers of excellence.

In time, then, we will see that what seems to remain true longest in the human scheme is that the young keep coming. And, in time, one or more of the young—always together with one or more of the old—will discover how to do what seems to escape us only to the time of its discovery. As long as the young keep coming, a surer conception of talent is foretold. As long as the young keep coming, so does the permanent agenda to search for superordinate ecologies of achievement.

My Path to This Study

Half a century of observation and study of school-communities have led toward the conclusions offered here about certain ways of stimulating students prone to science to expressing talent in its wide-ranging fields. During those years, I was fortunate in opportunities to study both scientists at work and scientists in the making. Generous latitude in time and resources for studies of methodologies in scientific research and for pertinent observation and testing of curriculum and instruction in schools, colleges, and universities allowed me to study in-depth programs and practices for the science prone and science talented.

Over a third of a century, making an average of 36 school visits per year of observation and investigation to about 1,000 schools, I clarified the conception that underlies this study of the *ecology of achievement* that is the result of the family-school-community ecosystem acting in mutualism with the cultural and university ecosystems. Further, through my study of 600 institutions representative of the broad spectrum of American schooling, I saw directly the disparities in resources and factors that affected curriculum and instruction, teaching and learning, within limiting and enabling environments.

In the planning and start-up operation of some 93 programs designed to evoke science talent, I refined my understanding of the major problems and first solutions in the conduct of family-school-community programs for the talented in the sciences and humanities in the United States and overseas.

A distillation of my studies and observations over 50 years comes together on these pages. Here are offered certain of the tested, revised curricular and instructional policies and practices useful in planning programs for developmental stage-shifts from general giftedness→science proneness→an early expression of science talent in the secondary school years.

References

American Association for the Advancement of Science. (1993). *Project 2061: Benchmarks for science literacy*. New York: Oxford University Press.

Bailyn, B. (1960). *Education in the forming of American society*. Chapel Hill, NC: University of North Carolina Press.

Bateson, G. (1979). *Mind and nature: A necessary unity*. New York: Bantam.

Bell, D. (1973). *The coming postindustrial society: A venture in social forecasting*. New York: Basic Books.

Borland, J. (1989). *Planning and implementing programs for the gifted*. New York: Teachers College Press.

Brandwein, P. F. (1947). The selection and training of future scientists. *Scientific Monthly, 545*, 247–252.

Brandwein, P. F. (1951). The selection and training of future scientists. II. Origin of science interests. *Science Education, 35*, 251–253.

Brandwein, P. F. (1955). *The gifted student as future scientist: The high school student and his commitment to science*. New York: Harcourt Brace. (1981 reprint, with a new preface) Los Angeles: National/State Leadership Training Institute on Gifted/Talented.

Brandwein, P. F. (1981). *Memorandum: On renewing schooling and education*. New York: Harcourt Brace Jovanovich.

Brandwein, P. F. (1988). Science talent: In an ecology of achievement. In P. F. Brandwein & A. H. Passow (Eds.), *Gifted young in science: Potential through performance* (pp. 73–103). Washington, DC: National Science Teachers Association.

Bridgman, P. W. (1945). Prospect for intelligence. *Yale Review, 34*, 450.

Bridgman, P. W. (1949, December). Scientific method. *The Teaching Scientist, 6*, 342.

Bruner, J. (1966). *Toward a theory of instruction*. Cambridge, MA: Harvard University Press.

Conant, J. B. (1952). *Modern science and modern man*. Garden City, NY: Doubleday Anchor.

Cremin, L. A. (1980). *American education: The national experience 1783–1876*. New York: Harper and Row.

Cremin, L. A. (1990). *Popular education and its discontents*. New York: Harper and Row.

Csikszentmihalyi, M., & Robinson, R. (1986). Culture, time, and the development of talent. In R. J. Sternberg & J. Davidson (Eds.), *Conceptions of giftedness* (pp. 264–284). New York: Cambridge University Press.

Feldman, D. H. (1974). Universal to unique: A developmental view of creativity and education. In S. Rosner & L. Abt (Eds.), *Essays in creativity*. Croton-on-Hudson, NY: North River Press.

Feldman, D. H. (Ed.). (1982). *Developmental approaches to giftedness and creativity*. San Francisco: Jossey-Bass.

Feldman, D. H. (with A. Benjamin). (1986). Giftedness as a developmentalist sees it. In R. J. Sternberg & J. Davidson (Eds.), *Conceptions of giftedness* (pp. 285–305). New York: Cambridge University Press.

Gould, S. J. (1981). *The mismeasure of man*. New York: W. W. Norton.

Gruber, H. (1986). The self-construction of the extraordinary. In R. J. Sternberg & J. Davidson (Eds.), *Conceptions of giftedness* (pp. 247–263). New York: Cambridge University Press.

Havighurst, R. (1972). *Developmental tasks and education* (3rd ed.). New York: McKay.

Kuhn, T. S. (1970). *The structure of scientific revolutions* (2nd ed.). Chicago: University of Chicago Press.

MacKinnon, D. W. (July 1962). The nature and nurture of creative talent. *American Psychologist, 17*, 484–495.

McClelland, D. C. (1973). Testing for competence rather than for intelligence. *American Psychologist, 28*, 1–14.

National Council of Teachers of Mathematics. (1991). *Professional standards for teaching mathematics*. Reston, VA: Author.

National Education Goals Panel. (1991). *Building a nation of learners*. Washington, DC: Author.

National Education Goals Panel. (1992). *Building a nation of learners*. Washington, DC: Author.

National Education Goals Panel. (1993). *Building a nation of learners*. Washington, DC: Author.

National Education Goals Panel. (1994). *National education goals report*. Washington, DC: Author.

National Science Teachers Association. (1992). *Scope, sequence, and coordination: Relevant research*. Washington, DC: Author.

Phares, T. K. (1990). *Seeking—and finding—science talent: A 50-year history of the Westinghouse Science Talent Search*. New York: Westinghouse Corporation.

Renzulli, J. S. (1977). *The enrichment triad model: A guide for developing defensible programs for the gifted and talented*. Mansfield Center, CT: Creative Learning Press.

Renzulli, J. S. (1978, November). What makes giftedness? Reexamining a definition. *Phi Delta Kappan, 60*(3), 180–184, 261.

Renzulli, J. S., Reis, S., & Smith, L. (1981). *Revolving door identification model*. Mansfield Center, CT: Creative Learning Press.

Roe, A. (1953). *The making of a scientist*. New York: Dodd, Mead.

Schultz, T. (1981). *Investing in people*. Berkeley, CA: University of California Press.

Siegler, R., & Kotovsky, K. (1986). Two levels of giftedness: Shall ever the twain meet? In R. J. Sternberg & J. Davidson (Eds.), *Conceptions of giftedness* (pp. 417–435). New York: Cambridge University Press.

Sternberg, R. J. (1985). *Beyond IQ: A triarchic theory of human intelligence*. New York: Cambridge University Press.

Tannenbaum, A. J. (1983). *Gifted children: Psychological and educational perspectives*. New York: MacMillan.

Tannenbaum, A. J. (1989). Probing giftedness/talent/creativity: Promise and fulfillment. In P. F. Brandwein & A. H. Passow (Eds.), (1989), *Gifted young in science: Potential through performance* (pp. 39–55). Washington, DC: National Science Teachers Association.

Toulmin, S. (1977, June 9). Back to nature. *New York Review of Books, 24*, 3–6.

VanTassel-Baska, J. (1984). The talent search as an identification model. *Gifted Child Quarterly, 28*(4), 172–176.

Wallach, M. A. (1976). Tests tell us little about talent. *American Scientist, 64*, 57–63.

Yaeger, R. E., Aldridge, B. E., & Penick, J. (1983). Current practices in school science today. In *Science teaching: A profession speaks* (National Science Teachers Association Yearbook) (pp. 3–11). Washington, DC: National Science Teachers Association.

Part III
The Surveys

Remembrances from More than a Half-Century Back: The Surveys

Deborah C. Fort

As I began to analyze the 26 surveys that form the basis for this report on people influenced by Paul F. Brandwein (mostly his former high school biology students), I found that our[1] original purpose—to try to discover from the surveys we have gathered whether Paul's theory about what might turn gifted students toward science makes sense—has produced some striking but not always consistent results.

Paul's influence on those of his former high school students we have been able to track down and survey was profound in many ways. Eighteen of these 24 high school students became either physicians or academic research scientists; 4 others went into educational administration and/or teaching, particularly at the primary (and in one case nursery school) level; and 2 chose nonscientific professional fields. (In addition, we made surveys available to some of Paul's colleagues and received two responses.)

Paul had hoped to be able to follow up on his students as they matured to see how many of them who in high school enjoyed the factors he defined in *The Gifted Student as Future Scientist* (reprinted in this volume) actually chose scientific or technological careers. In that book, he wrote,

> Three factors are considered as being significant in the development of future scientists: a Genetic Factor, with a primary base in heredity (general intelligence, numerical ability, and verbal ability); a Predisposing Factor, with a primary base in functions which are psychological in nature; an Activating Factor, with a primary base in the opportunities offered in school and in the special skills of the teacher. *High intelligence alone does not make a youngster a scientist.* [Italics added.] (1955/1981, p. xix)

A bit later, he added, "*All factors are generally necessary to the development of high-level ability in science; no one of the factors is sufficient in itself*" (1955/1981, p. 12).

In 1983, Paul tentatively concluded that his observations (1955/1981) confirm his many findings that these factors still need to be considered in assessing the schooling

---

[1] This project would not have been possible without the generous contributions of my colleagues Richard Lewontin, James P. Friend, and the late Walter G. Rosen, all of whom were Paul's students at Forest Hills High School a half-century ago. Deeply grateful acknowledgment for the help on this section proffered by my surviving collaborators James P. Friend and Richard Lewontin. In spite of their best efforts, mistakes no doubt survive. They are mine alone.

D.C. Fort, *One Legacy of Paul F. Brandwein*, Classics in Science Education 2, DOI 10.1007/978-90-481-2528-9_24, © Springer Science+Business Media B.V. 2010

and education that nurtures scientists-to-be. He further defined the predisposing factor as including "questing" (or a skeptical view of accepted positions, which also embodies what is generally called curiosity) and "persistence" (or a degree of independence that sustains effort) and the activating factor as a "key [environmental] factor" (or the provision of a space overseen by a sympathetic teacher and/or mentor, which offers the psychological safety and freedom necessary for experience in experimental work). True experimental work, which calls forth "questing" and "persistence" within an environment of "psychological safety and freedom," can indeed be useful in probing new knowledge (1983, pp. 53–54).

Later in the same essay—"Do We Expect School Science to Nurture Creativity?"—Paul wrote,

> In our continuing effort to secure a tenable hypothesis that might shed light on the self-selection of scientists-to-be, we note this point: Of a group of 624 students who participated in the science program at Forest Hills, New York (1955), 62 were selected for further observation as the experimental group (rating above 4 on an inventory of traits comparing them to working scientists); and 62 served as controls, rating below 3 on the inventory. The 62 experimentals and the 62 controls were matched in I.Q. and general scholastic average.
>
> At this point, some 30 years after the initial study, from [our] still-preliminary study of those who have been followed, it appears that 22 of the experimental group of 62 have committed themselves to scientific research; and 13 have committed to technological fields in the area of science. Twelve committed themselves to teaching science in the high schools. Among the 62 "controls" (tentatively selected as probably intending careers other than science or technology), 6 are in scientific research, 8 are in technological fields, 6 are in teaching science in the high schools. An unsurprising but useful finding is: Out of 67, I have been able to observe at work and to interview at this period in the continuing study, the 28 working scientists; and 11 out of the 19 engineers are people who in their schooling were persistent and almost indefatigable *in pursuing experimental work probing unsolved problems.* They used an inquiry approach on the highest levels.
>
> The operation called "doing an experiment" involved work in solving a problem for which a solution was *not* in the literature. This was so attested by working scientists. Further, in interviews, the 28 working scientists were still able to recall their early pleasure in pursuing experimental work. (This, of course, calls upon abilities and traits of personality different from those required by the usual laboratory exercise.) They also recalled the effect this had on their decision to pursue a scientific career. (1983, pp. 54–55)

So, some 30 years after high school, of the "experimental group" of 62 former high school students, by then in their 50s, some 47 were working in science, technology, or education,[2] compared to 20 of the controls in those fields. Paul's hypothesis appears sustained in that three-quarters of the study group were successful professionals in science and engineering while fewer than a third of the control group were.

In 2000 plus, we are again looking for the former Brandwein students, now in their 70s, even 80s. Of course, not everyone has survived, and they have been difficult to track down. Still, we had surveys (see Part IV, Appendix A, for the survey

---

[2] Although the original experimental group was not selected with an eye to finding future *educators,* given his own work as a teacher, teacher of teachers, and textbook writer, Paul would surely have been gratified to find that 12 of them had headed in that direction.

questions) returned from 25 Brandwein alumni, including one former graduate student (and the 26th respondent, a science-education colleague). Also, spouses and friends of deceased former students and colleagues contributed insights. (In the few cases I cite these, they will be clearly differentiated from the words of alumni formally surveyed.) Some nine alumni contributed both surveys and essays. The essays stand alone in Part II.

In any case, some of their insights, with attribution if so granted (as it was in all but a few cases), will be reproduced here. See the introduction to this volume for a summary of the essays from the others—scientists, science education professionals, editors, and teachers—who knew and were influenced by Paul either as their teacher, their mentor, their colleague, their friend, or a combination of all those relationships. Only a couple of the science educators, who were Paul's colleagues, filled out surveys, most preferring instead to contribute papers about their interactions.

Like Paul and many others (see, for example, Stephen Jay Gould's *The Mismeasure of Man* [1981] and Richard Lewontin's many works on the subject), I share a distrust of IQ as a measure of worth. In another context, Paul put it well: "Until you have *given* a gift, you are not gifted." In other words, potential *without* performance is not particularly valuable (if at all).

Paul did not distinguish between the contributions made by "scientists" and "technicians," whose work he saw as inextricably linked. In this position, he often found himself in the minority; it seems to me that as usual his view was and is sound.

Three Groups, Question by Question

The Students Who Became Scientists

The 18 members of this group (Forest Hills alumni who became scientists) we were able to track down and survey include William Bristow, MD, Eric Cassell, MD, Roberta F. Colman, PhD, Conrad Ellner, MD, James P. Friend, PhD, Jean Gansky, MS, Rhea Gendzier, PhD, Thomas Katz, PhD, James Ketchum, MD, Richard Lewontin, PhD, Walter G. Rosen, PhD, Andrew M. Sessler, PhD, Tom Schatzki, PhD, Lisa A. Steiner, MD, Herbert L. Strauss, PhD, Lubert Stryer, MD, Josephine Baron Raskind von Hippel, MD, and Bernard Weiss, PhD. Another Brandwein alumnus who went into science, Richard Goodman, PhD, contributed an essay rather than focusing on the survey. (See "Brief Encounters, Lasting Effects" in this volume.) The six students and colleagues who became teachers or educational administrators and the two alumnae who went into other fields but filled out surveys will be considered separately. (See below.)

Most of Paul's former students who became scientists, both researchers and practitioners, wrote in response to our first preliminary question—*I. In what context and when did you know Paul F. Brandwein?*—that they knew him as their high school biology teacher at Forest Hills High School (New York) in the 1940s and 1950s. Two encountered him not as their teacher but as an important figure in the high school's

266 D.C. Fort

science department, where he served as chair, and/or as adviser to the projects they submitted to the Westinghouse Science Talent Search.

A number of these former students mentioned the importance of identifying themselves as a member of Paul's chosen group, students who had access both to each other and to a room full of biological equipment available during and after school hours and with whom he on occasion walked from school to his apartment. Some of these respondents recalled setting up equipment for the next biology class's students, an activity helpful both for the teacher lucky enough to get their help and for the students themselves, some of whom would eventually become teachers. Eric Cassell, now an internist, wrote, "Paul was very encouraging to his students. But more, he made being his student a group activity so that we were all pleased to be included in 'his' group. That made a feeling of being special—and also that he was special. (Which he clearly was.)" For a psychiatrist from a dysfunctional family, Paul's group of students was also essential. He wrote,

> I had come from a rather stolid and suffocating family environment and had not yet spread my wings intellectually or emotionally until the time of getting into Dr. Brandwein's class and into the biology group. This group really became my first family in a very deep sense. I went through a disconcerting and disorienting sense of loss when we graduated in 1946.

While a number of respondents mentioned the importance to adolescents of group acceptance in a special place, which is by no means unique either to the mid-century or to teenagers with a scientific bent,[3] only Lewontin remembers trying to recreate this aspect of Paul's approach professionally in later years. He wrote,

> I have tried to make my teaching into a unique intellectual exploration of genetics, evolution and statistics, usually avoiding textbooks, and certainly disagreeing with them and organizing the knowledge in a completely different way than what is standard. My research has been dominated by skepticism of received knowledge and a demand for the greatest possible methodological rigor. One consequence is that I have done a lot of work in the philosophy of biology. I organized my workplace into a social space with a large central atrium off which all the offices of graduate students, postdoctoral fellows, visitors and me opened and with no closed doors.

Paul and most of the former high school students we located fell out of touch after the Forest Hills years. The two who remained in close contact are my co-researchers on this survey, the late Walter G. Rosen and Lewontin, both of whom worked on a Biological Sciences Curriculum Study project, which tried to uncover promising uninvestigated biology projects for secondary students (Brandwein, 1962). In addition, Paul remained in contact with Rosen while the latter was in graduate school. Lewontin's final encounter with Paul in the 1960s was not a happy one:

> ...After I became a professional biologist, we had a number of contacts, chiefly as a consequence of his work as an editor at Harcourt Brace. His editorship there involved working

---

[3] This statement I offer not only anecdotally (based on my own and others' memories of adolescence) but also with reference to many psychological analyses, such as those by Goldberg, Evans, and Hartman (2001), Chu (2005), and Swenson and Strough (2008). Of course, the phenomenon has also been treated fictionally, best in my experience in Margaret Atwood's *Cat's Eye* (1989), which also powerfully takes up the emotional toll involved in *not* belonging to the group of choice.

Remembrances from More than a Half-Century Back

with the Biological Sciences Curriculum Study, of which I was a member, in particular on the publication by Harcourt Brace of our collection of advanced experimental problems that high school students could work on. This connection resulted in the one serious disagreement I had with PB the last time we met in Chicago. It concerned the degree to which Harcourt Brace would concede to the pressure in states like Texas to leave out evolution from the high school texts. As a senior editor at Harcourt Brace, PB gave in to this pressure for commercial reasons to a much greater extent than I thought proper. This seemed to me particularly ironic since it was he who emphasized evolution in our high school biology class at a time when it was not a standard part of the New York State Regents curriculum.

It is also worth mentioning that Paul told me, sadly, that neither Florida nor Texas would buy a textbook that mentioned that the United States had in any way wronged Mexico by annexing large parts of it in the 1840s for Texas and California. As a publisher, he had, however reluctantly, to respect the bottom line. His excuse for compromise was that better students get *some* good science and history than *none*. He mourned that a "fine" text Harcourt published "lost *millions*."

Many respondents also answered our second question—*II. Would you be willing to share with us anecdote(s) about your interaction(s)?* James P. Friend's memory is particularly vivid:

On the second day of our biology honor class, Paul called on a girl to say what she had learned from the reading assignment. When she began talking in vague generalities and her discussion wandered from the topic, Paul stopped her by saying merely, "Please sit down." There were audible gasps from around the classroom and then dead silence. Paul went on to admonish the class to say only what you know to be true about a topic, to be able to say how you know it to be true (cite references), and finally to demonstrate why you know it is true (quality of the references).

Toward the end of the honor class, after we had studied and discussed heredity, evolution, and the races of mankind, Paul spoke often and passionately about the terrible tragedy of racism. It was the first I had ever even considered the topic, not having encountered any examples of it in my life up to that time. It was his emphatic discourse that impressed me so that during my lifetime, when I did encounter it, I always remembered Paul's urgings for tolerance and understanding.

I think it was around the last day of class that he said to us, "When you have something to say to an individual, ask yourself, 'Is it true? Is it kind? Is it necessary?' "[4] Those words have never left my memory and I try always to follow Paul's advice.

Paul taught a class in laboratory techniques with a lot of emphasis on careful observations. One day he had each member of the class bring in a few ccs of urine for testing. He had us then pour the urine into a large battery jar on the bench in front of the room. When that was done he stood next to the jar and said, "I want each of you to do exactly what I do." He put a finger into the urine and then he put his finger in his mouth. So we lined up in orderly fashion and one-by-one put a finger in the urine and then into our mouths. As may be imagined, each person gave some expression of revulsion. When everyone had finished, Paul then went back to the jar and said, "You didn't observe me carefully. I put this finger into the urine and this other finger into my mouth." It was an unforgettable lesson!

---

[4]Editor's note: A number of Paul's former students besides Friend remembered variations on this piece of advice. When Paul quoted it to me, he added these questions as introduction: "*How* do we know it? How *well* do we know it? . . ."

Bernard Weiss wrote,

I have three memories. (1) He advised us to prepare for the written Westinghouse exam by learning the meaning of all the words in the indexes of college science texts. It helped. I have to admit that even though you cannot learn science by merely learning its vocabulary, it did stimulate my curiosity. (2) I do not remember him looking at me or at any other person he was talking to. He seemed always to be looking off into space, lost in thought. (3) On the day that the seniors were allowed to take over the school, I volunteered to teach his biology class. I did not know what I was getting into: The class ran itself. He would sit in the back of the room, and each student would make a comment about the lesson and then call on another who had raised his or her hand. Whenever I attempted to stimulate discussion by alluding to a point that had not been raised, I discovered that it had been discussed in a previous class. This was the first time I had been exposed to the philosophy of setting the table and letting the students feed themselves. I was 15 at the time, and this lesson is one that has stayed with me for life.

Paul's former students had very little knowledge of his biography, about which he was consistently secretive. What they did offer was either fairly self-evident (he had a PhD in biology from New York University; he taught at Teachers College of Columbia University, for example), or wrong—one student guessed that he was an "American-assimilated Jew," although he went on to admit, "I have no direct evidence of his ancestry." In fact, Paul was of Austrian origin and characterized himself as a devout and "discordant" Catholic. He once told me, "I believe in God, and, therefore, I don't go to church," continuing, "I am discordant with all churches." To one of the former students who did not go into science, he did provide some background biographical details. On this, see the introduction to this volume.

Genetics?

Moving from these preliminary questions into the 10-question survey itself (see Part IV, Appendix A) brought some fuller responses, appropriate for scientists who had *evidence* about their own biographies rather than *speculation* about Paul's. Our first question had asked about genetic components that might have predisposed respondents to science—*1. Discuss in a general [nonstatistical?] way the genetic gifts that may have predisposed you to your career.* (This question fit closely with our fifth question—*From what kind of environment did you come?*—which went into the possible effects of home environments on future careers.) Most respondents brushed off, often with significant reservations, the genetics question, devoting nearly twice as much space to their family backgrounds. I will follow their lead, discussing these two central questions—heredity versus environment; nature versus nurture—together before returning to questions 2 through 4.

In examining our question about genetic gifts, many respondents referred to the accomplishments of their parents, one also referring to those of his grandparents, looking at parents as "the vessels through which genes are passed to grandchildren" and adding ironically that his "maternal grandmother was a real reader, [who] also made some very good cakes." Wrote Rosen, "I had 'intelligent' parents, whatever

Remembrances from More than a Half-Century Back

that means; good health and normal physique, no doubt at least in part a genetic inheritance." After citing some of her extended family's accomplishments, Rhea Gendzier concluded, carefully, "What any of this has to do with genetics I leave to those of you who are more comfortable speculating about that than I am." Population geneticist Lewontin wrote tartly, "There is no credible evidence of genetic 'gifts' of character or intelligence. The only genetic gifts I have were (1) excellent physical health, (2) my Y chromosome in a sexist world, and (3) my Caucasian features in a racist world." Approaching the question more concretely, neurobiologist Lubert Stryer, a recent winner of the National Medal of Science, admitted that another valuable quality for a scientist "is not needing that much sleep—six hours a night or less is what I need," continuing, "It helps to have physical stamina: Research can be like a marathon."[5]

Although a couple of respondents cited their high IQs and test scores, one important scientist wrote he had few genetic gifts, adding, "my brother has a higher IQ, and he never did anything in life." Several respondents noted their ease with mathematics, one reporting that other members of his family became professional mathematicians or went into related fields. On the other hand, Cassell described himself as "mathematically impaired," arguing that his "genetic endowment was the ability to theorize ... and the ability to connect to other people, which I now know is fundamental to love and essential to the care of the sick (as opposed to surgery, for example)." A couple of respondents found curiosity and a "positive response to challenges" more important qualities than intelligence. "People start telling you about themselves. You start probing and learning. Is this genetic or environmental?" another scientist asked, not rhetorically.

It is unsurprising that thoughtful practicing scientists often find it impossible to separate genetic traits from environmental ones. Their mentor agreed.

Growing Up Out of School

Respondents offered much fuller responses to our question about their home environments, appropriately because, even though they chose to do work beyond what was required and committed themselves to extracurricular activities, the family has much more influence in shaping the young than their comparatively few hours spent in school. Question 5—which in turn divided into several parts—

From what kind of environment did you come?
Were your parents professionals? In what fields?
Were they readers? What sort of materials did they choose?
Did your home environment offer

---

[5]Stryer was interested to learn that he shared this quality with Paul, who needed but 5 hours a night.

A. *Financial comfort—did you have to worry about money or sustenance?*
B. *Support of academic endeavor?*
C. *Treatment of your emotional needs?*

attempts to probe the importance of this essential background. Both parents of most respondents—11 of the 18—were employed professionals. In two families, only the father worked for pay (a nonworking mother is an oxymoron). The parents of three respondents, hampered by their lack of education, supported their families through less prestigious jobs; one of Paul's former students failed to indicate what kind of work his "extremely destructive" parents did between their bouts of "marital discord." That Josephine Baron Raskind von Hippel became a psychiatrist is hardly surprising given her "very intelligent and well educated" parents' professions—both were physicians, her mother, also a psychiatrist. Three of our respondents were WWII refugees. Two families had fled Hitler; one particularly unlucky family had to run from three horrendous 20th-century despots—from Germany's Hitler, from the USSR's Stalin, and from China's Mao—before finally finding refuge in the United States, where their teenage son Lubert (Stryer) became Paul's student at Forest Hills and grew up to write a widely read textbook on biochemistry, a text that, according to his publisher, sold close to a million copies in six editions, which he dedicated, "To my teachers, Paul F. Brandwein, Daniel L. Harris, Douglas E. Smith, Elkan R. Blout, and Edward M. Purcell."[6] Stryer noted that Blout, the only one of the five who survives today, "helped me develop into the scientist who I am. Knowing him is a wonderful experience, which continues. We maintain contact once a month when I get to Boston." Stryer also visited Paul's wife, Mary, in recent years before she, too, died.

Although the large majority of our respondents' parents owned books and read them, none of them read books on science or mathematics. Friend, among many other respondents, reports that his parents' libraries contained "many of the great works in literature ... [but] no books on hard science" except for (Eric Temple) Bell's *Men of Mathematics.*

Many of our respondents grew up during the Depression of the 1930s and were aware, as Rosen put it, that "money was tight." "We always scrimped," wrote another. In addition, refugees from WWII's horrors often arrived with almost nothing but the clothes on their backs and had to struggle not only with a depressed economy but also with an unfamiliar country and language. Scholarships helped put a number of respondents through college and graduate school. Reported Stryer, for example,

---

[6]Stryer's textbook *Biochemistry* is widely used. Its some million copies reached many students worldwide in six editions over about a third of a century. Said Stryer, "It contributed to our understanding of how vision begins." Stryer is sole author of the first four editions; in the last two he is the third author, joined by Jeremy M. Berg and John L. Tymoczko. A seventh edition is planned.

We had lots of worries in America. We were reasonably well-to-do in China; then came the hard years. I am grateful to Chicago and Harvard for scholarships and for allowing me to work at the University of Chicago Faculty Club as a waiter. A check came from Paul, my benefactor, for $300 freshman year. That would be like $2,300 today.

When we moved to Stanford, my wife and I made gifts to students and informed them of the tradition that Paul started, asking them not to repay us but to serve the next generation.

Working Hard?

According to Paul's *The Gifted Student as Future Scientist* (1955/1981), youngsters who may be science prone have a tendency to persist in difficult tasks. Accordingly, we asked respondents, *Would you characterize your approach to the tasks you decide to take on as "persistent" or not? Do you typically labor long to accomplish a goal, or do your achievements come easily?*

The adults the students became overwhelmingly characterized themselves as persistent. Wrote Rosen, for example, "My PhD supervisor said I was like a terrier going after its prey in a burrow or up a tree, refusing to quit." On the contrary, a couple of respondents, two of whom characterized themselves as "lazy," said they tended to choose projects whose solution did not promise to require much effort. Wrote Baron Raskind von Hippel, "I don't labor too much. Luckily my achievements have come easily; however, I was taught that work is a necessary part of life, and good grades were expected. I am more an 'idea' person than a persistent 'working through' person."

And several said that different tasks evoked different responses. Thomas Katz wrote, for instance,

> Persistence? In some ways "yes" and in many ways "no." I have made significant discoveries in a number of subfields of chemistry, and I am proud of the precision and detail of the papers describing them. Producing these papers required long, very hard work, both in the lab and in the library. Like most experimental organic chemists, I worked very long hours. However, once the important experiment had been done, I usually have pursued only few corollaries and did not continue to mine others. A couple of years ago, in an essay describing some of my work, a writer for the chemical trade magazine, *Chemical and Engineering News*, hit it right when she called me a chemical vagabond. I wouldn't call a vagabond a persistent type.

Many Other Mentors

Respondents were generous in thanking the many mentors (between two and nine) who guided and helped them after they entered college or university and, eventually, professions.[7] A couple of respondents found their first mentors in elementary

---

[7]Because of the difficulty of verifying the respondents' many mentors' names, specialties, and locations, I am listing almost none of them here by name.

school; several listed Paul as their first and most important mentor lifelong. One distinguished professor of chemistry and biology with a CV of almost 30 pages, including over 200 publications, put it simply: "Paul Brandwein was the most inspiring teacher. My family had no experience in science, and he was the one who stimulated me to continue." Many spoke of their subsequent mentors in university and college and graduate school. Wrote Weiss of his mentors, "The most valuable thing they taught me was how to ask questions: what questions could be approached experimentally and what small questions might produce big answers. I was also influenced by their infectious joy in what they were doing." Mentors need not be ensconced in elevated positions of power, however. For example, while paying tribute to nine mentors of whom "Paul was the first and most important and inspiring teacher," Friend remembered,

> Oddly enough, in doing my doctoral research in microwave spectrometry, I learned much more about the science and instrumentation from a graduate student named Jerome Kraitchman (known for the "Kraitchman equations" in the quantum mechanics of rigid rotors). We became good friends, and he seemed pleased to help me learn all that I needed to run the spectrometer. This included skills in understanding and making electronic equipment and working in a machine shop. He was eminently qualified to do this since he had been a radar technician in the Navy during WWII. It was Jerry who helped me get into the science of spectroscopy and encouraged me to join professional societies, go to their meetings, and the like. I consider him my real mentor in graduate school.

And when Paul, then in his early 60s, was serving as *my* mentor in learning about publishing in particular and living in general, he told me that one of the men he was also helping was about his age, another an illiterate neighbor in his 40s. Still another Brandwein alumnus paid tribute to his mentor, the chair of his department:

> He was a truly great man, whom I have never stopped idolizing. It also makes me tearful to write that. He was honest and generous—not such common traits in academic medicine. He taught me, by making it possible, that there was a place for scholars in medicine—for practicing and doing research and studying. I wasn't a good bench-worker. I'm not sure why; I tried hard enough. When I started writing about medicine itself and doing research on that (for example, on doctor-patient communication) he was very supportive, and when I started writing about death and the care of the dying, he let me know that I had begun to find my niche. I remember that when I published my first paper, which was a bit way out, I was so worried somebody else would realize what I had and get there before me. [My mentor] said not to worry, if the idea was really good, the problem wouldn't be anyone stealing it, the problem would be anyone accepting it.

But perhaps it is Stryer who put great mentors' roles most succinctly, acknowledging, "I've been so lucky with my mentors," and, having praised the five of them to whom his textbook is dedicated (including Paul) at some length, concluded, "They carried their learning lightly."

Passing on the Encouragement and Guidance

The great majority of respondents described their own efforts to serve as mentor to their students and other people in need. Only one interviewee chose not to answer the question; another approached the role intimately, saying succinctly that she had failed as a mentor because "None of my five children were ever interested in geology." All of the others took their mentoring roles seriously. Katz explained simply, "That's my job." Two engage in formal mentoring programs: one is directing a mentoring network for biomedical research at a major university "aimed at new college (university) faculty to encourage them to be engaged in biomedical research"; the other, now retired, explained, "I [was] ... a project leader at U.S. Department of Agriculture with a lab of about eight full-time people," continuing,

> I prefer this hands-on research and guidance job to the more administrative department head position I had previously. My oversight is pretty close, similar to [that of] a university professor. I have always felt mentoring my staff to be a major component of my job, trying to advance their careers. At the extreme, one of my staff, now getting her PhD in cognitive science at the University of California, Berkeley, I first hired out of high school and hired her every summer till her BS, then took her on as a junior scientist. I have sent a number of my junior staff to get their PhDs, one of them now has replaced me as project leader. During their PhD [work], I [served as a] member of their thesis committees, ... picked their thesis problem with them, and, in effect, ... was their research adviser. This program I developed myself ... is unusual at the Department of Agriculture. It is, incidentally, frequently the only way to ... find engineers who are citizens, a Department requirement.

Paths to Professions

In contrast to the basically unified response to our questions about having and becoming mentors, our fourth question—*How did you come to think of yourself as a particular sort of professional? Was your ability uncovered by formal tests or a course of study? Or did you "self-identify" (Brandwein's term)? Or a combination?*—evoked a variety of responses, difficult to classify. There was an element of self-identification in many of the choices respondents made, but other forces were often just as important. Katz best summarized the role of self-identification as a path to career:

> I loved the introductory course in organic chemistry and was seduced by the professor, who gave me opportunities to work in his lab. I was good at it, carried out many many experiments, took or audited every course in organic chemistry that was offered, had a home in the lab, and spent most of my time with the graduate students, postdoctoral students, and professors. I suppose one would say I self-identified. I would say I fell in love.

A couple fell into their fields simply because they were "good at [them]." For three respondents, tests were a roadblock; for one, her ease in excelling at them encouraged her to go into and stay in academia. For some, family pressure influenced them to go into fields they would not have freely chosen, though often they came to like where they were. For example, one respondent tried to resist familial pressure to become a physician, but ended up in the field anyway. Wrote James Ketchum,

> I wanted to be a scientist from age seven or eight and recall writing "I want to be a scientist and help struggling mankind" in a composition at age eight in fourth grade. My mother had me tested and found I was in the top 1 percent [of IQ] at age five. She had me take

private piano lessons, French, elocution, acting, art, swimming, and dance lessons from ages 5 to 16. She wanted me to be a doctor, and I started out in pre-med at Dartmouth but shifted briefly to philosophy and then to psychology before returning to pre-med in my junior year. I always resisted complying with my mother's ambitions for me, but ended up agreeing "in spite of" my recalcitrance. I chose psychiatry as a specialty while still in college, but after residency training was persuaded to move into psychopharmacology research for a decade while in the army.

Four of the practicing physicians who responded to our survey were psychiatrists, one an internist; three went into research in academia. Ellner left his "rather stolid and suffocating family environment" behind to help others through psychiatry. He reported that he felt that he "was behind the curve and that I had a certain valid and valuable way of pursuing my interests in science. My abilities really were self-identified more than recognized by others." Cassell reported that he "self-identified," but that he took a long time to do so:

I wasn't good at tests. I couldn't memorize like my medical school classmates. I had to read the same thing in three or four books until I understood, then I could answer questions on tests. And I didn't self-identify until I was in my 40s and I didn't get confident until my beginning 70s (I'm 75 now). But one reason I have gotten more confident and begun to theorize better (at least I think so) is because I don't practice anymore. So I can lose the deeply ingrained habit of mind that keeps physicians careful, always careful.

He concluded his answer by warning, "Swingers kill patients. Mostly, people don't like it when doctors kill patients."

Weiss, sharing Cassell's view of the difficulty of objectively studying people, went into research on nonhuman subjects. He noted,

I identified myself early as one who would pursue science. I thought I would like to do it through medicine, which I perceived as being more useful than pure basic science. I soon realized that this was I and that humans were terrible experimental animals. So, after medical school, I pursued basic research in biochemistry and molecular biology.

Lisa A. Steiner, also an MD, reports, however, that her path toward a professorship at the Massachusetts Institute of Technology was "not really" through self-identification. Coming from a supportive home, and aided by the women's movement, she "zigzagged quite a lot, going first to graduate school in math, then to medical school, finally into basic research. I wished to do something that would be helpful to 'humanity.'"

In contrast, explained one agricultural scientist, Paul's influence turned him slightly away from his family's pattern:

In trying to get me reoriented (directed is too strong a word here) to the biological sciences, Dr. Brandwein was bucking an old family tradition. My father was an aeronautical engineer, his father was a civil engineer. My son became a construction engineer, my daughter an applied mathematician; she married a chemical engineer (whose father was a systems engineer), and our 11-year old granddaughter is showing a strong bias toward mathematics. Whether all this is genetics or environment at home, I leave to you. There really wasn't much question that I would become a scientist or the like. The only guidance I got from home was that my father discouraged academia, and I have worked in industry or government most of my life. My mother's hope that I would become a doctor fell on deaf ears. The end result was that I became a research scientist, with strong inclination toward applied math and low temperature material physics (model building, I guess). This is not to say that Brandwein did not leave a strong impression on me, but it was more general than specific (maybe he did influence me away from straight applied science). His influence was certainly stronger than that of any other science teacher in high school; at least I remember his name along with Mr. Lazarus, our physics teacher (and force behind the radio club). My memory of Paul Brandwein is that of a dedicated scientist (a very unusual thing in a high school instructor and a real role model), knowledgeable in teaching and encouragement. I wonder whether the model was not far more important to me than the technique.

Rosen followed his own inclinations into science, writing that "In some inchoate way I think I began to self-identify [as a biologist] in high school. I know of no formal tests or course of study that did it for me. Indeed, many aspects of the formal course of study were quite boring."

While Andrew M. Sessler's path toward his position as the future director of the Lawrence Berkeley National Laboratory (1973–1980) at the University of California, Berkeley, was set pretty much from infancy—"I wanted to be a scientist from the earliest days I can remember (about fourth grade) probably due to my dad's influence (he was a high school biology teacher and knew Brandwein[8])"— Lewontin said his future as a population geneticist occupying the Alexander Agassiz Professorship of Zoology and Professor of Biology at Harvard was an "accident." He explained,

I was reasonably good at math and at simple lab work and as a senior at Harvard I worked in the mouse room of L.C. Dunn, a professor at Columbia who was visiting for a year, and then worked in the laboratory of a former student of [Theodosius] Dobzhansky. My undergraduate record was marred by a period of failing work, so when I was invited by Dunn and Dobzhansky to come to Columbia as a student of Dobzhansky's, I jumped at the chance, fearing there were no good alternatives. After I arrived, I tried to go to the lab of another professor who was a cytologist to do what I thought was a clever experiment. He agreed it was a good idea but told [me] to go back to Dobzhansky, who would never speak to him again if he stole a student. As it turned out this was lucky for me, as I would have been a lousy cytologist.

"Potential **Through** *Performance"?*

Question 6—*Did you discover your professional bent through working on a meaningful project? In Brandwein's phrasing, did you discover your "potential through performance"?*—follows up on the data respondents offered in the fourth question about paths to profession discussed in the preceding section. The question elaborated: *Brandwein proposed an "operational approach"—one which created an*

---

[8]Editor's note: Paul was also friendly with Barbara Wolff Searle and Robert Paul Wolff's father, chair of the biology department where Paul first worked. Both contributed to this volume, Searle, a survey as well as an essay in Part I, Wolff, a survey. Although both fell under Paul's influence, neither became a scientist.

environment to facilitate self-identification by offering opportunities for students to demonstrate their abilities, persistence, and questing through their performance and products, and one that would generate activating factors. Brandwein's analogy puts this operational approach in clear perspective:

> *The Operational Approach is more like a training camp for future baseball players. All those who love the game and think they can play it are admitted. Then they are given a chance to learn how to play well under the best "coaching" or guidance (however inadequate) which is available in the situation operating. No one is rejected who wants to play. There are those who learn to play very well. Whether they can play well (in the big leagues) is then determined after they have been accepted into the big leagues (college and graduate school). (p. 23) (1955/1981)*

Almost all the respondents, obviously not including Lewontin, quoted above, found, as Cassell wrote, "the analogy is a good one," although in his case "performance" did not involve a project.[9] A number of respondents noted the central role of the coach in the framework of the "operational approach": Friend wrote, for example, that "Paul was certainly the expert coach who gave me the most personal attention and encouragement to undertake a career in science," while Katz was uncertain:

> I am less sure about the role of coaching. I was very lucky to have many professors encourage me and provide space in their labs for me. I was very lucky to have graduate and postdoctoral students in the lab to give me advice. But it was the opportunity to be in the training camp, to have a professional lab, a superb library, eager players all around, to watch how professionals do their work—the environment—that I think counted more than direct coaching. But I may be underestimating the coaching role played by graduate and postdoctoral students.

After mostly agreeing about the role of actually teaching geology in leading her into her career, Jean Gansky wondered parenthetically, "(or did PFB push me?)" "Oh yes," said Stryer. "That's a very good description. The process reinforced the good things that I did while doing experiments—getting them going—and then teachers can help students learn virtuosity. Getting going, setting up experiments is important," he continued.

Baron Raskind von Hippel basically agreed, raising, however, as did Steiner, the peculiarly female aspects of career in relation to her decision. Steiner[10] was one of the beneficiaries of both the women's movement and of Paul's being in the forefront of those who recognized that excellence is without gender, that females can achieve as surely and as superiorly as men. Wrote Baron Raskind von Hippel, who had originally inclined to becoming a researcher in enzymology,

---

[9]Cassell went, however, on to note "how few of the superior medical students [he taught] went on to realize their potential. Mostly they have been crippled by emotional difficulties, vanity, family problems, desire to be famous, and failure to work hard enough. Maybe also, by the social desire to be like others."

[10]She wrote that "my teachers, both in high school and college [in the 1950s], always encouraged me and never raised any question that I could not pursue a career because I was female. It was not until the emergence of the women's movement in the 1970s that I realized that this was somewhat unusual."

Remembrances from More than a Half-Century Back

... in retrospect, if I had received good "mentoring" in graduate school I might well have opted for a career in research, or at least in "academic medicine."

I do think that it is more difficult for women to combine research and raising a family than to combine a medical career with family duties. Obviously, there are a few who can do this, but it is not easy, and if you put off your research career at all, you quickly fall far behind your colleagues in knowledge and technology.

Contributions

Asked to name what they thought was their greatest professional contribution, the large majority of respondents cited breakthroughs in scientific research and/or writing. A few cited administrative accomplishments; three noted their contributions to improving the lives of others; a couple cited their work as teachers; two were not sure that they had made any significant breakthroughs.

Their greatest personal contribution? A couple said they did not understand the question: one left it at that; the other went on to write, "I have tried to provide a model of a scientific workplace that puts the smallest possible emphasis on hierarchical relations and the greatest possible emphasis on the doing of science as an individual creative activity carried on in interaction with others." A number cited their families, nuclear and extended, Sessler adding, "My three kids are all in science. Two are professors and quite famous." Several cited mentoring activities, one woman mentioning in particular "encouraging women to enter and persist in science"; another summarized, "The training of other scientists, who are now my colleagues." Besides his scientist children, Sessler mentioned his support of arms control and human rights; another spoke of his work in university governance in addition to his scientific achievements. A number were proud of their work in helping ease human suffering. One cited her teaching; another mentioned his success from humble beginnings.

A small majority of respondents believed that their personal and professional achievements can be separated, Ketchum noting that the first calls for a "pattern of achievement," while the second often comprises "a single major achievement." Agreeing, Lewontin elaborated, "Scientists carry out successful science in an immense variety of personal and social environments including complete isolation from others, absolute dominance of others, and a variety of other social interactions, some of which are destructive and some nourishing to different individuals with different needs." Sessler also affirmed separation but admitted, "but I do move constantly between them (many times a day as I answer e-mails, etc.)."

Cassell and Stryer agreed about both the separation and the interaction, Cassell noting, "I'm listening to Beethoven sonatas as I write, and bliss wasn't his state. That is one of the great and interesting questions, the relation of the private to the public person." He continued, "Great physicians must be good persons. Not surgeons, neurologists, dermatologists, and etc. Probably pediatricians also need to be good, but I don't know anything about them . . . ." He concluded,

I do know that the pressure for me to change and grow personally and emotionally virtually always came about because I knew that if didn't change, I couldn't meet the needs of some particular patient. And I would do and did anything for a sick person. I would do anything—intellectually and emotionally—almost without constraint and almost no matter where it led, to learn more about medicine and the care of the sick.

Paul's Interaction with Educators

All but two of the teachers and educational administrators who completed our survey, like the scientists whose responses are analyzed above, first knew Paul as a teacher. Middle school science teacher Joanne Gallagher Corey, who also contributed a short essay,[11] took a class from him at Columbia Teachers College. The following educators studied with him at Forest Hills: Hannah Flegenheimer, EdD, a specialist in learning disabilities; Alfred Schutté, EdD, a professor, educational administrator, and consultant; Ann Baron Brodell Ward, MS, a Head Start director; and Robert Paul Wolff, PhD, professor of literature and philosophy at the University of Massachusetts Amherst.[12] John F. Disinger, PhD, a professor of natural resources at Ohio State University, first became acquainted with Paul through his work and personally when Paul accepted Disinger's invitation to speak at a conference on environmental education in the 1980s. Flegenheimer, a Brandwein biology protégée at Forest Hills, did not end up as a bench scientist, but, as in the case of Barbara Wolff Searle[13] and her brother Robert Paul Wolff, her scientific training served her well in another field.

Our second preliminary question—*II. Would you be willing to share with us anecdote(s) about your interaction(s)?*—elicited several responses from this group. Corey remembered Paul's attention to her emotional safety:

> He took me aside toward the end of the course and told me to be careful, that I was very sensitive and might be hurt emotionally in the teaching profession ... that I "threw myself into things too much." He said that was OK as long as I knew what the consequences would be. I thought that was wonderful advice and never forgot it. It was surprising, also, to get such a sensitive reading from him. ... He did tend to be very sarcastic and witty! A great man!

On the other hand, Disinger valued Paul's careful attention to his work, remembering a letter in particular, coming from someone he did not then know personally:

> [Paul] took the time and made the effort to write it. In his letter, he mentioned his concurrence with points made and positions taken in that paper and other papers I had written, how highly he thought of my work, and (of course) how much he agreed with me. He also encouraged me to continue my efforts, as well as suggesting how I might extend them, etc. I appreciated the compliments and encouragement, found the suggestions insightful and helpful, and very much admired his style.

---

[11](See "Close Encounters, Lasting Effects.")

[12]Wolff's interactions with Paul are explored fully in the essay by his sister, Barbara Wolff Searle, in this volume.

[13]See her essay in Part I.

Schutté remembered the presence of observers in the back of Paul's Forest Hills biology classes, commenting, "There were always lessons that had moments for testing hypotheses, noting probabilities of events, many uses of observations, followed up by reasoning, not accepting all events, but looking at contradictions. In general, Dr. B.'s lessons were always ones where you had to *think.*"

Disinger paraphrased three of Paul's statements that he found particularly significant:

- Call it "environmental education" rather than "conservation education," partly because the word "environmental" is more inclusive, and partly because the word "conservation" is not an adjective.
- High school science teachers work too hard. . . .
- About the third time you attempt to teach something to someone else, you begin to understand it yourself.

Like the scientists, the educators tended to discount our question on genetic predisposition. Three praised their well-rounded upbringings, one asking, "Is curiosity, insatiable interest in reading and almost everything genetic? My interest in everything remains to this day." Although another admitted that she was lucky in having good verbal and mathematical skills, she attributed her success to "generalized ability to take on the tasks required for whatever the chosen career might be." Citing his ignorance of matters genetic, Robert Paul Wolff responded as follows:

I must concentrate on my publishing career (twenty-one books in philosophy, education, economics, Afro-American Studies, political theory, etc.). I haven't a clue what genetic gifts might have predisposed me to the career I chose. Since my knowledge of these matters is entirely secondary and based on a few books I have read, I am in no position even to speculate.

All but one of our educators came from families where both parents had professional careers. And all were widely read. Five of the six described their financial circumstances as comfortable but not wealthy, and the same five reported parents who "strongly supported" academic achievement. With one exception, all[14] reported parents supportive of emotional needs. The exception found her way through her superior performance in high school: She wrote, "Certainly I was a lonely, somewhat sad child who did not fit very comfortably into the world of childhood fun and games and whose school performance was not cause for respect or admiration among my peers—until high school."

So, like the scientists, the educators tended to credit environment more than genetics for their achievements. For four of the six, those achievements come as the result of persistent effort, while for two they come fairly easily. Wolff, however, equivocated:

I guess I am persistent—at least enough so to finish the books I write. But I write mostly in my head, talking things out with imaginary audiences. When I start writing, I write very quickly—one book in eight weeks, another in three, another in a month. I am persistent in

---

[14]In each case, the missing respondent simply skipped the question.

this one sense: I will not accept an incomplete or metaphorical or sketchy explanation for something.

All six had numerous mentors. Disinger, who did not count Paul among his, seeing him rather as a valued colleague, provided a definition:

In every case above, the individual I am here calling a "mentor" was highly competent, highly experienced, highly personable, and highly desirous of making sure that I, as an individual, learned and improved my knowledge and abilities. A few years ago I received a university-wide distinguished teaching award at Ohio State, at which time I observed that all nine of those with whom I shared that award routinely demonstrated the same characteristics with all of their students and colleagues. The latter observation comes largely from reading their letters of nomination and recommendation, and from spending delightful time talking with them. So to my mind those are the characteristics, qualities, and approaches that "work."

One of the educators paid Paul a special tribute, which, like Stryer's current financial gifts to the young in memory of Paul's $300 to him in the 1950s, passed Paul's generous example to the next generation:

Throughout my academic career I think Paul Brandwein was probably the most inspiring teacher I encountered, and his effect on me, beyond what I have already described, was to convince me of the importance of good teaching in the lives of young people. I ended up in the field of education, working to develop and improve instructional programs for students with disabilities, an outcome that was probably not coincidental. There have been a few other excellent teachers in my life: Elizabeth Koffka at Smith, Margaret Jo Shepherd at Teachers College, but they served only to reinforce the initial impact.

In return, with one exception, respondents all now serve or have served as mentors themselves. Corey was part of a formal mentoring program, while Disinger said he was not one in the "formal sense of the term," however:

I do know that I have taken personal interest in and worked closely and intensely with many students (and more recently, younger professionals) over the years, and that some (but not all) of them have remained both personal and professional acquaintances. Most of them are "successful" in the sense that they do what they do very well, but I find it difficult to claim any credit for that. Paraphrasing the Hippocratic Oath, Paul Brandwein once said the main task of the teacher is to do no harm, and one is pleased to know that at least some of his former students are successful in their fields and their personal lives at least in part because I (and others) did not get in their way.

Wolff was more emphatic:

That is mostly what I have been doing as a teacher for the past forty-nine years. Why? Because they are like my children. How? By trying to get them to see, as I do, the structure and organization of ideas, and by helping them to do better work (as well as worrying about their lives, as one would for a child).

Five of the six educators answered the related questions 4—*How did you come to think of yourself as a particular sort of professional? Was your ability uncovered by formal tests or a course of study? Or did you "self-identify" (Brandwein's term)? Or a combination?*—and 6—*Did you discover your professional bent through working on a meaningful project? In Brandwein's phrasing, did you discover your "potential* through *performance"?*—that they "self-identified," that they discovered their potential through performance. Corey "did not prepare to be a teacher," but she fell into a middle school biology classroom before she knew "a lesson plan from a hole in the wall." Loving teaching, like Schutté, she stayed in classrooms. Disinger also self-identified: "No one ... 'pushed' me in one direction or the other." One educator self-identified through her interest in the learning processes of her own children. With one exception, all found their potential *through* performance; the formal training, if any, came after the educators walked through the doors. That exception was Wolff, who took advanced courses before choosing his profession:

> Not really. For as long as I can remember, I have been extraordinarily smart, and as a boy, I was on fire. I realize this is rather self-serving and not at all modest, but it is the truth. By the time I was 17, I was taking graduate mathematical logic courses from some of the leading logicians in America.

For all the educators, their on-the-job performance comprised their greatest professional contributions; for one it was also her greatest personal achievement. For four of the others, although their personal contributions were still part of their professional achievement, they focused on the personal interactions that were essential to their jobs. Disinger added "family-related 'personal' stuff," while Ward was personally proudest of having raised "four fine children." Wolff was on her page: "Fathering two wonderful sons: Patrick Gideon Wolff, one of America's leading chess players, and Tobias Barrington Wolff, a leading gay rights legal activist and professor of law at the University of Pennsylvania."

For three educators the two achievements are inseparable; for three, they are distinct.

Paul and Students Not Choosing Science

Two respondents to the survey did not go into science, yet both affirmed their admiration for Paul's contributions to their personal and professional lives. For although Paul revered science—he loved its German rendition, *wissenschaft* (the business of knowing)—"knowing" for Paul was not limited to science but embraced art, literature, politics, the world, the universe. . . .

Barbara Wolff Searle, PhD, eventually chose to work as ombudsman for the World Bank rather than to become a bench scientist. See her essay in Part I of this volume. Eleanore Berman, the class scientist according to another Forest Hills alumnus, became an artist, specifically a painter and printmaker. According to the same classmate, while working on her Westinghouse project, Berman rediscovered a lost species of ameba, *Chaos chaos* or *Pelomyxa*. Her classmate reported, "Everyone was heartbroken that Eleanore didn't win the Westinghouse, but Paul made it OK." Before she died in September of 2004, Berman completed our survey, but some questions remain. For example, while Searle came in contact with Paul both as his student and because he was a colleague of her father, who was also a high school biology teacher, I have not been able to learn whether Berman was Paul's student at Forest Hills or merely his Westinghouse advisee and admirer.

282   D.C. Fort

She opened her survey with the following tribute:

What can I write about Paul Brandwein that any of you have not experienced—to me, mentor, friend, adviser, loyal champion throughout his life? We all loved him in a special way.

As I grew older and turned to the arts (I always painted, and this was a pursuit that could continue as I raised a family of four), I realized how much my early interest in biology and fascination with the secrets of life fed my creative work. I always as a child had taken refuge in nature, in woods, on lakes and fields.

The work in the biology lab at Forest Hills High School fed those interests at a higher level, and was heaven to me—and Paul's field trips on the old World's Fair grounds lake and fields were a source of wonder and learning. I still look at plants that grow in my own beautiful garden—thinking of him and the work we did together in the lab. My research project on the growth of *Spirogyra* won a prize for a paper delivered at the Museum of Natural History in New York. Forever after, I have peered into streams and ponds, marveling at the life within.

My work as an artist is informed by all this richness—the twisting forms of vines and stems—the field of forces that give rise to life in the universe—the light and dark of the hedges in my garden.

Paul owned several of my pieces and arranged for William Jovanovich to have a large black and white painting called "Canto" as well as an etching called "Interlude." I was so grateful for his continued encouragement and his sage advice when I would bring my personal problems to him. He was a giving, devoted friend to many, and I admired his deep selfless commitment to peace and the children of the world—as he wrote to me, "always losing battles, but always to try again, always on the road, *siempre al camino.* Another favorite: *Per ardua, ad astra,* or "through hard work, we reach the stars."

Searle remembers that "when I showed up, [Paul] immediately gave me a job in the laboratory; I was in charge of the amebas and paramecium that were used for teaching purposes. He was a central figure in my entire high school career."

She continued,

Paul was one of three or four teachers who created an atmosphere of intrigue and specialness around themselves. They were all friends and competed among themselves to get the attention of the kids that they identified as bright. They loved teaching and made it clear that they were really interested in us.

Paul was a bit stuffier than the others and hence came in for some kidding and rough treatment (from the other teachers). I remember once when he came into his office to find all sorts of three-dimensional objects in his filing cabinets—things like the fish bowl and such. We all thought this was hilarious. I'm not sure Paul did.

In his biology class he scarcely used the standard textbook. He let us know that what he was teaching us was more up-to-date and complete than the book,[15] which he considered watered down. (Much less so than today, but not to his taste.) That was one of the ways he got our loyalty and interest.

He treated "us (the special kids)," she continued, as responsible adults who could be relied upon to meet deadlines and learn what needed to be learned "without being monitored or spoon-fed."

---

[15]Paul's rebellious stance during his teaching days was in ironic contrast to the censorship he had to face when he went into publishing. See Lewontin's criticism of Paul's compromise on evolution in this essay.

Berman's memories are less classroom-centered, more personal. For example, she wrote that "[Paul] used to brew tea in the lab on cold winter afternoons. He played piano for us in the music room and loved Beethoven's Pastoral Sixth Symphony. Over the ensuing years, we would often meet for dinner or tea, and he visited my home to meet my family."

While Berman skipped our question on genetics, moving quickly to her home environment (as will I, shortly), Searle responded as follows:

> I was obviously smart and verbal, which I actually think I inherited more from my mother than my father. My brother[16] had a much higher IQ than I, and one of his sons has been the U.S. chess champion, so whatever role genes play in conferring intelligence runs in the family.

Both women came from accomplished families. Berman's father was a physician, and "my mother had a deep love of the arts and literature. I always saw the world through the eyes of both a scientist and an artist. Both of my parents were avid readers and collectors of art." Searle's "father was a high school biology teacher. One of his brothers was a physics professor. My mother had to leave school early to support her family,[17] but I think she was smarter than my father. We were definitely upper middle class intellectual Jews."

Like Berman's, Searle's parents were readers, her mother bringing home library books weekly. In addition, her father "met regularly with men friends in a reading group. In high school he led a group of my friends for months in a discussion of S. I. Hayakawa's *Language in Action.* We were all readers!"

Both families were comfortable financially, and both encouraged their daughters to achieve academically, Searle noting, "In fact, I faced none of the differential expectations that so many of my women friends did." Apparently her family, like Paul, was ahead of their time in their expectations for achievement from females. Berman skipped our question of attention to emotional needs, while Searle answered, "I didn't think my emotional needs were adequately addressed, but in this I was probably a typical teenager."

While Berman self-identified as an artist early—"My continued interest and passion for painting has been part of me since I can remember. I was encouraged to exhibit in my 30s by a good friend and professional painter"—Searle was identified as a future biologist by Paul, with encouragement from her father. It is unsurprising that she left science, after which "I found an inner direction that guided me. Many people gave me assistance but the direction was all internal."

In spite of Paul's misidentification of her direction into biology, Searle, like Berman, "thrived under his 'operational approach!' In fact," she continued, "I have always (in my successive career changes) discovered my potential through

---

[16]Robert Paul Wolff, also a biology student of Paul's, became a professor of philosophy and literature. His contributions are discussed in the preceding section of this essay.

[17]Searle and Wolff's mother was employed as the managing editor of *Child Study Magazine* (working her way up from her start as a secretary).

performance." Berman did the same, "by trying, doing, and succeeding with the gifts I had and the joy it brought to me and others."

Both women had mentors other than Paul, Berman citing a Manhattan painting teacher "with whom I studied from the age of 13 on ... a man of great talent and wisdom [who] taught me great love and understanding of the arts." Commented Searle, in spite of Paul's attempt to turn her toward biology,

Paul was probably the most important mentor of my career. Other people have influenced me, but often more negatively than positively. That is, I spent a lot of time trying to maintain my independence, and trying to figure out what I wanted to do in the face of people who assumed that I was on a different path.

And both have also served as mentors to others, Searle "unofficially," Berman turning her attention to two groups:

... sometimes I give advice to young painters, but I believe I mentored my four children more. All four are artistic, and all have a deep sense of the universe, which I see in how they are raising their own children. That these traits have been passed down, and as well to my beautiful grandchildren, brings me great joy and hope.

Their greatest professional contribution? Berman wrote, "The creation of meaningful works from the depths of my heart," elaborating, "I have been painting professionally for many years; my works hang in prominent corporate collections and some museums, [and] I have raised four accomplished children." Searle said the decision was "really difficult" for her, but replied,

I ran a project for eight years (while at Stanford) that was very successful and served as a model for many other projects, some of which are still functioning. I also helped the Chinese government (the State Education Commission and the Chinese Academy of Sciences) develop a World Bank-financed project that created about 50 graduate science centers which could award "world-class" PhDs.

She added, however, that she thought that "the external and the personal can be separated. I think other people would evaluate my contributions differently than I would."

Both women find their most important personal contribution in their interrelationships with others, Searle citing in particular, "two wonderful kids, many people who I have helped over the years." Berman included her philosophical positions, summarizing, "Passing along my insight, love of life, nature, and nonviolence to my children, who in turn, have passed it to their children (my five grandchildren), all wonderful young people."

References

Atwood, Margaret. (1989). *Cat's eye*. New York: Doubleday.
Berg, Jeremy M., Tymoczko, John L., and Stryer, Lubert. (2002, 2007). *Biochemistry*. New York: W. H. Freeman. (See also Stryer below.)
Brandwein, Paul F. (1955). *The gifted student as future scientist: The high school student and his commitment to science*. New York: Harcourt, Brace.

Brandwein, Paul F. (1955/1981). *The gifted student as future scientist: The high school student and his commitment to science.* New York: Harcourt, Brace. (Reprinted in 1981, retitled *The gifted student as future scientist* and with a new preface, as Vol. 3 of *A perspective through a retrospective,* by the National/State Leadership Training Institute on the Gifted and the Talented, Los Angeles, CA).

Brandwein, Paul F. (1962). Beginnings in developing an art of investigation. In Jerome Metzner, Evelyn Morholt, Anne Roe, and Walter G. Rosen (Eds.), *Teaching high school biology: A guide to working with potential biologists* (pp. 43–60) [Biological Sciences Curriculum Study Bulletin No. 2] (Report No. SE00088). Washington, DC: American Institute of Biological Sciences. (ERIC Document Reproduction Service No. ED011002).

Brandwein, Paul F. (1983). Do we expect school science to nurture creativity? In *NSTA yearbook: Science teaching: A profession speaks* (pp. 52–56). Washington, DC: National Science Teachers Association.

Chu, Judy Y. (2005). Adolescent boys' friendships and peer group culture. *New Directions for Child and Adolescent Development, 107,* 7–22.

Goldberg, David, Evans, Peter, and Hartman, David. (2001). How adolescents in groups transform themselves by embodying institutional metaphors. *Clinical Child Psychology and Psychiatry, 6*(1), 93–107.

Morholt, Evelyn, and Brandwein, Paul F. (1986). *A sourcebook for the biological sciences* (3rd ed.). San Diego: Harcourt Brace Jovanovich.

Stryer, Lubert. (1975, 1981, 1988, 1995). *Biochemistry.* New York: W. H. Freeman. (See also Berg, Tymoczko, and Stryer, above.)

Swenson, Lisa M., and Strough, Jonell. (2008, September). Adolescents' collaboration in the classroom: Do peer relationships or gender matter? *Psychology in the Schools, 45*(8), 715–728.

Part IV
Appendixes

Appendix A: The Survey

Richard C. Lewontin, PhD
17 Maple Avenue #3
Cambridge, MA 02139

Walter G. Rosen, PhD
5 High Street
Brunswick, ME 04011

Deborah C. Fort, PhD
3706 Appleton Street, NW
Washington, DC 20016

James P. Friend, PhD
108 Righters Ferry Road
Bala Cynwyd, PA 19004

Date

Recipient

Address

Dear_____,

Deborah can be reached in Washington, D.C., mornings at (202) 363-1673, deborah.fort@starpower.net; Walt in Maine at (207) 725-6045, fax (207) 729-2633, waltrosen@suscom-maine.net, Dick in Cambridge at (617) 495-2419; Jim in Pennsylvania at (610) 667-5142, frienddjp@comcast.net. Our postal addresses are above.

Background

We four—Deborah Fort, Walt Rosen, Jim Friend, and Dick Lewontin—are trying to investigate the reasons why people go into particular fields. Everyone who receives this survey was in touch in some context with Paul F. Brandwein. This is what you and we have in common. Our survey tries to find out what, if anything, else you share in your personal and professional choices, as well as tracing what, if any, of Brandwein's techniques as a teacher and colleague influenced his associates. Walt, Jim, and Dick are Brandwein's old high school students, his alumni; Deborah encountered him through her editorial work at the National Science Teachers Association more than a decade ago.

Based on his own and other teachers' experience, Brandwein concluded in the mid-1950s in his small, unique book *The Gifted Student as Future Scientist* that *High-level ability in science is based on the interaction of several factors—Genetic, Predisposing, and Activating. All factors are generally necessary to the development of high-level ability in science; no one of the factors is sufficient in itself* (1955/1981, p. 12). Among the genetic factors were "high verbal and mathematical ability and adequate sensory and neuromuscular control." The predisposing factors included traits he called *persistence*—a willingness "to labor beyond a prescribed time ... to withstand discomfort ... to face failure"—and questing—"a dissatisfaction with present explanations of aspects of reality" (p. 10). Brandwein's activating factor included "opportunities for advanced training and contact with an inspirational teacher" (p. 11).

(If you are answering this survey on paper, please use separate sheets. See below if you wish to use an electronic approach, which would help us out.)

   I. In what context and when did you know Paul F. Brandwein?
  II. Would you be willing to share with us anecdote(s) about your interaction(s)? Be as specific as possible.
 III. Do you remember any of Brandwein's statements that continue to resonate?
 IV. Can you provide us with any biographical details about Paul F. Brandwein?

Survey

Let's test Brandwein's theory in the light of your career in science/math/publishing/education:

1. Discuss in a general (nonstatistical?) way the genetic gifts that may have predisposed you to your career.
2. Would you characterize your approach to the tasks you decide to take on as "persistent" or not? Do you typically labor long to accomplish a goal, or do your achievements come easily? Please give specific details.
3. Did you encounter in your youth or later career inspiring teachers or other mentors? Who were they and how did they influence you?
4. How did you come to think of yourself as a particular sort of professional? Was your ability uncovered by formal tests or a course of study? Or did you "self-identify" (Brandwein's term)? Or a combination?
5. From what kind of environment did you come? Were your parents professionals?

   In what fields?
   Were they readers? What sort of materials did they choose?
   Did your home environment offer

Appendix A: The Survey

> A. Financial comfort—did you have to worry about money or sustenance?
> B. Support of academic endeavor?
> C. Treatment of your emotional needs?

Most of these factors are not ones that lend themselves to precise measurement by paper-and-pencil tests and, consequently, Brandwein proposed an "operational approach"—one which created an environment to facilitate self-identification by offering opportunities for students to demonstrate their abilities, persistence, and questing through their performance and products, and one that would generate activating factors. Brandwein's analogy puts this operational approach in clear perspective:

> The Operational Approach is more like a training camp for future baseball players. All those who *love* the game and think they can play it are admitted. Then they are given a chance to *learn how to play well* under the best "coaching" or guidance (however inadequate) which is available in the situation operating. No one is rejected who wants to play. There are those who learn to play very well. Whether they can play well (in the big leagues) is then determined after they have been accepted into the big leagues (college and graduate school). (p. 23)

6. Is this how you found your path to your profession? That is, did you discover your professional bent through working on a meaningful project? In Brandwein's phrasing, did you discover your "potential *through* performance"?
7. Have you served or are you now serving as a mentor to others? Why and how?
8. What would you now look upon as your greatest professional contribution?
9. What would you now look upon as your greatest personal contribution?
10. Can 8 and 9 be separated?

You _____ publish my responses with attribution.
    (may/may not)

Signature_____
                                                      (date)

All quotations are from *The Gifted Student as Future Scientist* (New York: Harcourt, Brace, 1955; 1981 reprint, with a new preface [Los Angeles: National/State Leadership Training Institute on the Gifted and the Talented]).

Demographics

If this section is too time-consuming to tackle, please just attach a CV or résumé. If you could send any of this information on disk or via e-mail, that would help our study a great deal.

Name, address, title, professional affiliations (past and present)

E-mail_____

Date of birth_____

Retired? Since when?_____

Current activities
Summary of formal education (with dates)
Honors, with dates

If you are answering on paper, please return this form and extra sheets to Deborah C. Fort, 3706 Appleton Street, Northwest, Washington, DC 20016 by [*date*], if possible. If you are able to provide information electronically, please label the word-processing program and operating system on your disk. Or send information to Deborah via e-mail at deborah.fort@starpower.net.

Either electronic approach will save us time and money.

If you'd like to get in touch with any of us, please do so. Our contact information appears above.

If you have any of your old Forest Hills High School yearbooks, could you make photocopies of the pages that relate to students you think studied with Brandwein or to Brandwein himself? Please label the date and page numbers clearly.

Thank you for taking the time to gather these data. We will keep you apprised of what happens to them.

Jim, Walt, Dick, and Deborah

Appendix B: Bibliographies of the Works of Paul F. Brandwein

Deborah C. Fort and Suzanne Lieblich

Publications (Excluding Textbooks)

Paul Brandwein's name appeared on all of the following publications—whether as author or editor is not always clear; however, it is probably safe to assume that in most cases Brandwein was the primary or sole author. Names of coauthors and other collaborators are provided below wherever possible.

(1937, October). (And Morris Rabinowitz). Certain substitutes for *Paramaecium caudatum* in high-school and college biology. *Science Education, 21*, 156–158.

(1937, December). Teaching demonstration of the monohybrid cross. *School Science and Mathematics, 37*, 1103–1105.

(1939). (And Douglas Marsland). *Manual of biology.* New York: Henry Holt.

(1939, February). Demonstrations dealing with photosynthesis. *School Science and Mathematics, 39*, 160–161.

(1940). Infection studies on the covered smut of oats (Doctoral dissertation, New York University and Brooklyn Institute of Arts and Sciences, Brooklyn Botanic Garden, Lancaster, PA).[1]

(1942, March). Studies in the teaching of biology. *School Science and Mathematics, 42*, 243–250.

(1942, December). Science film as a demonstration. *High Points, 24*, 69–74.

(1944, January). Modern role of biology teaching. *Teachers College Record, 45*, 265–271.

(1944, February). Some comments on the annual science talent search. *Science Education, 28*, 47–49.

(1944, April). Useful modification of the test tube. *Science Education, 28*, 164–165.

(1944, June). School joins the fight against cancer. *High Points, 26*, 73–75.

(1945, February). Four years of science. *Science Education, 29*, 29–35.

(1945, November). Some suggestions for individualized work in general science and biology laboratories. *School Science and Mathematics, 45*, 704–712.

(1945, November). Triple-track curriculum in science. *High Points, 27*, 49–54.

(1946, April). Unscientific method in science teaching. *Science Education, 30*, 158–159.

(1946, December). Reorganizing biology to meet the needs and interests of youth. *The Science Teacher, 13*(4), 59–61, 86.

---

[1] The puzzling documentation on Brandwein's dissertation has been consistent in a number of sources—what Lancaster, Pennsylvania, has to do with the Brooklyn Botanic Garden is unclear—but with this citation, as with many others on this and the other bibliographic lists, we have not been able to lay hands on a physical copy of the document in question to resolve the apparent contradiction.

(1947, February). Reorganizing biology to meet the needs and interests of youth. *The Science Teacher, 14*(1), 22–23, 35–36.

(1947, March). The selection and training of future scientists. *Scientific Monthly, 64*(3), 247–252.

(1948, January). To the board of examiners. *High Points, 30,* 9–15.

(1948, February). A substitution for the term "experimental method" as used in investigations in science education. *Science Education, 32,* 15.

(1948, March 1). *Teaching conditions affecting the work of science teachers in the public high schools of New York State.* Unpublished report of a study by a committee of the National Science Teachers Association and the New York State Science Teachers Association.

(1948, June). The selection and training of future scientists. *High Points, 30,* 5–13.

(1948, October). How one teacher does it (techniques). *The Science Teacher, 15*(3), 126.

(1948, October). On a certain indignity. *High Points, 30,* 5–9.

(1948, November). On the road to a high school curriculum. *High Points, 30,* 30–41.

(1948, December). Self appraisal through recordings. *High Points, 30,* 64–65.

(1949, October). How one teacher does it. *The Science Teacher, 16*(3), 132.

(1949, December 17). Fables for teachers. *School and Society, 70,* 404.

(1950, April). How one teacher does it. *The Science Teacher, 17*(2), 79–80, 93.

(1950, October 28). Fables for teachers. *School and Society, 72,* 276.

(1951, January). Still on the road to a high-school curriculum. *High Points, 33,* 12–17.

(1951, December). The selection and training of future scientists: II. Origin of science interests. *Science Education, 35,* 251–253.

(1952, February). The selection and training of future scientists: III. Hypotheses on the nature of "science talent." *Science Education, 36,* 25–26.

(1952, April). College Board's science tests. *The Science Teacher, 19*(3), 107–113.

(1953, January 1). Trends in high school general science. *National Association of Secondary School Principals Bulletin, 37*(191), 50–55.

(1953, April). The selection and training of future scientists, IV: Developed aptitude in science and mathematics. *The Science Teacher, 20*(3), 111–114.

(1953, April). Signposts toward the revision of high school chemistry. *School Science and Mathematics, 53,* 313–315.

(1955/1981). *The gifted student as future scientist: The high school student and his commitment to science.* New York: Harcourt, Brace. (Reprinted in 1981, retitled *The gifted student as future scientist* and with a new preface, as Vol. 3 of *A perspective through a retrospective,* by the National/State Leadership Training Institute on the Gifted and the Talented, Los Angeles, CA)

(1956). (And Nathan S. Washton, Brenda Lansdown, William Goins Jr., Abraham Raskin, and Harold S. Spielman). What should be the subject-matter competency of science teachers? *Science Education, 40,* 392–395.

(1956, November). New patterns in the education of gifted children in mathematics and science. *Kentucky School Journal, 35,* 20–21.

(1958). (And Fletcher G. Watson and Paul E. Blackwood). *Teaching high school science: A book of methods.* New York: Harcourt, Brace.

(1958). (And Evelyn Morholt and Alexander Joseph). *Teaching high school science: A sourcebook for the biological sciences.* New York: Harcourt, Brace.

(1958). (And Morris Meister). *Your future in science: The challenging opportunities awaiting you in exciting new fields.* Chicago: Science Research Associates.

(1960). (And Maxwell Reed and Wilfrid S. Bronson). *The sea for Sam* (Rev. ed.). New York: Harcourt, Brace, and World. (Paul F. Brandwein revised, edited, and supplemented with photographs the authors' 1935 original.)

(1960). (And Martha E. Munzer). *Teaching science through conservation.* New York: McGraw-Hill.

(1961). (And Alexander Joseph, Evelyn Morholt, Harvey Pollack, and Joseph F. Castka). *Teaching high school science: A sourcebook for the physical sciences.* New York: Harcourt, Brace, and World.

Appendix B: Bibliographies of the Works of Paul F. Brandwein 295

(1962). Beginnings in developing an art of investigation. In P. F. Brandwein, J. Metzner, E. Morholt, A. Roe, and W. G. Rosen (Eds.), *Teaching high school biology: A guide to working with potential biologists* (pp. 43–60) [Biological Sciences Curriculum Study Bulletin No. 2] (Report No. SE00088). Washington, DC: American Institute of Biological Sciences. (ERIC Document Reproduction Service No. ED011002)

(1962). Elements in a strategy for teaching science in the elementary school. In *The teaching of science* (pp. 107–144). Cambridge, MA: Harvard University Press. (Reprinted by Harcourt, Brace, and World, New York)

(1962). (And Elizabeth B. Hone, Alexander Joseph, and Edward Victor). *Teaching elementary science: A sourcebook for elementary science.* New York: Harcourt, Brace, and World.

(1962). (And Jerome Metzner, Evelyn Morholt, Anne Roe, and Walter G. Rosen). *Teaching high school biology: A guide to working with potential biologists* [Biological Sciences Curriculum Study Bulletin No. 2] (Report No. SE00088). Washington, DC: American Institute of Biological Sciences. (ERIC Document Reproduction Service No. ED011002)

(1965). *Substance, structure, and style in the teaching of science.* New York: Harcourt, Brace, and World.

(1966). *Notes toward a general theory of teaching.* New York: Harcourt, Brace, and World.

(1966). (And Evelyn Morholt and Alexander Joseph). *A sourcebook for the biological sciences* (2nd ed.). New York: Harcourt, Brace, and World.

(1966). Techniques of teaching conservation. In *Conference on Techniques of Teaching Conservation, October 10–12, 1966.* Milford, PA: Pinchot Institute for Conservation Studies.

(1967). *Building curricular structures for science, with special reference to the junior high school.* Washington, DC: National Science Teachers Association. (ERIC Document Reproduction Service No. 015134)

(1968). *Substance, structure, and style in the teaching of science* (Rev. ed.). New York: Harcourt, Brace, and World.

(1969). *Notes on teaching the social sciences: Concepts and values.* New York: Harcourt, Brace, and World.

(1969). Skills of compassion and competence. In L. J. Rubin (Ed.), *ASCD Yearbook 1969* (pp. 131–151). Washington, DC: Association for Supervision and Curriculum Development.

(1969). *Toward a discipline of responsible consent: Elements in a strategy for teaching the social sciences in the elementary school.* New York: Harcourt, Brace, and World.

(1969, February). Observations on teaching: Overload and "the methods of intelligence." *The Science Teacher, 36*(2), 38–40.

(1970). (And Hy Ruchlis). *Invitations to investigate: An introduction to scientific exploration.* New York: Harcourt Brace Jovanovich.

(1970, June). Early education and the conservation of sanative environments. *Theory into Practice, 9*(3), 178–186.

(1970, October). *Propositions toward the survival of a self-endangered species: The fundamental question.* Paper presented at the American Association of Colleges for Teacher Education–Organization of the American States [AACTE-OAS] Conference on Education and the Environment in the Americas (Report No. SP004788). Washington, DC: U.S. Department of Health, Education, and Welfare, Office of Education. (ERIC Document Reproduction Service No. ED049175)

(1971). *Ekistics: A handbook for curriculum development in conservation and education* (Report No. SO 002497). Sacramento, CA: California State Department of Education Bureau of Elementary and Secondary Education. (ERIC Document Reproduction Service No. 064196)

(1971). *The permanent agenda of man: The humanities: A tactic and strategy for teaching the humanities in the elementary school.* New York: Harcourt Brace Jovanovich. (Also available as an audio recording [1973], Washington, DC: National Education Association; Tempe, AZ: Arizona State University Library; Clarksville, TN: Austin Peay State University Library; and [2000], Washington, DC: George Washington University Library.)

(1971). (And Elizabeth B. Hone, Alexander Joseph, and Edward Victor). *A sourcebook for elementary science* (2nd ed.). New York: Harcourt Brace Jovanovich.

(1971). *Substance, structure, and style in the teaching of science* (Rev. ed.). New York: Harcourt Brace Jovanovich. (ERIC Document Reproduction Service No. ED108969)

(1971). (And Brenda Lansdown and Paul E. Blackwood). *Teaching elementary science through investigation and colloquium.* New York: Harcourt Brace Jovanovich.

(1971, March). Man's cumulative record—and his methods of intelligence. *The Science Teacher, 38*(3), 26–28.

(1971, March). On the easy satisfactions of the easy critics. *Science and Children, 8,* 9.

(1973). *Ekistics: A guide for the development of an interdisciplinary environmental education curriculum.* Sacramento, CA: California State Department of Education.

(1975). *Teaching gifted children science in grades seven through twelve.* Sacramento: California State Department of Education.

(1976). (And Evelyn Morholt). *Teaching learning strategies: Biology: Patterns in living things.* New York: Harcourt Brace Jovanovich.

(1976, November 3). *A useful tactic in stimulating curriculum change* (Report No. SP014096). Washington, DC: National Institute of Education. (ERIC Document Reproduction Service No. ED170244)

(1977). *The reduction of complexity: Substance, structure, and style in curriculum.* New York: International Center for Educational Advancement.

(1977, January). (And Robert Ornstein). The duality of mind. *Instructor, 86*(5), 54–59.

(1977, October). What is curriculum? *National Elementary Principal, 57*(1), 10–11.

(1978). Science and technology, humane purpose and human prospect. In J. J. Jelinek (Ed.), *Improving the human condition: A curricular response to critical realities* (pp. 94–128) [The Association for Supervision and Curriculum Development 1978 Yearbook]. Washington, DC: Association for Supervision and Curriculum Development.

(1978, May). (And Isaac Asimov, Carl Sagan, F. James Rutherford, Edward M. Kennedy, Randolph W. Bromery, Donald Fredrickson, and Donald Kennedy). Convention '78 revisited. *The Science Teacher, 45*(5), 27–34.

(1979, July). A general theory of instruction: With reference to science. *Science Education, 63,* 285–297.

(1980, January). Thoughts on individualization. *Instructor, 89*(6), 36.

(1980, February). Thoughts on a direction in teaching. *Instructor, 89*(7), 34, 36.

(1980, March). Thoughts on compassion and competence. *Instructor, 89*(8), 28.

(1981). *Memorandum: On renewing schooling and education.* New York: Harcourt Brace Jovanovich.

(1983). Do we expect school science to nurture creativity? In *NSTA yearbook: Science teaching: A profession speaks* (pp. 52–56). Washington, DC: National Science Teachers Association.

(1983). *Horizon committee report.* Washington, DC: National Science Teachers Association.

(1983, Fall). Schooling and education in a postindustrial era: The coming development of the first educational system. *School Library Media Quarterly, 12*(1), 10–19.

(1983). *Notes toward a renewal in the teaching of science.* New York: Coronado Publishers.

(1986). (And Evelyn Morholt). *Redefining the gifted: A new paradigm for teachers and mentors.* Los Angeles: National/State Leadership Training Institute on the Gifted and the Talented.

(1986). (And Evelyn Morholt). *A sourcebook for the biological sciences* (3rd ed.). San Diego: Harcourt Brace Jovanovich.

(1986, May). A portrait of gifted young with science talent. *Roeper Review, 8*(4), 235–243.

(1987, September). On avenues to kindling wide interests in the elementary school: Knowledges and values. *Roeper Review, 10*(1), 32–40.

(1988). Conservation: Its permanent agenda in America. In *A permanent agenda for conservation: Proceedings of the 35th Annual Conference* (pp. 1–17; Report No. SE 059264). Madison, WI: Conservation Education Association. (ERIC Document Reproduction Service No. ED317381)

(1989). (And A. Harry Passow [Ed.], Gerald Skoog [Contributing Ed.], and Deborah C. Fort [Association Ed.]). *Gifted young in science: Potential through performance.* Washington, DC: National Science Teachers Association.

Appendix B: Bibliographies of the Works of Paul F. Brandwein 297

(1989, November). Toward a permanent agenda for schooling. *Education and Urban Society,* *22*(1), 83–94.

(1991, March). (And Lynn W. Glass). A permanent agenda for science teachers, part I: International innovations. *The Science Teacher, 58*(3), 42–46.

(1991, April). (And Lynn W. Glass). A permanent agenda for science teachers, part II: What is good science teaching? *The Science Teacher, 58*(4), 36–46.

(1991, May). (And Lynn W. Glass). A permanent agenda for science teachers, part III: Implications for curriculum as part of a permanent agenda. *The Science Teacher, 58*(5), 22–25.

(1992, April). Science talent: The play of exemplar and paradigm in the science education of science-prone young. *Science Education, 76,* 121–123.

(1995). *Science talent in the young expressed within ecologies of achievement* (Report No. EC 305208). Storrs, CT: National Research Center on the Gifted and Talented. (ERIC Document Reproduction Service No. ED402700)

Textbooks and Series

Paul Brandwein's name appeared on all of the following publications—whether as author or editor is not always clear. Names of coauthors and other collaborators are provided below wherever possible. For several reasons, it has often proved impossible to track down all elements of a series. Many, perhaps most, bibliographies and libraries do not list textbooks. Used copies available through Internet sources sometimes help, often expensively, to fill some of the gaps. Finally, most of Brandwein's texts were published by Harcourt, Brace; Harcourt, Brace, and World; or Harcourt Brace Jovanovich. The latter was sold in the late 1980s. Thus, the publishing history of these enormously influential texts is sometimes murky.

Notes on Textbooks and Series[2]

Science for Better Living, first published in 1950 by Harcourt, Brace, was a series of science textbooks for the junior high school grades (grades seven through nine).

Concepts in Science, first published in 1966 by Harcourt, Brace, and World, later by Harcourt Brace Jovanovich, began as a series of science textbooks for grades one through six, each book with an accompanying teacher's edition, and other resources including laboratory kits (equipment and materials) with accompanying teacher's manuals, activity books for grades four through six, teaching tests for grades three through six, and a boxed set of ungraded investigation cards. Sometime between 1966 and 1968, materials for the "beginning" (kindergarten) level were added. (The chronology of this development is ambiguous.) Starting with the second edition, textbooks were color-coded to indicate grade level (yellow for kindergarten and blue, red, green, orange, purple, and brown for grades one through six, respectively).

---

[2] Available audiovisual materials for *Concepts in Science* are listed in Part IV, Appendix B, "Audiovisual Materials."

In 1968, the series added textbooks on life, energy, and matter for use in grades seven through nine. These textbooks also had corresponding teacher's manuals, workbooks, and comprehensive tests.

As the series evolved, the textbooks were revised several times. The information in each edition differed significantly because of, as Brandwein put it, "ever-increasing new data and rapidly growing knowledge" (personal communication to Deborah C. Fort, 1989).

Some *Concepts in Science* textbooks were available in paperback, large-print, and/or Braille editions. The various editions and materials are listed here as available information permits; there were undoubtedly others.

Science and Technology, published in 1985 by Coronado Press (San Diego), was a series of science textbooks for the elementary grades. Additional volumes for the middle and upper grades were planned but apparently never published.

(1950). (And Leland G. Hollingworth, Alfred D. Beck, and Anna E. Burgess). *Science for better living: Complete course.* New York: Harcourt, Brace.

(1952). (And Leland G. Hollingworth, Alfred D. Beck, and Anna E. Burgess). *Science for better living: Complete course.* New York: Harcourt, Brace.

(1953). (And Leland G. Hollingworth, Alfred D. Beck, and Anna E. Burgess). *You and your inheritance: Science for better living.* New York: Harcourt, Brace.

(1953). (And Leland G. Hollingworth, Alfred D. Beck, and Anna E. Burgess). *You and your world: Science for better living.* New York: Harcourt, Brace.

(1955). (And Leland G. Hollingworth, Alfred D. Beck, and Anna E. Burgess). *You and science: Science for better living.* New York: Harcourt, Brace.

(1956). (And Paul E. Blackwood and Hyman Ruchlis). *Discoveries in magnetism and junior scientist's kit.* Chicago: Science Research Associates.

(1956). *You and your inheritance: Science for better living.* New York: Harcourt, Brace.

(1956). *You and your world: Science for better living.* New York: Harcourt, Brace.

(1960). (And Leland G. Hollingworth, Alfred D. Beck, Anna E. Burgess, and Violet Strahler). *You and science: Science for better living* (Rev. ed.). New York: Harcourt, Brace.

(1960). (And Leland G. Hollingworth, Alfred D. Beck, and Anna E. Burgess). *You and your inheritance: Science for better living.* New York: Harcourt, Brace.

(1960). (And Leland G. Hollingworth, Alfred D. Beck, Anna E. Burgess, and Violet Strahler). *You and your resources: Science for better living* (3rd ed.). New York: Harcourt, Brace.

(1960). *You and your world: Science for better living* (Rev. ed.). New York: Harcourt, Brace.

(1964). (And Jerome J. Notkin, Paul E. Blackwood, and Herbert Drapkin). *Exploring the sciences.* New York: Harcourt, Brace, and World.

(1964). (And Alfred D. Beck, Violet Strahler, Leland G. Hollingworth, and Matthew J. Brennan). *The world of living things.* New York: Harcourt, Brace, and World.

(1964). (And Alfred D. Beck, Violet Strahler, Leland G. Hollingworth, and Matthew J. Brennan). *The world of matter: Energy.* New York: Harcourt, Brace, and World.

(1965). *Science for better living: Complete course.* New York: Harcourt, Brace, and World.

(1966). (And Arthur W. Greenstone, Frank X. Sutman, and Leland G. Hollingworth). *Concepts in chemistry: Laboratory manual.* New York: Harcourt, Brace, and World.

(1966). *Concepts in science* series.[3] New York: Harcourt, Brace, and World.

---

[3]Our source, the *Concepts in Science 2* teacher's edition, confirms the publication in 1966 of the textbooks, teacher's manuals, laboratory kits, workbooks, teaching tests, and laboratory cards for all levels indicated here except for kindergarten. The record is ambiguous and even contradictory on when materials for the kindergarten level were first published.

Appendix B: Bibliographies of the Works of Paul F. Brandwein 299

(And Elizabeth K. Cooper, Paul E. Blackwood, and Elizabeth B. Hone). *Concepts in science K; Concepts in science 1; Concepts in science 2; Concepts in science 3; Concepts in science 4; Concepts in science 5; Concepts in science 6.*[4]

(And Elizabeth K. Cooper, Paul E. Blackwood, and Elizabeth B. Hone). *Concepts in science K: Teacher's edition; Concepts in science 1: Teacher's edition; Concepts in science 2: Teacher's edition; Concepts in science 3: Teacher's edition; Concepts in science 4: Teacher's edition; Concepts in science 5: Teacher's edition; Concepts in science 6: Teacher's edition.*

(And Hyman Ruchlis). *Concepts in science: Classroom laboratory K; Concepts in science: Classroom laboratory 1; Concepts in science: Classroom laboratory 2; Concepts in science: Classroom laboratory 3; Concepts in science: Classroom laboratory 4; Concepts in science: Classroom laboratory 5; Concepts in science: Classroom laboratory 6* [Kits that include a teacher's manual and materials for 6 pupils].

Concepts in science 4: A workbook: Experiences; Concepts in science 5: A workbook: Experiences; Concepts in science 6: A workbook: Experiences.

Concepts in Science 3: Science teaching tests; Concepts in Science 4: Science teaching tests; Concepts in Science 5: Science teaching tests; Concepts in Science 6: Science teaching tests.

(And Hyman Ruchlis). *100 invitations to investigate* [Set of ungraded laboratory cards].

(1967). (And Elizabeth K. Cooper). *Concepts in science K.*[5] New York: Harcourt, Brace, and World.

(1968). (And Hyman Ruchlis). *Concepts in science: Classroom laboratory K* [Teacher's manual].[6] New York: Harcourt, Brace, and World.

(1968). (And Robert Stollberg and R. Will Burnett). *Energy: Its forms and changes (Concepts in science).* New York: Harcourt, Brace, and World.

(1968). (And Clifford R. Nelson). *Energy: Its forms and changes: Teacher's manual (Concepts in science).* New York: Harcourt, Brace, and World.

(1968). (And Robert Stollberg and R. Will Burnett). *Life: Its forms and changes (Concepts in science).* New York: Harcourt, Brace, and World.

(1968). (And Herbert Drapkin). *Life: Its forms and changes: Teacher's manual (Concepts in science).* New York: Harcourt, Brace, and World.

(1968). (And Robert Stollberg and R. Will Burnett). *Matter: Its forms and changes (Concepts in science).* New York: Harcourt, Brace, and World.

(1968). (And Violet R. Strahler). *Matter: Its forms and changes: Teacher's manual (Concepts in science).* New York: Harcourt, Brace, and World.

(1970). *Concepts in science* series (2nd ed.). New York: Harcourt, Brace, and World.

(And Elizabeth K. Cooper, Paul E. Blackwood, and Elizabeth B. Hone). *Concepts in science* [Separate color-coded book for each grade, K–6].[7]

(And Elizabeth K. Cooper, Paul E. Blackwood, and Elizabeth B. Hone). *Concepts in science: Teacher's edition* [Separate color-coded book for each grade, K–6].

Concepts in science: Experience book.[8]

---

[4]A separate edition of the *Concepts in Science* textbooks was published by the California State Department of Education, Sacramento, in 1966 and 1967. (Volumes for grades one and two were found.)

[5]One bibliography gave this date and authorship.

[6]One source listed just the teacher's manual and gave this date.

[7]One can assume that the kindergarten level was published, although no such volume was found, because that level was included in earlier and later editions.

[8]By the time the *Experience Book* was published, Harcourt, Brace, and World had apparently changed its name to Harcourt Brace Jovanovich. Our source did not indicate the grade level, but the levels are presumably the same as those for the first edition.

Concepts in science: Science teaching tests.[9]

(1970). (And Nancy W. Bauer and others). *The social sciences: Concepts and values* [Series of textbooks for the elementary grades]. New York: Harcourt, Brace, and World.

(1972). *Concepts in science* series (3rd ed.). New York: Harcourt Brace Jovanovich.

(And Elizabeth K. Cooper, Paul E. Blackwood, Elizabeth B. Hone, and Thomas P. Fraser). *Concepts in science* [Separate color-coded book for each grade, K–6].

Concepts in science: Teacher's edition [Separate color-coded book for each grade, K–6].

Concepts in science: Classroom laboratory [Kits that include a teacher's manual and materials for 6 pupils; separate color-coded kit for each grade, K–6].

(And Elizabeth K. Cooper, Paul E. Blackwood, Elizabeth B. Hone, and Thomas P. Fraser). *Concepts in science: Experience book.*[10]

Concepts in science: Science teaching tests.[11]

(1972). (And Robert Stollberg, Arthur W. Greenstone, Warren E. Yasso, and Daniel J. Brovey). *Energy: Its forms and changes (Concepts in science,* 2nd ed.). New York: Harcourt Brace Jovanovich.

(1972). (And Robert Stollberg, Arthur W. Greenstone, Warren E. Yasso, and Daniel J. Brovey). *Life: Its forms and changes (Concepts in science,* 2nd ed.). New York: Harcourt Brace Jovanovich.

(1972). (And Robert Stollberg, Arthur W. Greenstone, Warren E. Yasso, and Daniel J. Brovey). *Matter: Its forms and changes (Concepts in science,* 2nd ed.). New York: Harcourt Brace Jovanovich.

(1972). (And Violet R. Strahler). *Matter: Its forms and changes: Teacher's manual.* New York: Harcourt Brace Jovanovich.

(1974). (And Arthur W. Greenstone). *A Searchbook: Physical science.* New York: Harcourt Brace Jovanovich.

(1974). (And Arthur W. Greenstone and Violet R. Strahler). *A Searchbook: Physical science: Teacher's manual.* New York: Harcourt Brace Jovanovich.

(1975). *Concepts in science* series (Newton ed.). New York: Harcourt Brace Jovanovich.

(And Elizabeth K. Cooper, Paul E. Blackwood, Elizabeth B. Hone, Margaret Cottom-Winslow, Thomas P. Fraser, and Morsely G. Giddings). *Concepts in science* [Separate color-coded book for each grade, K–6].[12]

Concepts in science: Teacher's edition [Separate color-coded book for each grade, K–6].

Concepts in science: Classroom laboratory [Kits that include a teacher's manual and materials for 6 pupils; separate color-coded kit for each grade, K–6].

Concepts in science: Experience book.[13]

Concepts in science: Science teaching tests.[14]

(1975). (And Daniel J. Brovey, Arthur W. Greenstone, and Warren E. Yasso). *Energy: A physical science: A searchbook (Concepts in science).* New York: Harcourt Brace Jovanovich.

(1975). *Life: A biological science: A searchbook (Concepts in science).* New York: Harcourt Brace Jovanovich.

---

[9]No volumes were found, but they probably were available, because they existed for the 1966, 1975, and 1980 editions.

[10]Only the brown level was found, but the levels are presumably the same as for the first edition.

[11]No volumes were found, but they probably were published, since they existed for the 1966, 1975, and 1980 editions.

[12]Despite a lack of evidence, we assume the kindergarten level was included, because it appeared in the previous and following editions.

[13]Our source, *Experience Book [Brown]*, did not indicate the other levels, but they are probably the same as those for the first edition.

[14]Our source did not give the levels, but they are presumably for grades three through six.

Appendix B: Bibliographies of the Works of Paul F. Brandwein 301

(1975). (And Sigmund Abeles, Arthur W. Greenstone, and David Kraus). *Life: A biological science: A searchbook: Teacher's manual*. New York: Harcourt Brace Jovanovich.

(1975). *Matter: An Earth science: A searchbook (Concepts in science)*. New York: Harcourt Brace Jovanovich.

(1975). (And Sigmund Abeles, Arthur W. Greenstone, and David Kraus). *Matter: An Earth science: A searchbook: Science teaching tests (Concepts in science)*. New York: Harcourt Brace Jovanovich.

(1975). (And Violet R. Strahler). *Matter: An Earth science: A searchbook: Teacher's manual (Concepts in science)*. New York: Harcourt Brace Jovanovich.

(1980). (And Nancy W. Bauer). *The community: Living in our world*. New York: Harcourt Brace Jovanovich.

(1980). *Concepts in science* series[15] (Curie ed.). New York: Harcourt Brace Jovanovich.

> (And Elizabeth K. Cooper, Paul E. Blackwood, Margaret Cottom-Winslow, John A. Boeschen, Morsely G. Giddings, Frank Romero, and Arthur A. Carin). *Concepts in science* [Separate color-coded book for each grade,K–6].
>
> (And Elizabeth K. Cooper, Paul E. Blackwood, Arthur A. Carin, and Frank Romero). *Concepts in science: Teacher's edition* [Separate color-coded book for each grade, K–6].
>
> (And Elizabeth K. Cooper and Paul E. Blackwood). *Concepts in science: Classroom laboratory* [Kits that include a teacher's manual and materials for 6 pupils; separate color-coded kit for each grade, K–6].
>
> *Concepts in science: Activity book.*[16]
>
> *Concepts in science: Science teaching tests.*[17]
>
> *Concepts in science* [Blue; Braille].
>
> *Concepts in science* [Red, orange, and purple; large print].
>
> *Concepts in science: Classroom laboratory* [Purple; pupil's manual, Braille].
>
> (And Elizabeth K. Cooper, Paul E. Blackwood, Margaret Cottom-Winslow, John A. Boeschen, Morsely G. Giddings, Frank Romero, and Arthur A. Carin.) *Concepts in science* [Special ed.; paperback].

(1980). (And Nancy W. Bauer). *The Earth: Living in our world*. New York: Harcourt Brace Jovanovich.

(1980). (And Warren E. Yasso and Daniel J. Brovey). *Energy: A physical science (Concepts in science*, Curie ed.). New York: Harcourt Brace Jovanovich.

(1980). (And Violet R. Strahler). *Energy: A physical science: Teacher's manual (Concepts in science*, Curie ed.). New York: Harcourt Brace Jovanovich.

(1980). (And Nancy W. Bauer and Erika Mahoney). *The family: Living in our world*. New York: Harcourt Brace Jovanovich.

(1980). (And Warren E. Yasso and Daniel J. Brovey). *Life: A biological science (Concepts in science*, Curie ed.). New York: Harcourt Brace Jovanovich.

(1980). (And Warren E. Yasso and Daniel J. Brovey). *Matter: An Earth science (Concepts in science*, Curie ed.). New York: Harcourt Brace Jovanovich.

(1980). (And Nancy W. Bauer, Burton R. Clark, and others). *The United States: Living in our world*. San Francisco: Center for the Study of Instruction; New York: Harcourt Brace Jovanovich.

(1980). (And Nancy W. Bauer). *The world: Living in our world*. Harcourt Brace Jovanovich.

---

[15]It was impossible to determine how many levels have been produced in paperback, large-print, Braille, and other special editions.

[16]The levels are presumably the same as for the first edition.

[17]Our source did not give the levels, but they are presumably for grades three through six.

(1985). *Science and technology* series. San Diego: Coronado Publishers.
(And Burnett Cross and Sylvia S. Neivert). *Things around us* [Grade 1]; *Changes around us* [Grade 2]; *Changes we make* [Grade 3]; *On planet Earth* [Grade 4]; *Planet Earth in space* [Grade 5]; *In a new age* [Grade 6].

Audiovisual Materials

Although all of these nonprint contributions appear in bibliographies under Brandwein's name, it has been impossible to determine the extent of his involvement in their production. For convenience, we have divided them into three groups:

I. Filmstrips, mostly for the primary and middle school grades, composed in the 1970s and 1980s. All were published in New York by Harcourt Brace Jovanovich.
II. Textbook-based audiovisual materials.
III. Other audio recordings.

Filmstrips

Back and forth: A probe into energy transfer.
Back to plants: A probe into a source of energy.
Balanced diet: A probe into a web of life.
Bending light: Making rain.
Beyond Earth: A probe into space.
Bread: A probe into packaging prize genes.
The cat: A probe into inborn and learned behavior.
Changing: A probe into solids, liquids.
Colors in white light: City plants.
Cube and clip: A probe into changes in matter.
Day or night? A probe into Earth's motion.
Dinosaur: A probe into fossils.
Discovery of bacteria: Radio waves from space.
Finding out: A probe into push pull.
Fitness: A probe into adaptation.
Fitness to live: Oil, gas underground.
From space: A probe into a major resource.
Green energy: A probe into capturing energy.
Growing up: A probe into likenesses.
Invisible push: A probe into the pressure of air.
Living colors: A probe into adaptation by structure.
Making change: A probe into changing environment.
Moon shapes: A probe into the phases of the Moon.
Moving water: A probe into the water cycle.

Appendix B: Bibliographies of the Works of Paul F. Brandwein

New plants from seeds: Where rain comes from.
A push is a push: A probe into action and reaction.
Past and present: A probe into change.
Records from the past: A probe into fossil evidence.
Seeing the unseen: A probe into the structure of living things.
Shadows: A probe into how light behaves.
Sink or swim: A probe into conserving soil.
Sound effects: A probe into sound waves.
Splash: A probe into water for living things.
Stretch and shrink: A probe into molecular motion.
Telescope in space: Changes in traits.
Tilt: A probe into multiplying force.
Traveling seeds: A probe into special adaptations.
Under water: A probe into an environment.
Uses of dirt: A probe into conservation.
Vanishing act: A probe into chemical change.
The wandering fossils: A probe into the geologic cycle.
Where does it fit? A probe into common structure.

Textbook-Based Audiovisual Materials

Concepts in science [Brown (level 6); audio recording]. (Based on the textbook series, Curie ed.). Southfield, MI: Readings for the Blind.

Concepts in science [Green (level 3); teaching tests; audio recording]. (Based on the textbook series, Curie ed.). Milwaukee, WI: Volunteer Services for the Visually Handicapped.

Concepts in science [Green (level 3); teaching tests; audio recording]. (Based on the textbook series). New York: Harcourt Brace Jovanovich.

Concepts in science [Purple (level 5); filmstrip]. (Based on the textbook series). New York: Harcourt Brace Jovanovich.

Concepts in science [Purple (level 5); teaching tests; audio recording]. (Based on the textbook series, Newton ed.). Milwaukee, WI: Volunteer Services for the Visually Handicapped.

Concepts in science [Red (level 2); audio recording]. (Based on the textbook series). New York: Harcourt Brace Jovanovich.

Concepts in science [Red (level 2); audio recording, duplicating masters, and teacher's notes]. (Based on the textbook series, Curie ed.). New York: Harcourt Brace Jovanovich. Available in Anchorage School District Library, Anchorage, AK.[18]

---

[18]Materials (7 audiocassettes, 37 duplicating masters, and teacher's notes) were presumably created for grades K–6, but only grade two appears to be available.

Concepts in science [Red (level 2); filmstrip]. (Based on the textbook series). New York: Harcourt Brace Jovanovich. Available in Anchorage School District Library, Anchorage, AK.

Life: A biological science (Concepts in science) [Test booklet; audio recording]. (Based on the textbook series, Curie ed.). Alamogordo, NM: New Mexico School for the Visually Handicapped.

Science and technology [Grade 5; audio recording]. (Based on the textbook series). Vacaville, CA: Volunteers of Vacaville.

Other Audio Recordings

Now hear this! [Audio recording of speech by Paul F. Brandwein delivered at the dedication of the Oakland Museum, September 26, 1969]. Berkeley, CA: Pacifica Tape Library.

The permanent agenda of man: The humanities: A tactic and strategy for teaching the humanities in the elementary school [Audio recording based on a book by Paul F. Brandwein]. Washington, DC: National Education Association; Tempe, AZ: Arizona State University Library; and Clarksville, TN: Austin Peay State University Library (all 1973); also (2000), Washington, DC: George Washington University Library. (Also available as a 1971 publication [New York: Harcourt Brace Jovanovich].)

Index

A

Abstract Reasoning, 167, 174
Activating factor, 77–78, 136, 139, 150, 164,
  166, 169, 171–174, 183, 187–188,
  224–225, 232, 263–264, 276,
  290–291
Activities, end of first term (college), 152
Advanced science class, 155–156, 170
  teacher's function, 155–156
Affection, 22, 199, 235
Aldridge, Bill G., 62–63, 66
Amnesty International, 7
Army Chemical Center, 24
Army General Classification Test, 144
Association of Biology Teachers of New York,
  155, 184
Atmospheric chemist
  Crutzen, Paul (Nobelist), 26
  Molina, Mario (Nobelist), 26
  Rowland, Sherwood (Nobelist), 26
Augmented environment, 224–225, 228
  high school, design of, 249
Author (writer), 87

B

Behavior characteristics, gifted students
  attendance, 177–178
  leadership, 177
  planning ability, 177
  responsibility, 175–176
  trends in personality traits, 178–180
  ways of work, stages, 181–182
Benedict, Ruth, 18
Benjamin Franklin High School, 148
Bennett, Dean B., 98, 120, 122
Bennett, Hugh Hammond, 92
Berliner, Ernst, 11
Berman, Eleanore, 281, 283–284

Biological Sciences Curriculum Study (BSCS),
  17, 61, 82–83, 122, 229, 266–267
Biologist, 10–11, 41, 148, 236, 266, 269, 275,
  283
Brandwein, Mary, 119–123, 128–130
Busse, Thomas W., 224
Butterfield, Earl C., 227

C

California State Department of Education, 101,
  299
Carleton, Robert H., 61, 63–64
Carman, Harry J., 156
Charisma, 22, 38, 58
Chaucer, 48
Chauncey, Henry (president of the Educational
  Testing Service at Princeton), 138
College Entrance Examination Board, 171, 194
College-preparatory model, 251
Columbia Teachers College, 51, 278
Columbia University, 7, 20, 24–25, 144, 213,
  268
Commitment to science, 79, 145, 158–160,
  163–164, 167, 234
  subsidiary notions, 159
Conant, James B., 143, 148, 253
Concepts in Science, 76, 95, 297–299
Conceptual scheme, 73–76, 139, 142–143,
  146–148, 248–249
Conflict resolution, 137
Conservation (conservationist), 85, 91–99,
  102–104, 108, 118–130
Conservation Learning Summit, 125–130
Cornell University, 10, 91, 131
Critical thinking, 71, 103, 225–226, 253
Critical Years Ahead in Science Teaching, 187
Cross-cultural, 106–107, 110–112, 114–115
Curriculum, 73, 75–76
  and instructional designs, 250

D

Dartmouth College, 12
Datta, Lois-Ellen G., 223
Depression, effects of, 48, 270
de Solla Price, Derek, 219
"Developed ability," 140, 159, 166, 171–173, 175, 178, 196–198
DeWall, Marily, 61–67, 119–131
Dewey, John, 96, 105
Differential Aptitude Inventory of the Psychological Corporation, New York City, 167
Disadvantaged, 66, 218, 235, 238
Dobzhansky, 20, 275
Doxiadis, Constantinos A., 102
Drexel University, 26–27
Druger, Marvin, 65
Dyad: genome interacting with environment, 221

E

Earth Day, 93
Ecology, 12, 92, 94, 101, 107, 111, 121–122, 124–125, 211
Ecology(ies) of achievement, 57, 66, 78, 86, 215–239, 243–257
  contrasting ecologies, 234–235
  cultural heredity, 217
  early environment of child, 215
  effects of, 231–233
  environment-driven factors, 218
  first thesis, 215–221
  gene- and environment-driven factors, 217
  need for talented in science, 236–237
  in school, 232
  school environment, as dyad, 221–226
  schooling, effective, 217
  scientific knowledge, 219
  student as "performing scientist," 226–228
Ecosystem(s), 57, 79, 94, 114, 121–122, 217–218, 243, 245, 247, 256–257
Educational Resources Information Center (ERIC), 83–85
Education (educational), 9, 16–18, 23–24, 30, 33, 41, 57–58, 62–63, 69, 72, 78, 81–88, 91–99, 101–104, 105–117, 119–130, 135–136, 138, 141–142, 144–145, 151, 153–156, 158, 166, 183, 217–219, 222, 224, 231–232, 234–236, 243–246, 250–251, 263–265, 270, 278–280, 284
  system, 138, 243

Ekistics: A Guide for the Development of Environmental Education Curriculum, 101–104
Eisenhower (President), 138
"Elements in a Strategy for Teaching Science in the Elementary School," 55–56, 83
Empowerment (empower), 116, 117
Environmental (environmentalism, environmental education, environmental literacy), 92–93, 94, 96
  and heredity, relationship of, 246
Equal access to opportunity, xxii
ERIC, *see* Educational Resources Information Center
Evoking science talent, intereffective elements, 253
Evolution (Darwinism, Darwin), 266–267
Experiential learning (approach), 110
Extracurricular science work, 152

F

Family-school-community program, 245
Field-based, 122, 124–125, 129
Field-specific hypothesis, 225
First Gulf War, xvii–xviii
First World War (WW I), 217
Forest Hills High School, 5, 9, 16, 20, 22, 24, 29–30, 37, 41–42, 43, 47, 55, 61, 145, 151, 157, 169, 174, 184, 190, 205, 226, 231, 282
Framework, 96–98
Friend, James P.
  accomplishments at Isotopes, Inc. (later part of the Teledyne Corporation), 25
  Work at the National Academy of Sciences on
    the chlorofluorocarbon problem (1981–1982), 27
    the nuclear winter problem (1983–1985), 27
    the stratospheric flight problem (1975–1980), 27

G

Gallagher, James J, 219
Gardiner, Mary, 11
Gender equity, 276–277
  gender discrimination, 22
  gender imbalance, 32
General education, program of, 153

Index

Genetic Factor, 77, 139, 148, 166–170, 173–174, 182–183, 223–224, 263, 290
   in relation to science potential, 168
Genetics (genetic), 268–269
Genetic Studies of Genius, 159–160, 179
George Washington High School, 55, 148, 166, 252
George Washington University, 11–12
Getzels, Jacob W., 137, 224–225
The Gifted Child, 160, 179, 189
Gifted (giftedness, gift), 9, 24, 37, 47, 57, 61, 78–79, 135–139, 143, 166, 178, 215, 219, 221–222, 224–228, 237, 245, 255, 265
The Gifted Student as Future Scientist, 55, 78, 81, 135–213, 221, 263, 290–291
   behaviors
      attendance, 177–178
      leadership, 177
      planning ability, 177
      responsibility, 175–176
      trends in personality traits, 178–179
      ways of work, 181–182
   "conceptual scheme", 143
   identification by testing, 166–174
   proposals on local/national levels, 189–195
   self-identification, 151–165
   teachers, 184–188
   working hypothesis, 146–151
   raw material for the study, 145–146
Gifted Young in Science: Potential Through Performance, 57, 61, 66, 88
Global Rivers Environmental Education Network (GREEN), 98, 105–106, 120
Golay, Marcel, 25–26
Goodrich, Hubert B., 150, 169, 183, 186, 224
Gore, Al, 94
Gould, Stephen Jay, 66, 256, 265

H

Hammerskjöld, Dag, xx
Harcourt Brace Jovanovich, xxi, xxvii, 42, 56, 63
Harvard University, 7, 20, 56, 62, 83
Heredity and environment, relationship of, 246
Heterogeneous schools, 229, 233–235
"High level ability in science," 77, 150, 159, 223, 254, 290
   students, categories, 144
High Schools in Westinghouse Science Talent Search (1942–1988), 230, 233

Home environment, 269–270, 283
Human rights, 7, 106, 277

I

"Identical genomes," 216, 221
Identification by testing (gifted students)
   genetic factor, 166–168
      aptitude test, 167
      honor classes after the ninth grade, 167
      in relation to science potential, 168
   predisposing factor, 168–169
   studying the students, 169–174
      Man-to-Man Rating Scale, 169–170
      productivity, 170
      "reliability" criterion, 170
      test of developed ability, 171–172
      Westinghouse National Science Talent Search Examination, 171–174
"Independence training," 223
"Instructed learning," 78, 244, 247–249
   equal opportunity, 248
   inquiry-oriented teaching, 248
Integrated problem solving, 111
Intelligence, 9, 77, 96, 139, 159, 187, 217–218, 220–223, 226, 228, 245, 253–255, 269, 283
   See also IQ
Interdisciplinary approach, 51, 111
Inventory of predisposing factors, 163–164, 169, 171, 175, 200–202
IQ, 265, 269
   See also Intelligence

J

Jackson, Nancy E., 227
Jackson, Philip W., 224
Johns Hopkins University, 228
Jovanovich, William, 282

K

Knapp, Robert H., 150, 169, 223
Kolb, David A., 96
Kotovsky, Kenneth, 227–228, 248

L

Laissez-faire policy, 176–177
Lawrence Berkeley National Laboratory, 7, 275
Leadership Institute(s), 124–125
Lecture, 5, 10, 32, 37, 42, 52, 56, 62–63, 65, 70, 106, 119, 122, 244, 249
Leopold, Aldo, 92
Lewin, Kurt, 96, 116

Livermore, Norman, 101, 103
Lucan-Burton-Newton model, 220

M

MacCurdy, Robert D., 178, 229
"The Magic Synthesis," 137
Man-to-Man Rating Scale, 140, 163, 169–171, 175, 198, 201
Mann, Paul B., 184
Mansfield, Richard S., 224
Massachusetts Institute of Technology (MIT), 11, 24, 39–40, 63, 82, 274
Mathematical ability, 77, 136, 148, 166–167, 183, 290
Mathematics, 7, 17, 42, 111, 136, 138, 148, 153, 155–156, 158–159, 161–163, 168, 190–192, 227–229, 236, 244–245, 248–251, 254–255, 269–270
Meister, Morris, 160, 179
Meitner, Lisa, 10
Mentor (mentoring, mentored, mentorship), 19–20, 52, 271–273, 280, 284
"Methods of intelligence," 220–221, 226, 253, 255
Michels, Walter, 11
"Model" to guide (environment preparation), 222
Modest (humble, humility), 61–62
Morholt, Evelyn, 42, 55, 57, 88, 119, 146, 204

N

National Academy of Engineering, 37
National Academy of Sciences, 7, 27, 46, 220
National Association for Gifted Children, 61
National Council of Teachers of Mathematics, 250
National Education Goals, 251
National Research Center on the Gifted and Talented, 57, 79
National Science Foundation, 32, 59, 62, 131, 138, 190, 193, 245
National Science Teachers Association (NSTA), 52, 57, 61–67, 120, 122, 127, 129–131, 249, 290
Scope, Sequence, and Coordination, 249
yearbook, summary, 244
National/State Leadership Training Institute on the Gifted and the Talented, 135
Nature (natural), 5, 9, 12, 15–16, 20, 43, 52, 63, 82, 92–96, 99, 101–102, 106–107, 112, 114, 117, 125–127, 139, 143, 145, 148, 150, 160, 163,

166, 169, 171, 176–178, 181, 191, 204–205, 224, 231, 243, 268
Need for talented in science, 234–235
Network(s), 109–110, 112–113
New York University, 26, 45, 58, 63, 268
North American Association for Environmental Education, 61, 130
Numerical ability, 139, 163–164, 167, 174, 263

O

Oat smut, 10, 45
Observation, careful, importance of, 87, 267
Occupational Program Undergirding Science (OPUS), 223
Operational Approach (O.A.), 276
  benefits of O.A./other observations, 164–165
  in blueprint, 151–161
  gifted students, 147
  major differences in students, 147
  observations on, 161–165
  place of interest, 162–163
  raw materials, 161
  selection through self-identification, 161–162
Oppenheimer, Jane, 41
Oppenheimer, J. Robert, 10–11
Original work, 11, 39, 51, 56, 75
Origins of American Scientists, 186
Oxtoby, Toby, 144

P

Padalino (John "Jack"), 62, 119–130
Parloff, Morris B., 222
Passow, A. Harry, ix, 57, 61, 66, 88, 226
Paul B. Mann Biology Congress, 184
The Paul F-Brandwein Institute, 119–130
Peer groups, importance of, 6, 9, 10, 15, 16, 18–19, 30, 45, 266, 271, 282
"Performing scientists," 215, 226–228
Persistence, 77, 149–150, 164, 169, 173, 187, 200–202, 223, 226–227, 254–255, 264, 290–291
Phenix, Philip, 82
Phi Delta Kappan, 228
Piaget, Jean, 65, 96
Piano, 37, 56, 63, 65, 83, 274, 283
Pinchot, Giffort, 92
Pinchot Institute for Conservation Studies at Grey Towers, 92, 105, 122, 125
Place-based learning, 99
Plato, 95
Pogo (Walt Kelly), 99

Index

A Policy for Scientific and Professional Manpower, 144–145
Precollege education, 245, 250
Predisposing Factor, 77–78, 136, 139–140, 149–150, 163–164, 168–176, 178, 182–183, 187, 223–224, 227, 263–264, 290
Presidential Award for Excellence in Science Teaching for New York State, 52
Primary Mental Abilities form A (verbal-meaning/reasoning, number/word fluency), 167
Proposals (gifted students)
  on local level, 189–193
    biology registration, 191–192
    increased registrations in physics/chemistry, reasons, 190
    proposals dealing with basic science courses, 190–192
    special groups, solutions, 192–193
  on local and national levels, 193–195
    colleges, 194
    public schools, 193–194
    teachers' organizations, 194–195
Protégé(es) as mentors, 272–273
Psychiatry, 11–13, 149, 274
Psychological Corporation tests, 178

Q

Questing, 77, 149–150, 164, 169, 182, 187, 222–223, 227, 264, 290

R

Rabkin, Yakon M., 219
The Races of Mankind, 18, 22, 69
Racism, 18, 22
Radcliffe, 32–33, 47
  raw material for study, 145–146
Reflections of a Physicist, 181
Refugé(es), 43, 270
"Remedial experiencing," 95
Rivers, 106–108, 110–115
Roosevelt, Theodore (President), 91–92
Rutgers Creek Wildlife Conservancy, 119–120, 128–129

S

Santer, Ursula Victor, 41
Sato, Irving, 135
Schools and schooling
  Benjamin Franklin High School, 148
  Bronx High School of Science, 5, 160, 173–174, 189, 228

ecology(ies) of achievement, 232
  effective, 217
  school environment, as dyad, 221–226
  elementary and secondary school science education, 87–88
"Elements in a Strategy for Teaching Science in the Elementary School," 55–56, 83
Family-school-community program, 245
Forest Hills High School, xxvi, 5, 9, 16, 20, 22, 24, 29–30, 37, 41–42, 43, 47, 55, 61, 145, 151, 157, 169, 174, 184, 190, 205, 226, 231, 282
George Washington High School, xxi, 55, 148, 166, 252
  heterogeneous, 229, 233–235
Harvard Medical school, 11
  high school, after, 32–33
  high school, design of, 249
  nonschool environment, importance of, 269–271
  proposals for serving gifted students, 193–195
  rural, 85
  select and heterogeneous, 234–235
  special science schools, 254
Stuyvesant High School, 174
  teaching science in junior high school, 51, 55, 82
Schwartz, George, 11, 17, 41, 204
Schweitzer, Alfred, 48
Science for All Americans, 249
Science curriculum, 46, 73, 87, 189, 226, 248
Science education, 23, 56, 58, 62–63, 67, 79, 81–88, 91–92, 96, 102, 124–125, 244, 250–251, 265
  responsibilities in, 153–154
Science educators, 61–63, 66, 265
Science K–12, 249
Science News Letter, 153
Science prone student(s), 77–78, 86
Science shy, 66–67
Science Since Babylon, 219
"*The Science Sponsor's Handbook*," 189
Science talent, 143
  in young expressed within ecologies of achievement
    conception of, 254–257
    curricular approach, 248–251
    early self-identification of, 252–254
    instructional approach, 247–248
    limiting environments, 244–246

portent of, 255–256
Search, 228
"Science Talent: In an Ecology of Achievement," 215–239
Science Talent in the Young Expressed Within Ecologies of Achievement, 57, 66, 79, 243–259
The Scientific American, 153
"Scientific civilization," 219
Scientific literacy, 250
Scientific method, 153, 253
Scientific Monthly, 153
Second World War (WW II), xvii, xx, 16, 39, 67
Selected Science Teaching Ideas of 1952, 160
Self-identification? (self-identified self-identify, self-identifying), 75, 273, 281, 283
Self-identification (gifted students)
  observations on operational approach, 161–165
  operational approach in blueprint, 151–161
Sherburne, E. G., 229
Siegler, Robert S., 227–228, 248
Silber, Robert, 61
Skepticism, 47, 75
Skoog, Gerald, 57, 61, 66–67, 88
"Slow" learner, 66, 141, 151
Small Futures, 218
"Social invention," 159–160, 257
A Sourcebook for the Biological Sciences, 42, 82
Special education, program of, 154
Special science schools, 254
Sputnik, 236, 250, 253
Stanley, Julian C., 228–229
Stapp, William B., 92, 98, 105–117, 129
State University of New York at Buffalo, 46
Students (learners), 66, 69, 70, 72–76, 78, 94–95, 98, 123, 136, 139, 147, 215, 218–219, 221–222, 224, 226, 235–238, 243–257, 269, 280
Student as teaching assistant/lab preparer, 6, 23, 31, 282
Study of Mathematically Precocious Youth at Johns Hopkins University, 228
Swarthmore, 29, 32–33

T
Teacher (teach, teaching, teacher of teachers), 30, 57, 70–72, 105–106
  features of stimulating teachers, 185–186

stimulating youngsters to enter science, 184–185
Teaching the Gifted Child, 219
Team effort, 145
Technological Innovation in Science: Adoption of Infrared Spectroscopy, 220
Teller, Edward, 10
Terman, Lewis M., 9, 142, 159–160, 178
Test of Developed Ability in Science (TDAS), 175, 196–198
  Forms A and B, 171–172
Test(s), 6, 27, 41, 51, 55–56, 66, 71, 76, 109, 111, 137, 140, 148, 150–151, 168, 171–172, 175, 178, 196–198, 222, 225–229, 232, 235–237
Testing Approach, 147
Tolba, Mostafa K., 107
Torrance, E. Paul, 224
"Transformative power," 256
Truman, President and Mrs. (Bess), 10

U
Udall, Stewart I., 126
Union of Concerned Scientists, 7
University of California, Berkeley, 37, 44, 273, 275
University of Connecticut, 57, 80, 91, 122
University of Iowa, 45, 63
University of Michigan, 39, 105, 109, 113, 122, 129
University of Oregon, 12
University of Wisconsin, xxvi
U.S. Agency for International Development, 33
U.S. Department of Agriculture, 40, 92, 131, 273

V
VanTassel-Baska, 253
Verbal Ability, 77, 139, 148, 166, 263
Verbal Reasoning, 167, 174

W
War(s), xvii–xviii, 5, 16, 18, 30, 39, 67, 217, 220
Watershed, 105–117
Watson, Fletcher G., 63, 82–83, 148, 173
"Weed and seed" approach (Government-University-Industry Research Roundtable), 245
Weltfish, Gene, 18
Western Regional Environmental Education Council (WREEC), 102, 104

Index

Westinghouse (now Intel) Science
Talent Search (the Search,
the Westinghouse, Westinghouse),
10, 19, 23, 29, 31–32, 43, 47, 165,
178, 229–230, 233, 253, 255, 266
Wheeler, Keith A., 105, 119–131
Witty, Paul, 160, 166, 168, 189
Wolfle, Dae, 144
Women's Medical College of Pennsylvania, 11
Working hypothesis, gifted students, 146–151

World Bank, 33, 281, 284
World Council for Gifted and Talented
Childreden, ix

Y

Yager, Robert, 63–64, 111

Z

Zinsser, Hans, 48